The Complete Guide to Photorealism for Visual Effects, Visualization, and Games

This new edition offers the most up-to-date and comprehensive guide to accomplishing and perfecting a photorealistic look in digital content across visual effects, architectural and product visualization, and games.

Emmy award-winning VFX supervisor Eran Dinur offers readers a deeper understanding of the complex interplay of light, surfaces, atmospherics, and optical effects, and then discusses techniques to achieve this complexity in the digital realm, covering 3D, 2D, and generative AI methodologies. In addition, the book features artwork, case studies, and interviews with leading artists in the fields of VFX, visualization, and games. Exploring color, integration, light and surface behavior, atmospherics, shading, texturing, physically-based rendering, procedural modeling, compositing, matte painting, lens/camera effects, and much more, Dinur offers a compelling, elegant guide to achieving photorealism in digital media and creating imagery that is seamless from real footage. In this second edition, the book explores generative AI as a tool for producing photoreal content. It focuses on various workflows and techniques for controlling the generative process and guiding AI models toward the desired results.

The book's broad perspective makes this detailed guide suitable for VFX, visualization, and game artists and students, as well as directors, architects, designers, and anyone who strives to achieve convincing, believable visuals in digital media.

Eran Dinur is an Emmy and VES award-winning VFX supervisor, artist, and author of *The Filmmaker's Guide to Visual Effects* (2024). His film and TV work includes *Marty Supreme*, *The Trial of the Chicago 7*, *Hereditary*, *The Greatest Showman*, *Uncut Gems*, *The Wolf of Wall Street*, *Boardwalk Empire*, *Star Trek*, and *Iron Man*. He is an adjunct professor at the School of Visual Arts and the author of several popular VFX courses at fxphd.com.

The Complete Guide to Photorealism

for Visual Effects, Visualization, and Games

Second Edition

Eran Dinur

Routledge
Taylor & Francis Group
NEW YORK AND LONDON

Designed cover image: Top: *Fir Forest* by Gökhan Karadayı. © Gökhan Karadayı and ScansLibrary
Bottom left: *Simple Interior* by Javier Wainstein. © Javier Wainstein
Bottom right: *Hanging Panda* by Massimo Righi. © Massimo Righi

Second edition published 2026
by Routledge
605 Third Avenue, New York, NY 10158

and by Routledge
4 Park Square, Milton Park, Abingdon, Oxon, OX14 4RN

Routledge is an imprint of the Taylor & Francis Group, an informa business

First edition published by Routledge 2021

Library of Congress Cataloging-in-Publication Data
Names: Dinur, Eran author
Title: The complete guide to photorealism for visual effects visualization and games / Eran Dinur.
Description: Second edition. | New York, NY : Routledge, 2026. | Includes bibliographical references and index.
Identifiers: LCCN 2025050314 (print) | LCCN 2025050315 (ebook) | ISBN 9781032966557 paperback | ISBN 9781032966564 hardback | ISBN 9781003590293 ebook
Subjects: LCSH: Rendering (Computer graphics) | Photo-realism | Digital images—Editing | Digital cinematography | Architectural rendering—Technique | Video games—Design | Cinematography—Special effects
Classification: LCC T385 .D585 2026 (print) | LCC T385 (ebook)
LC record available at https://lccn.loc.gov/2025050314
LC ebook record available at https://lccn.loc.gov/2025050315

ISBN: 978-1-032-96656-4 (hbk)
ISBN: 978-1-032-96655-7 (pbk)
ISBN: 978-1-003-59029-3 (ebk)

DOI: 10.4324/9781003590293

Typeset in Avenir
by Apex CoVantage, LLC

CONTENTS

PREFACE TO THE SECOND EDITION

In the few years since the release of this book's first edition, generative AI swept the digital visual world in a storm. New developments happen at a dizzying speed. Yesterday's hype is today's old news as AI companies constantly tout their latest model as they vie for attention (and subscribers). VFX, design, and visualization studios find themselves in a mad dash to keep up with the rushing current, and artists feel the pressure to continuously learn and adapt. Add to this the lack of clarity around copyrights and ethics, and the overhanging concerns about the unpredictable changes in our industries. It's the Wild West out there, and it's as exciting as it is intimidating.

For this edition, I wrote an entirely new chapter on generative AI. I chose to concentrate only on aspects that are relevant to this book – namely, generative AI as a tool for producing photoreal visual content. It is a powerful tool, no doubt, but it lacks the sort of precise control and predictability that 3D and 2D artists are accustomed to. In this new chapter, I look at the various techniques that artists can employ to guide AI generation toward the desired results, from prompt syntax and inpainting, to ControlNet models and IP adapters. I decided to focus on Stable Diffusion's open-source world because it is readily available to everyone and offers a remarkable level of flexibility in designing custom workflows. And, in keeping with the goals of this book, I strived to explain the "why" and not just the "how".

Of course, traditional 3D and 2D are still the core of this book, and for this edition, I updated and revised various sections to reflect the latest changes. I also added new case studies that cover topics such as realistic plastic shaders, photoreal nature in architectural visualizations, PBR lighting techniques, hybrid 2D/3D matte painting and more. Finally, this edition is filled with new and inspiring photoreal artwork by VFX, visualization and games artists, using 3D, 2D, and gen AI techniques. I find it particularly exciting to showcase such a diverse group of artists from around the world.

ACKNOWLEDGMENTS

Writing this book was a true journey for me, and I would like to thank all the people who supported me along the way, and shared their expertise, insight, and talent.

Many thanks to Dr. Amodsen Chotia of the University of Paris for his invaluable help in clarifying complex physical concepts of light and light-surface interaction, and to my friend Dr. Ariel Lindner for connecting us.

A big fat thank you to my colleague Ben Zylberman for creating the book's superb illustrations, and for being ever so patient toward my never-ending requests.

I would like to thank all the talented artists from around the world who kindly contributed some of their brilliant artwork to this book: Angelina Pronina, Caleb Spenrath, Cornelius Dämmrich, David Baylis, Gökhan Karadayı, Hadi Karimi, Javier Wainstein, Jose Iuit, Julian Sadokha, Massimo Righi, Mohamed Raof, Piotr Krynski, Sergey Vasnev, Suburbia Studio, Valeryi Orlov, and Waleed Nabil.

Thanks also to Brainstorm Digital's artists and founders Richard Friedlander and Glenn Allen for allowing me to use images from our demo reels, and to John Montgomery and Eduardo Abon for letting me use footage from my fxphd compositing course.

I owe special gratitude to a unique group of artists who offered not just their artwork, but also fascinating words of wisdom that shed light on their own creative process: Dax Pandhi, Irena Šmitáková, Kevin Samar, Mikhail Sizou, and Raphael Rau.

Many thanks to the Routledge editorial and production team: Daniel Kershaw, Danielle Zarfati, Alison MacFarlane and Kate Fornadel.

To my amazing family – how can I ever thank you enough? Karin, my partner in life, and my children. Ayala and Yotam, thank you for your patience, wisdom, and understanding. I love you all so much!

I dedicate this book to the memory of my parents.

Introduction

Photorealism is a vital component in digital visual media: in visual effects, seamless integration with live-action footage is crucial for telling a believable visual story, and photorealism lies at the core of all successful VFX magic. In architectural and product visualization, photoreal CG renders are now the norm, and often replace photography in advertisements, presentations, and catalogs. Photorealism in video games is constantly evolving along real-time rendering technologies, with new games consistently raising the bar. The dramatic rise of generative AI opens up a completely new frontier for photorealistic visuals.

Yet despite the technological advancements and the wealth and quality of tools available to digital artists, achieving photorealism in digital media remains a complex, challenging, and often elusive goal. Numerous factors come into play, and these factors stretch across different crafts, from modeling, texturing, and shading, to lighting, matte painting, and compositing. Think about creating photoreal content, and numerous questions come to mind: how do you generate imagery that is indistinguishable from real photographs? How do you successfully emulate light and its interactions with surfaces? How much detail is needed to make digital visual content convincingly real? How to avoid the dreaded "CG look"? How to achieve realistic daylight and atmospheric depth? How to get the best out of physically-based renderers? What are BRDF shaders? How can render passes help in compositing? How to control generative AI and guide it toward the desired results? And so on, and so on.

This book provides a comprehensive guide to achieving photorealism in digital media. It approaches the subject through a broad perspective that spans various crafts and caters to digital artists working in a variety of fields. The key to photorealism lies in plain sight: it is everywhere around us. To produce convincingly realistic digital imagery, artists must expand their knowledge beyond the specific software tools and techniques of their craft and study the real world: the way light interacts with different materials; how Rayleigh and Mie scattering in the atmosphere transform the light of the sun; the way dirt and grime form on surfaces; the differences between human vision and cameras; the complex structure of plants; how the aperture of a lens affects defocus; or the different characteristics of metallic and dielectric surfaces (to mention just a few of the subjects discussed here).

In writing this book, I set myself two goals: the first is to provide an in-depth study of real-world light, the physical interaction of light and surfaces, material properties, atmosphere, depth, and cameras. These topics are vital for digital artists, yet much of the available information about them is explained in hardcore scientific jargon and through complex equations. In this book, I approach

DOI: 10.4324/9781003590293-1

these subjects from an artist's point of view – focusing on their visual (rather than mathematical) aspects.

The second goal is to address the tools and techniques that digital artists can use to visually emulate reality and produce photoreal content. This includes CG approaches (path tracing, BRDF shaders, PBR texturing, procedural modeling, etc.), 2D workflows (seamless integration, extraction techniques, working with color, render passes, etc.), and generative AI techniques (inpainting, ControlNet, prompt syntax, IP adapters, etc.). In most cases, I have intentionally avoided software-specific discussions, knowing that readers rely on a wide variety of different applications and tools. So, whether you use Maya or Unreal Engine, V-ray or Redshift, Substance Painter or Mari, After Effects or Nuke, ComfyUI or InvokeAI, this book provides the general knowledge that can complement software-specific books, courses, or tutorials.

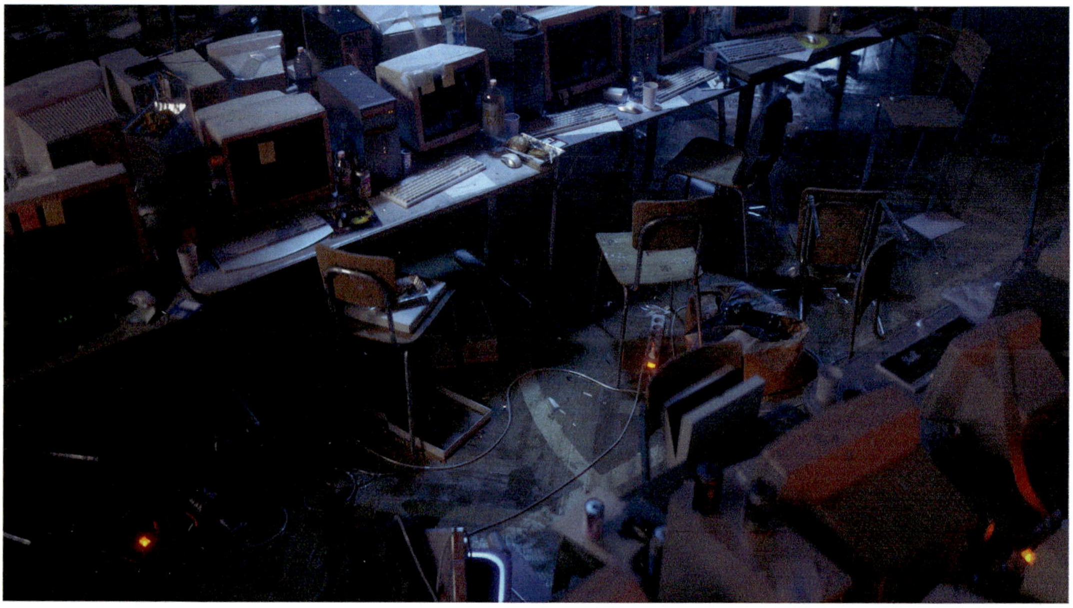

2003 by Cornelius Dämmrich. Created in Cinema 4D.

A note about Animation

I am of course completely aware of the importance of animation in the context of realism. Clearly, even the most photoreal modeling, texturing, shading, and lighting cannot salvage a poorly animated character, a lack of a proper sense of weight and inertia, or bad timing and gesturing. But the subject of animation is too vast to be covered in this book in any meaningful way, and it stands independent from other subjects discussed here. Moreover, there are already many excellent books devoted to animation and motion capture. Therefore, this book remains focused on the visual rather than kinetic characteristics of realism (with only a few minor exceptions in discussing motion blur and lens flares). In other words, this book is about how things **look**, and not about how things **move.**

A Quick Overview of the Book

The book is divided into four parts: **Part 1** (Core Concepts) addresses fundamental aspects of photorealism and color; **Part 2** (The Real World) is an artist-friendly exploration of the physical characteristics of light, materials, the atmosphere, and lenses; **Part 3** (The CG World) examines photoreal CG topics in reverse order, starting from rendering and lighting, then shading, texturing, and modeling; **Part 4** (The 2D World) is dedicated to the 2D aspects of photoreal content in compositing, matte painting, image manipulation, and generative AI.

Here is a short description of each chapter:

Chapter 1: Reality and Photorealism

A discussion of the fundamental aspects of photorealism, such as the difference between human vision and photography, detail and imperfections in simulated reality, the uncanny valley and more.

Chapter 2: Photorealism in Digital Media

This chapter compares the different roles, methodologies, aesthetics and challenges of photorealism in three main categories: visual effects, games, and visualizations.

Chapter 3: Color

Understanding color is crucial for achieving photorealism – and this chapter offers a methodological look at essential color subjects like color perception, additive color theory and operations, black and white levels, and dynamic range.

Chapter 4: Light Essentials

The fundamental physical properties of light are explained here from a visual perspective. Subjects covered include the particle/wave duality, photons and the electromagnetic spectrum, light decay, and direct/indirect light.

Chapter 5: Light Interaction

What happens to photons when they hit an obstacle? This chapter explores the principal light/surface interactions: diffuse and specular reflection, absorption and transmission.

Chapter 6: Daylight

Daylight is explained by studying the interaction of sunlight with atmospheric elements and the effect of Rayleigh and Mie scattering. This chapter also covers the principles of aerial perspective.

Chapter 7: Nighttime and Artificial Lighting

An overview of common emitters and fixtures, luminance and temperature measurements, and natural nighttime light.

Chapter 8: Shadows

Analysis of shadows: softness and color, nested and overlapping shadows, contact, and proximity shadows.

Chapter 9: Basic Material Properties

This chapter expands the subject light-surface interaction by examining material properties like diffuse/specular balance, large and small-scale roughness, the Fresnel effect, and the difference between metallic and dielectric materials.

Chapter 10: Lens and Camera Characteristics

This chapter provides a discussion of depth of field, Bokeh, lens distortion, chromatic aberrations, lens flares, lens blooms, motion blur, and grain.

Chapter 11: Rendering and Lighting

A brief overview of rendering and global illumination methods followed by a close look at path tracing and physically-based rendering. This chapter also focuses on various CG lighting tools like area lights and image-based lighting and discusses effective lighting practices for varying scenarios.

Chapter 12: Shading

This chapter starts with a glance at the evolution of shader technologies, then moves on to the various attributes of BRDF and how they contribute to photoreal renders. Additional shaders like hair/fur and volumetric are also discussed.

Chapter 13: Texturing

This chapter offers an overview of the PBR texturing workflow and discusses best practices and techniques for image-based and procedural textures.

Chapter 14: Modeling

The focus in this chapter is on aspects of modeling that relate to photorealism, like modeling for light and procedural generation of complex natural elements such as plants and terrains.

Chapter 15: Integrating 2D Elements

This chapter provides a thorough examination of different techniques for 2D integration and edge treatment, such as color matching, edge keying, matte-less extractions, and motion blur reconstruction.

Chapter 16: Integrating CG Elements

An overview of lighting and utility render passes and their use for improving the integration and realism of rendered elements, followed by additional techniques for "sweetening" the CG look.

Chapter 17: Lighting in 2D

This chapter examines different techniques of using color to modify the lighting, create atmospheric depth, shadows, and reflections in 2D.

Chapter 18: Lens and Camera Effects

In this chapter we look at tools and techniques for simulating lens and camera effects like depth of field, distortion, chromatic aberrations, and grain.

Chapter 19: Photorealism with Generative AI

The immense potential of generative AI for photoreal content creation is often outweighed by the lack of precise control and predictable output. In this chapter we explore various techniques and tools for guiding and controlling Stable Diffusion models, like ControlNet models, Inpainting, effective prompt syntax, and more.

Epilogue

A few words about the future of photorealism.

PART 1

CORE CONCEPTS

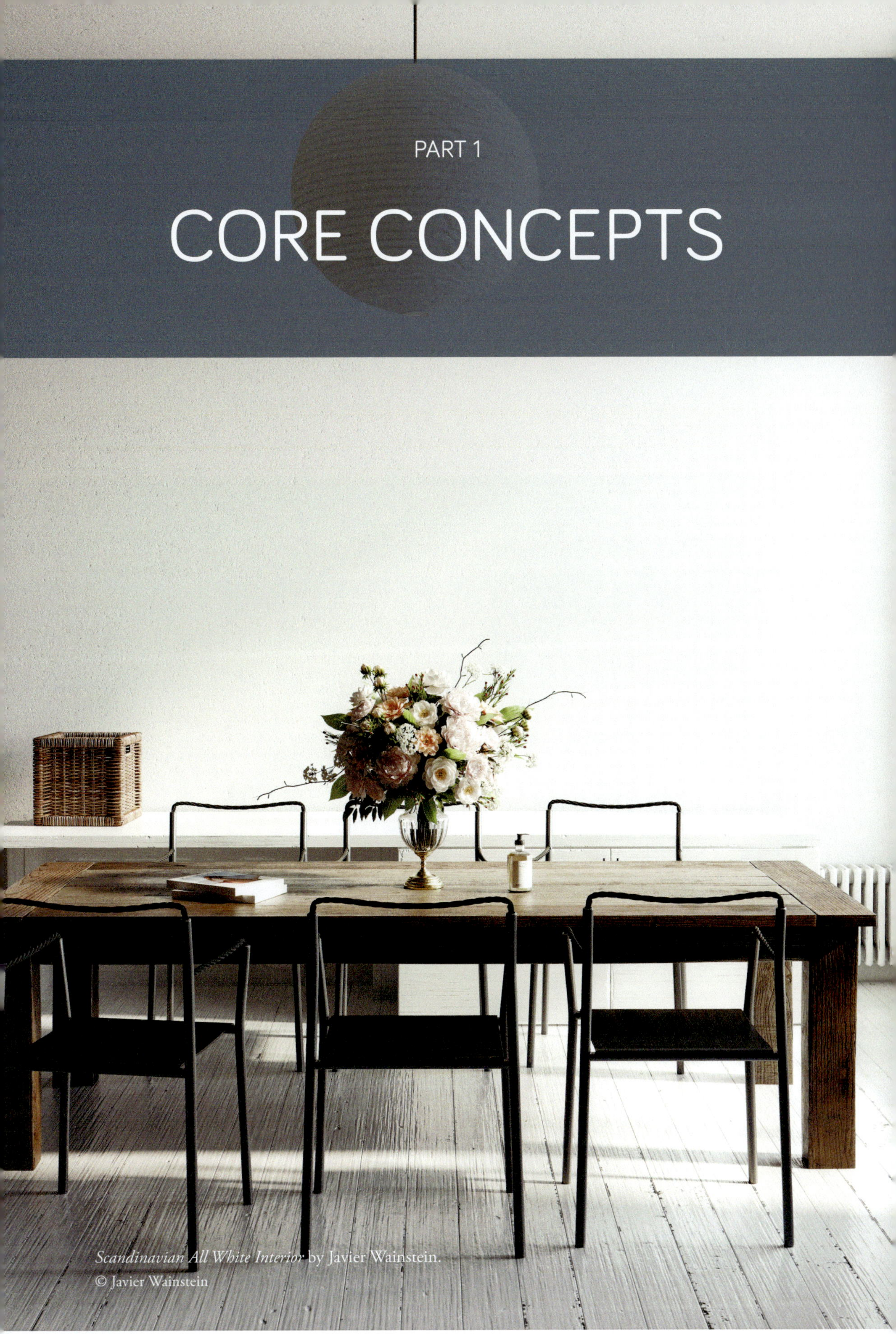

Scandinavian All White Interior by Javier Wainstein.

Chapter 1

Reality and Photorealism

Human Vision and Cameras

What is the definition of photorealism? Here is a common one: "If it looks like a photo, it is photoreal". This is a short, clear, and straight-forward answer, but it raises two essential questions: first, what does a photo "look like"? And second, why "look like a photo"? Why not just "look real"? Let me start by addressing the second question. The reality we see through our own eyes cannot be recorded and played back. We have no way (not yet, at least) to connect a person's brain to a monitor and see the world through that person's eyes. Everything we see on a screen or a monitor that we consider real (as opposed to animation, cartoons, or motion graphics) must first be captured by a camera. From the earliest printed photos of the 19th century to today's phones and tablets, the only way we can watch "transported reality" (any reality that's different from what we see around us) is through the mediating assistance of a camera. Our perception of reality in movies, videos and still photography is very tightly tied to a "photographic look", because essentially, it is the only look we have ever seen on a screen. Therefore, if we wish to create convincingly realistic digital content, we need to emulate the characteristics of photography and cinematography, rather than the way we see the world with our own eyes. This leads to the first question: what does a photo look like? In other words, if we could literally hook up a person's brain to a monitor and compare that person's vision to the camera's, would they look different? And in which ways?

This is of course a tricky question, exactly because we cannot perform this hypothetical comparison. If we think of "seeing" as a single process, then viewing photographs (or movies) is a double process: we are seeing through our own eyes reality that was "seen" (captured) by the camera. Take, for example, focus. Just like a camera, our eyes can dynamically shift focus. But we cannot possibly "view" our vision's out-of-focus areas. On the other hand, a photo on paper or on screen allows us to view areas that are in or out of focus equally well. So, while human defocus forever lies at the periphery of our vision, photographic defocus is an observable, meaningful part of the picture and a crucial artistic and narrative aspect of photography and cinematography. This is but one example that shows why, without the ability to observe our vision in the same way we observe footage, the comparison is somewhat flawed. Yet despite these limitations, it is still worth exploring, if only for the purpose of identifying those specific characteristics of photography that are distinctly different from human vision. Understanding those characteristics is crucial to making our digital content "look like a photo".

DOI: 10.4324/9781003590293-3

The Similarities

On a superficial level (and without digging too deep into anatomy or optometry) our eyes operate very much like a camera. The cornea and lens form the optical assembly that is the biological equivalent of a camera lens. The cornea acts like the static outer glass, while the inner lens is flexible and controllable to allow focus changes (much like the inner glass in a camera lens, which moves back and forth when the focus is adjusted). Our iris functions as the diaphragm, and controls the size of the opening (the aperture) through which light comes in. The purpose is the same as in a camera: to restrict the amount of light that reaches the retina in a brightly lit environment and let more light in when the environment is dark. Although the retina is structurally different from a camera sensor (it is curved, and its receptive cells are not evenly distributed along its surface) it still performs a similar task, sensing different light frequencies and intensities, and passing that information through the optic nerve to the brain. However, despite the initial similarities, there are some critical differences between cameras and human vision, and those differences define the characteristics of photographed material.

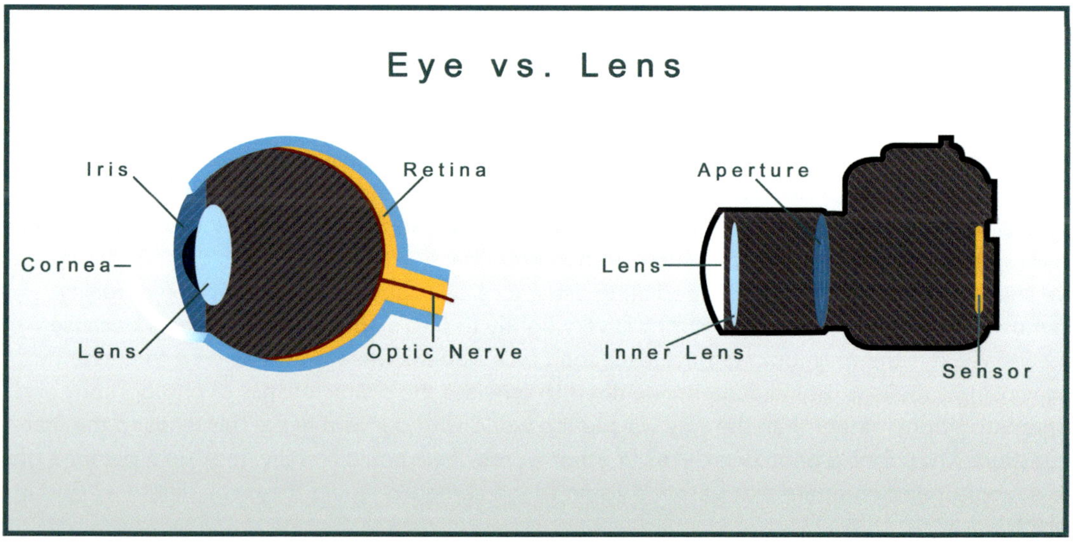

On a fundamental level, our eyes and the camera are very similar.

Field of View

The sensitivity of a camera sensor is balanced across its entire surface, and good lenses maintain a consistent sharpness and quality from edge to edge. The correlation between the focal length and the field of view is thus easy to measure and define. Human vision, on the other hand, has a distinct sharpness falloff from the focal center toward the fringes of the viewing field. If you gaze forward while waving your hand near your left ear, you can still see the hand – but it is quite blurry. To see the hand in focus, you must either bring it closer to the center of your vision, or instead turn your eyes hard left (which then makes everything at the front blurry). The nature of our peripheral vision means that unlike lenses, there is no precise definition of our field of view. Should we measure the entire width of the vision field, or only the sharpest area? And where exactly do we draw the line between sharp and blurry? So, while precise numbers are debatable, it is generally agreed that our overall field of view is roughly comparable to a 22–24mm lens on a full-sensor camera, and our effective (relatively sharp) field of view is comparable to a 43mm lens. Medium lenses are thus traditionally used to convey human vision, although they clearly do not mimic our wider peripheral vision.

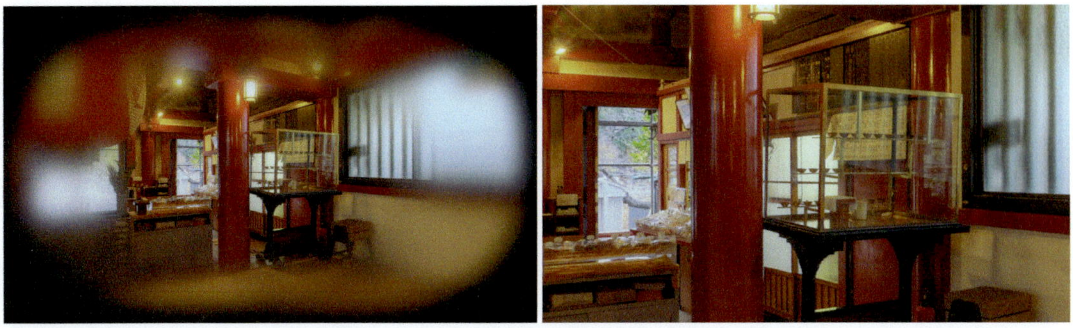

Left: an approximate depiction of human vision (roughly 22mm center-focused). Notice how odd and "unreal" it feels when viewed as a photograph. Right: a 43mm equivalent of our effective viewing area.

The most important difference in relation to field of view, however, is the fact that we cannot change our focal length. In photographic terms, we are forever limited to a single prime lens – no switching lenses, no adjustable zooms. This does make sense if you think of our eyes as a spatial navigation tool (it would be very hard to properly assess distance if we constantly changed our focal length), but it also means that the wide variety of perspectives that characterizes photography, from the exaggerated depth of wide lenses to the compressed perspective of long lenses, is simply not available to us. Yet even though changeable perspective is foreign to our vision, we completely embrace the rich perspective vocabulary of photography and cinematography and accept it as "real". We have no problem watching footage shot with a 12mm or 250mm lens, even though both lenses show the world in a way that we can never see with our own eyes. To achieve photorealism, we must treat field of view in photographic terms, and all but ignore the fixed-perspective, peripheral nature of our own eyes. Interestingly, tilt-shift effects and other photographic techniques that somewhat simulate human vision are often perceived as artificial and strange. This is but one example that shows how accustomed we have gotten to the language of photography and to viewing the world through a lens.

Seeing with the Mind

We think, and cameras don't – this is an important difference between human vision and camera, and an essential concept for photorealism. To illustrate this difference, let us examine two scenarios that should feel familiar to anyone who's ever used a camera:

1) On a trip to the mountains, you are standing on a high overlook, admiring the breathtaking landscape of tall peaks and deep valleys. Inspired by the beautiful view, you pull out the camera and take some photos. But to your disappointment, the photos convey none of the sense of depth and grandeur you experienced.

2) While on an apartment hunt, you find a place you really like – spacious, well-lit, and with a nice view outside. You take some photos, but they don't do the apartment justice. If you expose for the interior, the windows blow out and you cannot see the view outside, and if you expose for the windows, the interior looks dark and depressing. Either way, the pictures of the apartment do not look and feel like what you've seen with your own eyes.

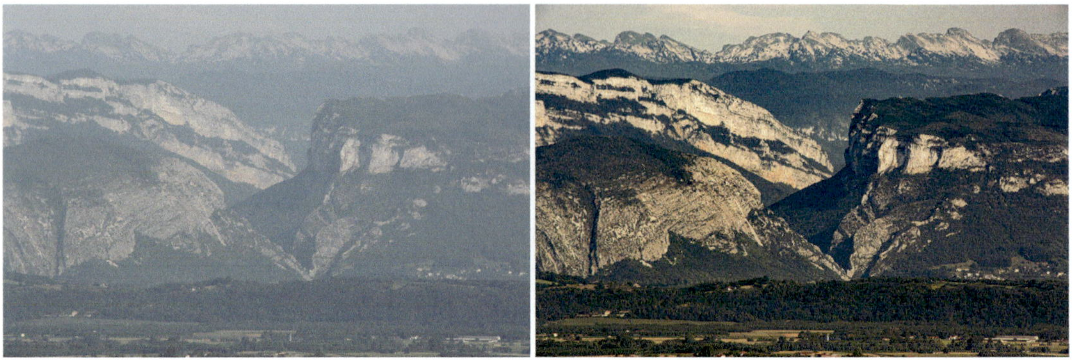

In the actual photograph (left), the camera records light exactly as it reaches the sensor, including haze and humidity that severely flatten the image. But we "see through" the haze with our mind, intuitively reconstructing detail and depth (right).

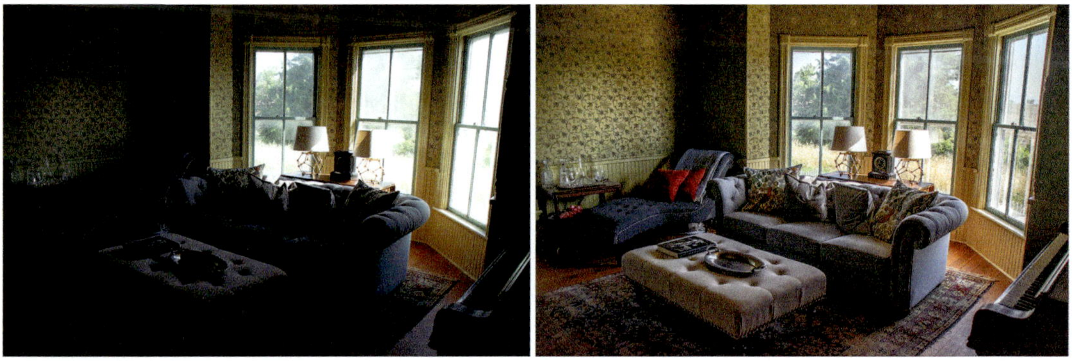

The extreme luminance difference between the interior and exterior are captured as muddy blacks and blown-out highlights by the camera (left). But our mind balances the exposure, lifting the shadows and toning down the highlights.

The camera merely records light that enters the lens and hits the sensor, and produces an objective snapshot of the environment. Whether that snapshot conveys anything meaningful is completely up to the photographer – indeed, the art of photography and cinematography is explained, debated, and taught in thousands of books, courses, and tutorials, because it involves so much more than just point and click. But there are no books about the "art of seeing" – maybe because we are all unrecognized masters of this art. While our eyes record light just like the camera, our brain constantly interprets and enhances that recording. We cannot separate the two processes, and they happen simultaneously (and mostly subconsciously). In the first example, you experience the beauty and grandeur of the landscape because you have knowledge and understanding of the scale and the depth of the view. You know that the mountains are tall, and the valleys are deep. You can see through the haze and make out the tiny houses of the distant villages. You are taking in the grand view while at the same time zooming in (mentally, not physically) on the smallest detail. The subjective snapshot captured by our eyes is simultaneously augmented by our memories, knowledge, experience and emotions, and this augmentation becomes an inseparable part of the viewing process.

In the second example, as you observe the apartment, your mind instantly compensates for the extremities of bright light and darkness. It brings out detail in the shadows and at the same time tones down the highlights, which allows you to experience both the interior and the exterior with an optimal exposure. This constant real-time exposure balancing act is simply not possible with a camera and can only be achieved as a post process. However (and this is the important part) when we look at photographs or footage, we seem to be more at ease with the typical single-exposure look of photography, then with any attempt to emulate the way we see. Take, for example, HDR (High Dynamic Range) photography: the process of merging several different exposures generates an image that mimics our vision quite closely. Yet HDR photography never quite caught up as a popular mainstream style. Most likely because the typical HDR look, with its highly detailed sky, lifted shadows, and toned highlights, has an eerily strange feel to it. Even though it simulates our vision, it feels "unreal" because it lacks the deep shadows or blown-out highlights we are so used to see in photography. Again, just as with tilt-shift effects, the attempts to emulate human vision in photography have not been very successful, which supports the notion that photoreal digital content should look like a photo (That said, HDR images have an essential role as a CG lighting tool which is unrelated to HDR as a photography genre. This role is discussed in Chapter 11).

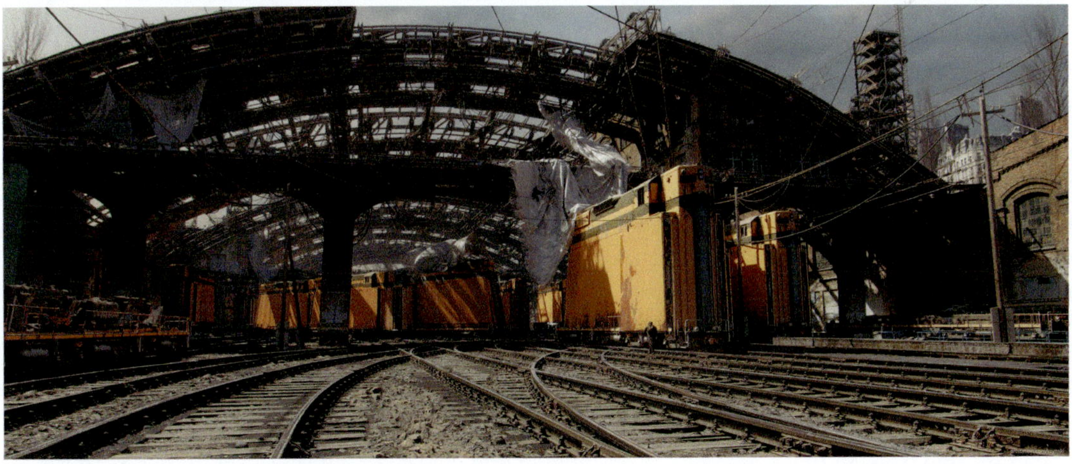

Le Ciapog Station by Piotr Krynski.

Created and rendered in Blender. © Piotr Krynski.

The Uncanny Valley

The term uncanny valley was originally coined in the 1970's in relation to humanoid robots, but was later adopted by the media in discussions about CG characters in movies. The "valley" denotes a dip in positive emotional reaction to a human-like character, particularly when the appearance and behavior of that character feels almost (but not quite) real. We seem to react very positively to CG characters when they are distant enough from real humans (for example, the main characters in Toy Story). The uncanny valley, that eerie, spooky sensation, is our reaction to characters that look and behave very much like real humans, yet something, just "something", doesn't quite feel human in them. Movies like *Final Fantasy: The Spirits Within* (2001) or Robert Zemeckis' 2004 *The Polar Express* drew considerable criticism regarding this subject. Both represented a milestone in motion capture techniques, and the CG humans in those movies displayed a higher level of realism than anything seen before. But precisely because of that, they drew negative reactions. In his review of *Final Fantasy*, Peter Travers noted "a coldness in the eyes, a mechanical quality in the movements",[1] while Marc Savlov of the Austin Chronicle wrote that "Watching the humans in *The Polar Express* is like watching people through a smeary car windscreen – the realism is there, but there's something a tad skewed and surreal about the whole affair".[2] The term uncanny valley was used numerous times since to describe similar reactions to CG characters, even as hyper-real CG characters continued to evolve and improve.

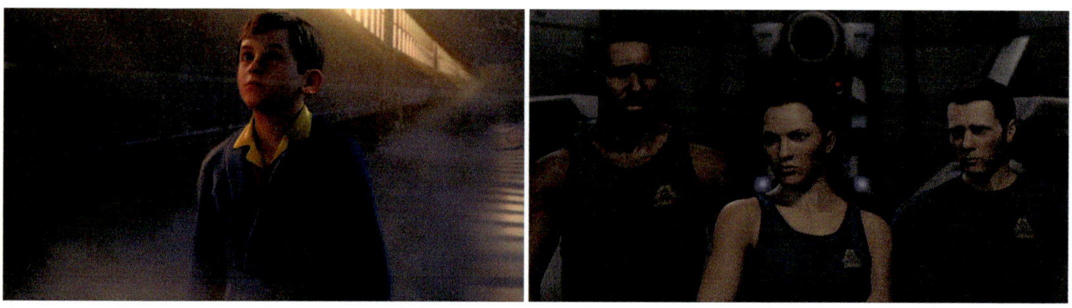

The Polar Express (left) and *Final Fantasy the Spirit Within* (right) generated much debate on the subject of the uncanny valley.

This is not an animation book, neither does it focus specifically on CG characters. But I think that the uncanny valley can be observed from a much wider perspective, one that applies to any element, animated or static, as well as to entire environments. Although inanimate objects do not draw the same emotional response from the viewer as human characters, a milder version of that uncanny sensation does happen whenever we view digital visuals that are almost photoreal, but still lack that "something" to look truly believable. For example, when watching a movie where the visual effects are impressive, yet still feel "off", or when viewing an architectural render that is impressively realistic, except that something, just "something" in it feels too sterile perhaps? Or too perfect? When we watch CG or matte-painted environments that are deliberately not photoreal, we are essentially in the same comfort zone as traditional 2D animation, at a safe distance from the uncanny zone. It is not surprising that successful CG animated features (like most Pixar movies, for example) intentionally keep their environments (detailed and beautifully lit as they are) within a non-photoreal region. But any attempt at creating photoreal imagery, whether it is for VFX, games, or visualizations, carries with it the inherent risk of the uncanny valley – that little "something" that kills the illusion of realism. That something is often quite elusive. As a VFX supervisor I find myself frequently chasing that intangible cause as I review a shot and try to figure out why exactly it does not feel quite real. The key here is to be aware of a wide range of factors that can affect photorealism. This book examines these factors and provides guidance that can help bridge the uncanny valley.

As generative AI becomes an essential part of practically every aspect of digital visual creation, we are facing a new type of uncanny valley – on one hand, a phenomenal ability to mimic photography and generate quasi-unlimited amount of detail, and on the other hand, a tendency to make errors in perspective, shadows, anatomy, that are not characteristic to CG or 2D work. In Chapter 19 we will look at various ways to guide and control generative AI and avoid this new type of uncanny valley.

The Detail Conundrum

Nothing embodies the challenge of creating believable imagery more than the huge gap between the real world's level of detail and the constraints of computer graphics. Imagine one of those "infinite" camera moves that starts on a single blade of grass and gradually pulls back to reveal the clearing, the forest, the hill, and finally the entire mountain range. In the real world, every single element that makes up the environment continues to exist regardless of whether it is distinguishable in the frame or not. But from a practical point of view, only some of the detail is necessary at every stage of the camera move. When the camera frames a single blade of grass, we probably only need to create the neighboring plants, and maybe a rock or two in the back. There is certainly no reason to create the forest, cliffs, or the entire hill. Conversely, when our camera is pulled back all the way and frames the entire mountain range, it does not make any sense to create every blade of grass, every rock, and every tree. Although we are looking at trillions of grass blades and leaves and millions of trees and rocks, we cannot, from that distance, see any of them individually.

We can thus place an imaginary threshold (let's call it "the bare necessity line") that defines the amount of crucial detail at various distances from the subject. This variable level of detail is approached differently in VFX, visualization, and games. In visual effects, the position and movement of the camera is usually already known, which makes it easier to define where the bare necessity line lies, and create exactly the amount of detail that's needed for a specific camera in a specific shot. However, when elements or environments are created as an asset that is intended to be used in multiple shots (and be viewed from different distances), enough detail must be created to cover all scenarios. That is normally the case in games, where the camera is roaming freely, and real-time performance is paramount. The bare necessity line in games is almost always handled through a system of **dynamic level of detail** (**LOD**). In such a system, elements have multiple versions with varied amount of detail, and the game engine automatically switches between the versions depending on the distance from the camera.

It seems then, that the detail conundrum can be solved quite successfully with some careful consideration of camera distance and framing, or an effective LOD system. But this is only a partial solution. In the context of photoreal digital imagery, the true challenge is not creating the right amount of detail for a certain camera, it is accounting for the detail that is left out. Let me explain: in the real world, all detail is always there, every little bit of it. While the grass, rocks and leaves in our camera pullback example may not be visible to the camera when viewed from half a mile away, they still contribute in a million tiny different ways to the perceivable look of that hill. The incredible visual complexity of the world around us is the accumulation of all the massive amount of detail at any scale level. When we look at a brick wall from a hundred feet away, we cannot possibly see the detail on every single brick. But the cumulative result of all these minuscule features forms a larger assembly that affects the overall appearance of the brick wall and gives it that intangible photoreal richness that is so difficult to replicate digitally.

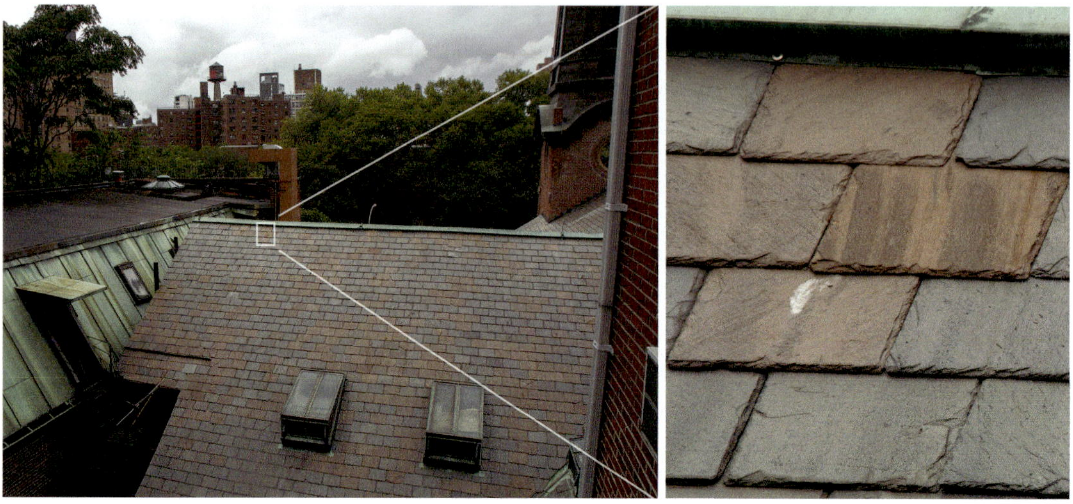

The close-up detail (right) cannot be distinguished when zoomed out (left). But it is still there – giving this shingled roof the elusive richness of real-world textures.

Emulating that richness without committing to generating every tiny feature on every scale level is a continuous challenge. Moreover, it is a challenge that runs through all of the different crafts: modeling, texturing, shading, lighting, matte painting, and compositing. At every step of the way, the necessity of adding enough detail is balanced against the limitations of hardware, schedule, and budget. And at every step of the way, there are methods, shortcuts, and tricks to keep that balance at a good place while avoiding an artificially digital look. The concept of detail is thus at the core of photorealism and runs as a thread through the entire length of this book.

The Role of Imperfections

Nature is chaotic and infinitely varied. There are rarely any straight lines, no two leaves look exactly the same, no two rocks are identical, every tree is unique, and every cloud is a different shape. Surfaces are twisted, broken, muddy, covered with moss, covered with dirt, stained with humidity, or decay. Man-made objects are of course much more structured and organized: straight lines, perfect angles, polished surfaces, uniform colors. But even precisely manufactured objects are not perfect: weather, dust, gravity, human interaction, and a whole range of factors allow at least some of nature's chaos to invade our creations. Computer graphics is a virtual medium born out of pure mathematics and is inherently perfect. In the world of CG, nothing is easier than creating a perfectly accurate sphere with a perfectly shiny, spotless surface. And of course, nothing feels more fake. Photorealism therefore requires a persistent "battle" against the default flawlessness of CG. This is done at any level – from breaking symmetry in modeling through adding grime and dirt to textures and all the way to introducing lens aberrations, flares, and grain at the compositing stage.

Case Study: Detail and Imperfections in Man-Made Objects

Man-made objects are deceptively difficult to portray realistically in 3D. Compared to the complex, irregular, and chaotic appearance of natural elements, the simple geometry, smooth surfaces and consistent materials may, initially, seem easier to emulate. But this is in fact the essence of the challenge – lack of detail is CG's nemesis, and detail in man-made objects is usually sparse and subtle. It's a truly fine balance – not enough detail, and the result feels digital and artificial, too much, and it easily loses believability.

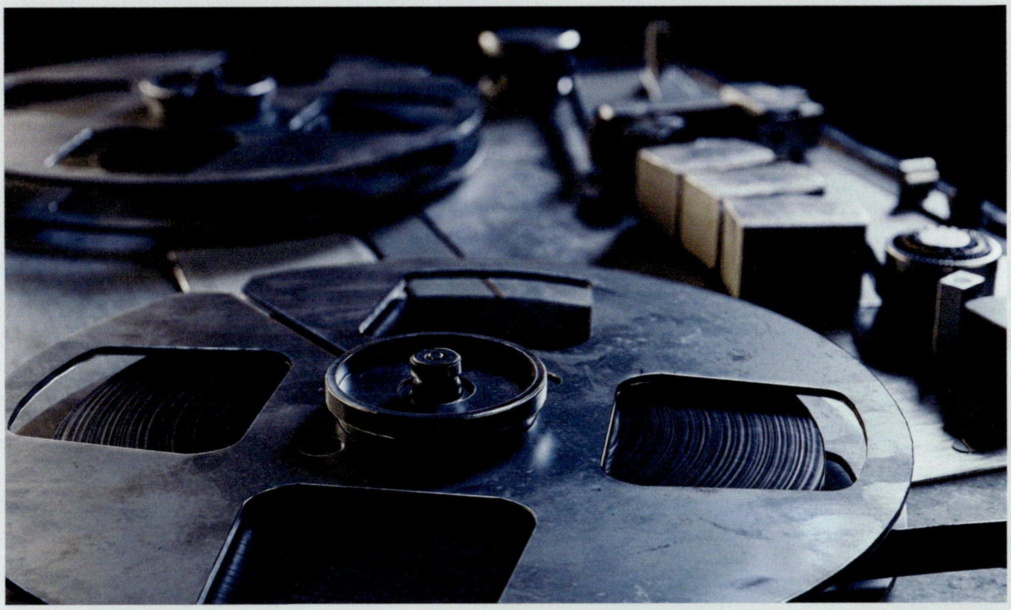

Nagra Tape Recorder by Angelina Pronina. Maya, ZBrush, RizomUV, Substance Painter, Marmoset Toolbag. © Angelina Pronina.

This rendition of the venerable Nagra tape recorder by environment and prop artist Angelina Pronina is a good example of keeping that fine balance. Angelina had very little to work with – the shapes are mostly simple cubes and cylinders, the material just unpainted metal. Yet this render works because of careful attention to a few essential aspects of modeling, lighting, texturing, and lens emulation (each of these aspects will be discussed in depth in the following chapters):

- Subtle bevels along the edges to catch the light and generate rim highlights, a crucial modeling feature in man-made objects (specular reflections discussed in Chapter 5, modeling for lighting in Chapter 14).
- Physically accurate metallic shaders that react to light with proper reflective characteristics without feeling overly shiny (shading discussed in Chapter 12).
- Curated textures that drive various shading properties and suggest subtle surface inconsistencies like scratches and dust (texturing covered in Chapter 13).
- A simple lighting setup that couples a cooler source with a warmer fill (real-world and CG lighting discussed in Chapters 6 and 11, respectively).
- Carefully adjusted depth of field that emulates real-world photography and lens

characteristic (DOF discussions in Chapters 10 and 18).

When creating photoreal renditions of man-made objects, close attention to detail in modeling and lighting and finely balanced textures and shaders are crucial for escaping the CG look and making even the simplest elements feel rich and realistic.

The Reality of the Unreal

While architectural and product visualizations are generally rooted within the confines of reality, VFX and games often depict imaginary worlds: alien planets, futuristic cities, fantastic creatures, or environments that do not abide by earthly laws. What role does photorealism serve in the context of fantasy and science fiction? Is it still an important factor, considering that the visuals often depict things we have never seen or experienced? Photorealism is achieved by studying and mimicking the world around us – light, surfaces, atmosphere . . . but what if these behave very differently in the imaginary environment we want to portray? Can we still maintain believability while creating otherworldly environments? How can we make the unreal look real?

With all the digital tools we have at our disposal, we can certainly break every rule. We can create a world where physical laws are upside down, a world that is so different from anything we've ever seen, that it makes no sense visually. But it is hard to draw the viewer into such a world. If you examine prominent sci-fi and fantasy movies, you'll notice that the most visually believable ones do not break all the rules, but instead maintain certain earthly aspects. Take, for example, *Avatar*'s iconic Hallelujah Mountains: while their most striking aspect (giant forested cliffs floating in mid-air) is certainly unreal, specific features like the jungle vegetation, rock surfaces, lighting, clouds, and atmospheric depth all feel familiar and recognizable. This careful balance between real and unreal, combined with meticulous attention to detail, places an otherwise impossible environment deep within the sphere of believability. Handled differently, the concept of floating mountains could have easily succumbed to overly-fantastic visuals or poorly-done VFX, causing the viewers to distance themselves from the environment. This kind of thoughtful balance between unreal and photoreal elements contributes to the success of many other sci-fi/fantasy visuals, for example: the folding Paris streets in *Inception*, the motionless floating alien spaceship in *Arrival*, the lobster-roach aliens in *District 9*, or the environments in games like *Horizon Zero Dawn* and *Crysis*.

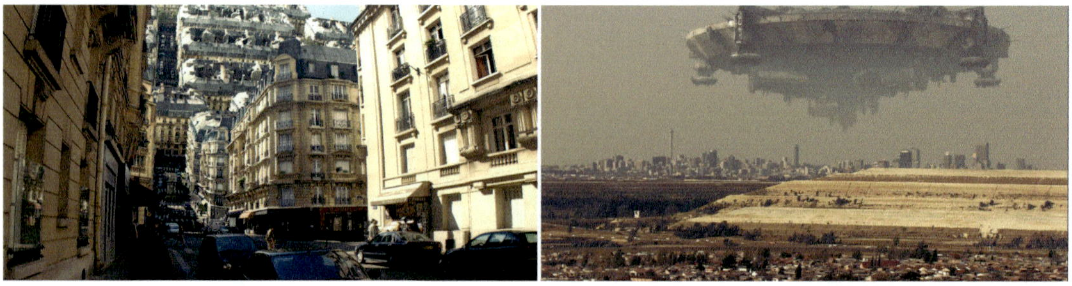

Inception's folding Paris streets (top left) and *District 9*'s alien spaceship (top right) are great examples of a successful visual balance between fantasy and familiar realism.

Image Quality and Photorealism

We live in a world where visual quality is highly venerated. Camera manufacturers tout their products' megapixels, phones are judged by their screen quality and resolution, and ultra-HD TV is the new norm. While some of this "bigger is better" hype is more marketing than substance (more megapixels will not necessarily give you better photographs), one cannot ignore the impressive technological advances in digital imaging and display, and the striking improvements in quality over the last decade. It is important to note, however, that photorealism has little to do with the quality of digital capture and display. The essence of photorealism has not really changed since the first photographs of the 19th century. A grainy, black and white, low-quality 300x200 pixels photo can still look more photoreal than a shiny 8K render. Resolution, number of pixels, bit depth, dynamic range – all these features are critical for creating, capturing, and displaying high-quality, high-definition imagery, but they do not necessarily augment the photoreal quality of the digital content.

This is a notion that I often find hard to convey to students. Much attention is devoted nowadays to technical specs, yet the essential qualities required for achieving photorealism are not necessarily tied to such specs. Those qualities include the subtle interaction of light and surface, the richness of textures, the sense of depth, the characteristics of optical lenses, and many other aspects that are discussed here. While factors like bit depth and high dynamic range do play an important role in the actual process of creating photoreal imagery, successful visual recreation of reality is independent of specs like resolution and color depth, and the results should look as convincing at the lowest quality as they would at the highest.

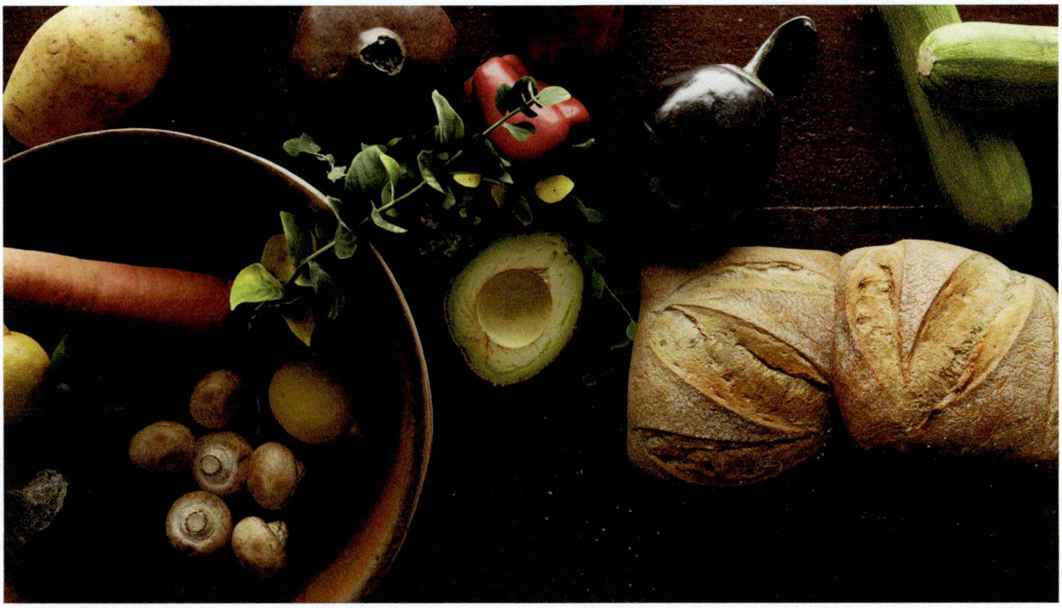

Image by Waleed Nabil. Rendered in Unreal Engine.

2D and 3D Workflows

In contemporary digital content creation, the distinction between 2D and 3D is often blurry. Visual effects in particular are driven by a pragmatic "use whatever method works best" approach. 3D and 2D workflows are combined in matte painting and compositing (in fact, 2D compositing applications like Nuke or After Effects offer quite an extensive array of 3D tools). Architectural visualizations also often require a combination of 3D and 2D, and while games are inherently 3D, matte painting and compositing techniques are increasingly used to augment their cinematic quality.

From the photorealism perspective, 2D and 3D seem, initially, like very different workflows. Ask a 3D generalist and a matte painter to create an image of a castle on a hill, for example. The first will model and texture the structure and terrain in 3D and then illuminate the entire scene to produce the final render. The matte painter will gather pieces of images from various sources and meticulously combine them into a coherent environment using coloring, masking, and painting. Photorealism is achieved in 3D through the complex mathematical process of simulating the interaction of light and surfaces, and in 2D by manipulating real photographic material.

Yet 2D and 3D workflows also converge at multiple points (for example, depth of field and motion blur can be part of the 3D render, or applied separately in 2D through depth and vector passes) and complement each other (2D techniques are used extensively for texture map creation, while 3D lighting and surface properties are often re-adjusted in compositing). So, while the separation of 3D and 2D in this book is indeed somewhat artificial, it allows the reader to examine photorealism from two perspectives: that of the CG artist in Part 3 and that of the compositor/matte painter in Part 4.

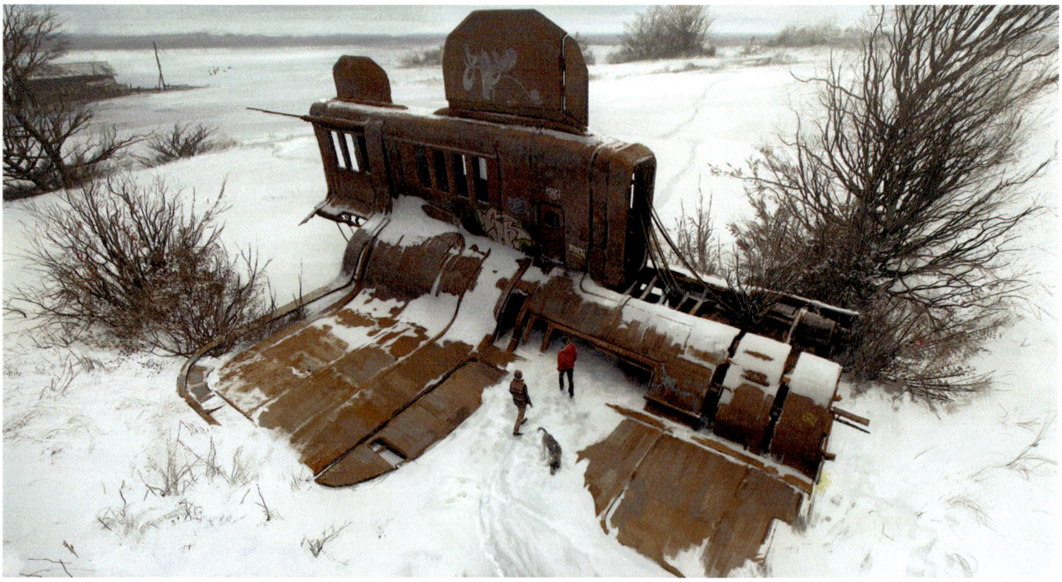

Winter Walk by Sergey Vasnev is a good example of combining CG elements with 2D matte painting techniques using Blender and Photoshop.

Generative AI models create images and videos in a purely 2D process. They are trained on 2D material and generate images and videos by iteratively rearranging random pixels into coherent imagery. There are no volumes, surfaces, or light rays to trace. Yet here too, 2D and 3D converge: 3D models and scenes are used to control and guide the generative AI process, and specialized algorithms can extract 3D information such as depth, normal, or human poses from 2D images. In addition, AI models are becoming better at creating 3D models from 2D material and can even generate UV maps and textures. As will be discussed in Chapter 19, digital artists can achieve more with generative AI by combining it with 2D and 3D workflows.

Notes

1 Peter Travers, 'Final Fantasy', *Rolling Stone*, July 06, 2001.
2 Marc Savlov, 'The Polar Express', *The Austin Chronicle*, November 12, 2004.

Chapter 2

Photorealism in Digital Media

The principles of photorealism are rooted in the physical interaction of light and matter, and thus apply equally to the various fields of visual digital media. However, each field represents different aesthetics, expectations, challenges, and methodologies. In this chapter we will look at the role of photorealism and how it is approached in three main categories: visual effects, games, and visualization.

Visual Effects

I doubt that there is a VFX artist out there who likes to be told that their work "doesn't look real" or "looks too CG". Creating believable "real" visual elements is at the heart of VFX work because those elements are combined with live-action footage. When visual effects do not share the same photoreal qualities of the footage, they damage the believability of the film and the flow of its storytelling. Even when a specific shot is completely done in CG and involves no integration with live action, it still sits in the cut among other live-action shots and thus must maintain a comparable photoreal quality.

The importance of photorealism and integration in VFX dictates certain approaches that put greater emphasis on plausibility than on aesthetics or style. A pretty sky may end up looking rather ugly when overexposed and blown out, for example, but this is often necessary to integrate the sky with the footage. A beautifully detailed matte painting may end up completely out of focus if the camera is focused on the foreground in a practical shot. Such compromises on beauty for the sake of realism are arguably more common in VFX than in games or visualization. Yet even within the world of VFX, approaches vary. In advertisement, for example, there's greater emphasis on presenting the product or concept at its best over preserving photorealism, and some movies or TV shows may intentionally opt for a non-photoreal look as a stylistic choice. Yet overall, such artistic deviations are more an exception then than the norm, and VFX artists generally strive for the highest level of photorealism.

Since VFX production happens mostly in post, after the film has been already shot and cut (or at least roughly cut), the work is usually performed for specific camera angles, moves, and framings. This allows the VFX team to focus the resources only on areas that are visible to the lens and ignore everything that is outside the frame. Matte paintings are rarely done out of context of a specific shot or sequence, as is the case with most 2D compositing shots. Some assets serve

DOI: 10.4324/9781003590293-4

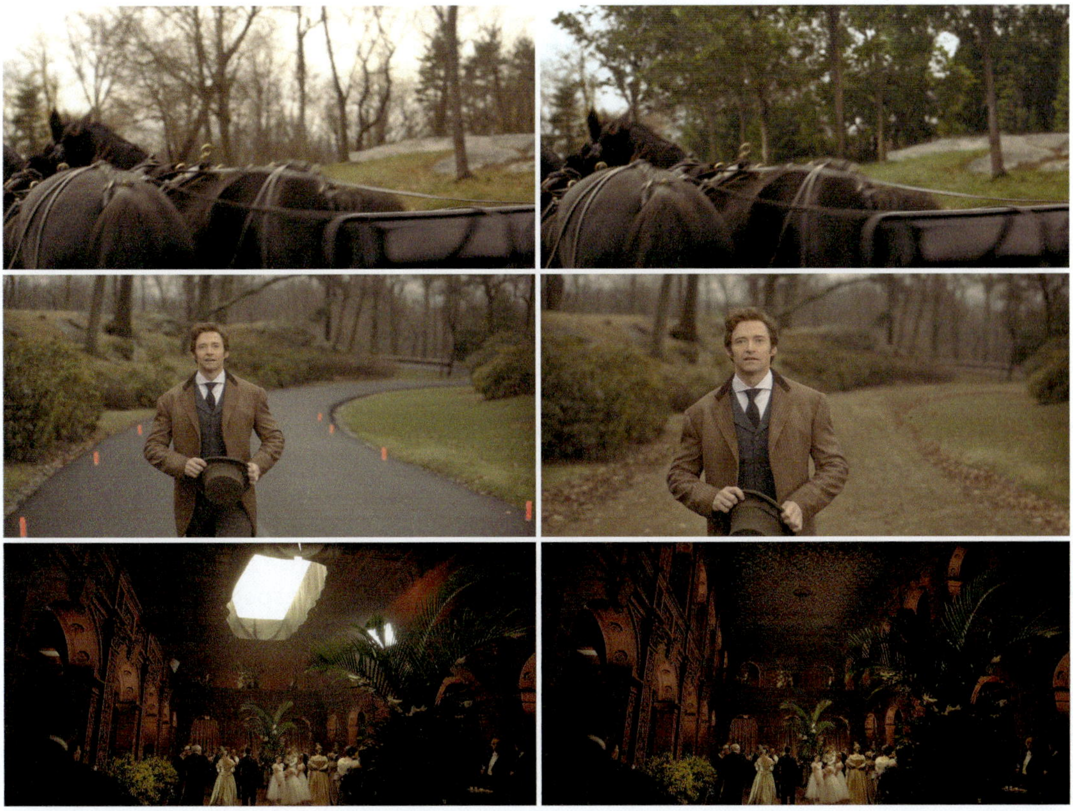

Changing the season, fixing the road, and adding a ceiling in *The Greatest Showman*. Even such seemingly simple VFX tasks require great attention to detail and integration, and a thorough photoreal approach.

© Twentieth Century Fox. Visual effects by Brainstorm Digital.

multiple shots (like game environments and characters) and must look good from multiple angles, but the final lighting and compositing tweaks only take place when the asset is integrated with the footage. That last 10% of the work has a critical role in establishing the level of photorealism and believability in VFX.

That said, more and more film and TV productions are using LED volume stages. There are obvious advantages – no ugly green screens, no need for rotoscoping or compositing, true in-camera integration, physically accurate light spill and light wrap. But this process drastically changes the VFX pipeline. Not only does it require that all material (2D and 3D) be completed in pre-production – the advantage of creating VFX only for the camera is gone, and environments generally need to encompass 180 or even 360 degrees, just like in games.

While this has increased the use of real-time rendering in the VFX industry, the core VFX work still relies on non-real time (**offline rendering**), which allows for incredibly high CG complexity and rendering quality. Those massive VFX scenes we see in high-budget films require equally massive simulation and rendering times, not to mention large teams of artists performing a variety of tasks, from modeling and texturing to lighting, rotoscoping, and compositing. Yet such big resources are not necessarily a guarantee for photoreal results. Visual effects for films and TV are not done in limbo – there are strict (and not always realistic) schedule and budget constraints, while various factors such as the aesthetic taste of the VFX supervisor, the director's vision, or the studio's agenda can also affect the output. It would be naïve to assume that every film with a big VFX budget is the embodiment of photorealism, or that small-budget movies or TV programs cannot reach comparable levels of realism. Examples of stunningly photoreal VFX can be found throughout the entire budget range, and while it certainly helps to have larger resources, what really matters at the end is the artist's understanding of light and color and their keen eye for detail. The concepts, principles, and techniques in this book should serve any VFX artist, whether they are working on a tentpole blockbuster or a student project.

Games

One can go back 30 years and still find some incredibly convincing visual effects (*Jurassic Park*, for example). But when it comes to games, the 1990's feel like the Stone Age. Compare the 1991 adventure game *Indiana Jones and the Last Crusade*, to a contemporary equivalent such as *Uncharted 4* (2016): the difference between the two-dimensional pixelated graphics of the first and the rich, cinematic visuals of the second is so immense that it is hard to believe these two games are only 25 years apart. Indeed, no other aspect of visual digital content has seen such a remarkable progress as real time video games graphics. In the four decades since the Atari console, hardware and software developments have completely transformed the visual experience of video games. As technology grew ever more powerful, the once unachievable dream of creating immersive, photoreal games became feasible. And game engines continue to evolve faster than ever: at the time of writing this book, Epic announced Unreal 5, a revolutionary update that includes real-time global illumination and unlimited poly count.

Game artists today are often expected to design, model, texture and light environments, props, and characters that are as detailed and believable as their counterparts in cinematic visual effects. To be clear, photorealism is certainly not a prerequisite in games. There are many popular game genres that proudly and happily remain within the realm of stylized or cartoonish graphics (for example, most Nintendo games or the majority of indie games). But there is certainly a growing number of titles that strive to offer gamers the most realistic experience possible, and their level of photorealism reaches higher grounds every year.

Unlike VFX, games graphics are rarely combined with live-action footage and are predominantly computer generated. This alleviates the challenges of integration that characterize VFX and allows for more stylistic freedom. But the road to photorealism in video games requires overcoming some tough obstacles: first, most 3D game environments cannot be designed for a specific camera framing or movement. All playable environments must accommodate a freely moving camera and look believable from every angle. Common VFX methodologies like 2D or 2.5D projections can only be used outside the playable area (distant backgrounds or skies, for example). Environments in open-world games, in particular, must be very large, and maintain quality and level of detail throughout their extent. In addition, everything needs to perform smoothly in real time through the game engine. This requires constant attention to polygon count and texture sizes, a robust LOD system, and adherence to hardware and software rules and restrictions.

These key challenges show why VFX is still ahead of video games in terms of photorealism. Working for specific cameras and without the restrictions of real-time rendering (and adherence to a specific engine) gives VFX a distinct advantage. But as is evident by new games and all the latest developments, things are changing very fast. The restrictions are diminishing while hardware capabilities increase, and game engines now offer truly cinematic features like depth of field, realistic lens flares, or discreet color grading. It will not take much longer before the gap between VFX and games is bridged. Already, game engines and real-time technologies are used on movie sets and in VFX, and games look more and more like movies. To that end, all of the principles of photorealism and photoreal CG techniques discussed here, as well as many key 2D aspects, apply to contemporary games creation as much as they apply to VFX.

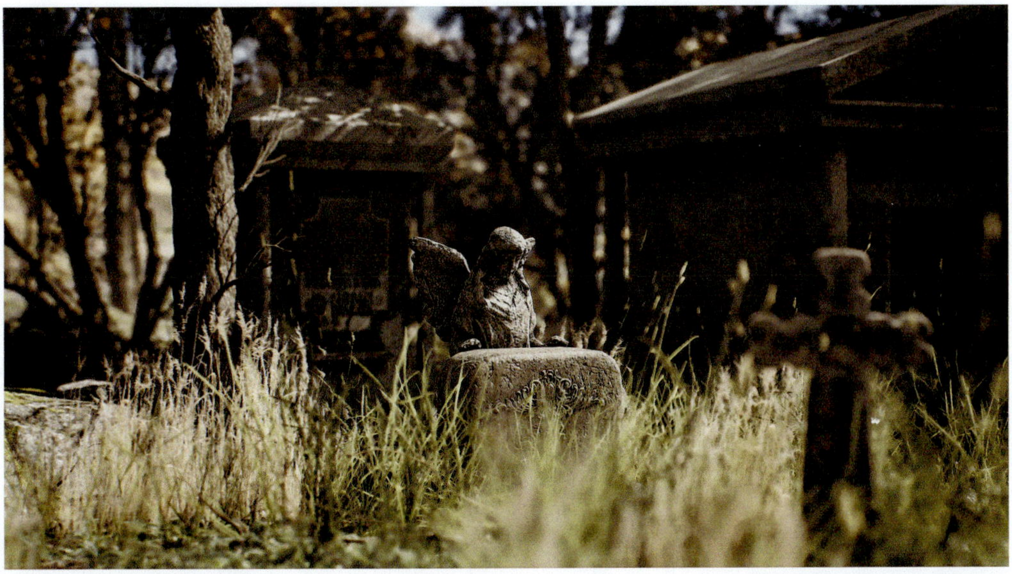

A frame from a game-inspired real-time animation by Pasquale Scionti, created and rendered in Unreal Engine. © Pasquale Scionti.

Visualization

The visualization field is vast and varied. Not every type of visualization requires a photoreal look, and in some cases such a look may be unnecessary, or even counterproductive. In medical, scientific, or forensic visualizations, for example, precise graphical representation of concepts, procedures, and events often takes precedence over photorealism. A scientist examining a visualization of a certain type of protein is interested in the molecular structure and is quite content with simple colored shaders. A forensic visualization needs to precisely depict an event, so effort is invested in animation and timing rather than photoreal lighting or peripheral detail. But photorealism is an essential ingredient whenever the goal of a visualization is to present the subject in the most faithful and true-to-life way possible. This is often the case in architectural and product visualizations.

Architectural Visualization

The need to visually present a yet-unbuilt building has of course existed long before computer graphics. The art of architectural illustrations goes all the way back to ancient Greece and has reached the highest level of sophistication through the industrial revolution and into the 20th

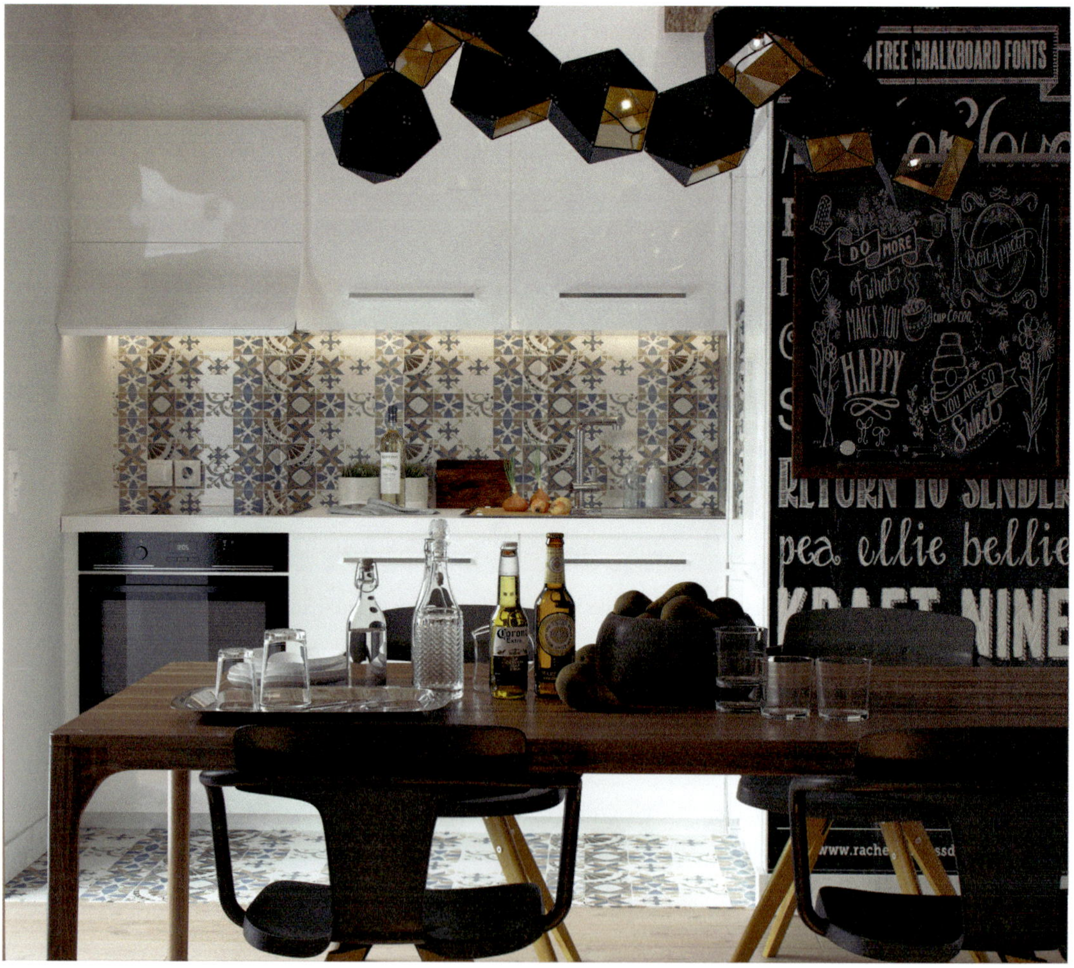

Les Faisses De Cantecor by Julian Sadokha. 3ds Max, Corona Render.

©Julian Sadokha, @Planet Studio.

century. Unlike technical working drawings (floor and elevation plans, sections, schedules) which are intended for contractors, illustrations strive to convey the aesthetics and visual impact of the architecture rather than its technical details. Artists creating such illustrations need to not only master the technical aspects of this craft, but also the artistic skills required to present the subject in a dynamic and enticing way. Indeed, many such illustrations are considered true works of art, and go well beyond their practical purpose.

The art of architectural illustrations changed drastically when 3D computer graphics entered stage. As the process of conceiving, planning and drafting moved almost entirely to CAD software, architects discovered the potential of 3D models, detailed textures and realistic lighting. CG renders became the standard in architectural visualizations because they effectively combine the traditional mediums of illustration and miniature models by allowing viewing and rendering from any position and angle. Virtual CG architecture can be used for walk-throughs or fly-overs, and offers the invaluable flexibility of changing interior lighting and time of day, or swapping textures and colors for quick design prototyping.

In VFX and games, the image is the final product. But visualization is only a waypoint to the final product, it is the means and not the goal. In architectural visualization, the imagery is

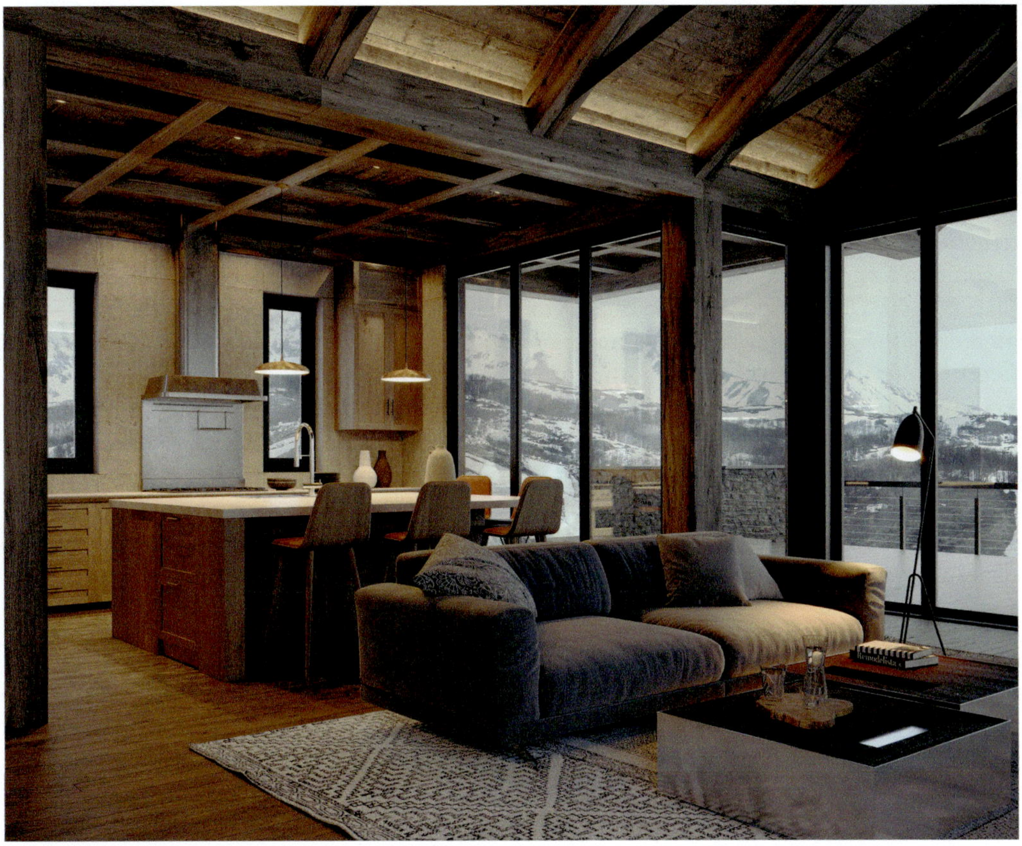

Aspen House by Julian Sadokha. 3ds Max, Corona Render.

© Julian Sadokha, ©Suburbia Studio, ©SWAG Architecture.

rendered long before the final product has materialized. This leads to a certain inevitable tension between photorealism and abstraction: developers and clients want to see photoreal renders, to get the most accurate representation of the yet-to-be-built building or interior. Architects prefer more abstract renders – after all, what if the final product fails the image that predicted it in some way? Photorealism could in fact be a liability. There is yet a different tension that lies at the boundary between art and commerce: in a craft so steeped with tradition as architecture, it is only natural that CG rendering stirred resistance at its onset and is still a subject of some controversy. Opponents of the medium point to its computerized precision and lack of human touch, its soulless objectivity (as opposed to the subjective flow of an artist's drawing), or (as is sometimes the case) the drab and unexciting appearance of computer-generated images.

Indeed, the world of architectural visualizations has its share of generic and uninspiring renders. But when done well, such renders become a truly powerful medium for presenting architecture as an immersive and moving experience. This is where photorealism comes into play. Unlike traditional illustrations, where a few strong lines are often more effective than fine detail, CG architectural renders do not fare well when they lack convincing photoreal quality and detail. It is not surprising that many of the early forays into physically-based rendering (radiosity, photon mapping, IES lights) were adopted first by ArchViz artists.

The extent of photorealism in a specific architectural project depends on the intended purpose of the visualization and the style of the architect and the artist. For example, many ArchViz artists go to great lengths in creating the most detailed textures and lighting, but completely disregard optical lens properties like depth of field, producing images that are unrealistically sharp across the entire depth. This is often done intentionally to allow viewing the entire structure sharp and clear, even at the price of some loss of photorealism. Another example: consider the ubiquitous use of libraries of generic 2D humans. From my own perspective as a VFX supervisor, I am sometimes perplexed by the apparent dissonance that rises from populating a meticulously built and rendered environment with hard-edged 2D cutouts that clearly don't match the lighting (and sometimes not even the perspective). In a movie, such a dissonance would certainly be shrill, but in the world and language of architecture, the added human cutouts are viewed more as an embellishment rather than co-equal elements. As such, their apparent lack of realism does not necessarily hamper the impact of the main subject. Here is yet another example: in both VFX and games, great care is taken to add the right amount of dirt, grime, and imperfections to avoid a synthetic, CG look. But dirt and imperfections are not necessarily the architect's highest priorities. The common aesthetic and stylistic approach to ArchViz naturally emphasizes a clean, fresh look, and what may feel overly sterile form a VFX and games points of view is the norm in ArchViz. To compensate for the lack of grime and imperfections, artists add highly detailed auxiliary elements such as furniture, upholstery, carpets, curtains, wall decorations, or potted plants.

Contemporary architectural visualizations range from basic renders and quasi-illustrated presentations to elaborate clips that feel more like Hollywood-produced movies. These go beyond the standard CG rendering aspects of ArchViz and usually involve animation and compositing techniques that are more characteristic to visual effects. In addition, game engines like Unreal and VR technologies are increasingly being used to present architectural projects within an interactive, real-time environment. Such visualizations share the same challenges and limitations as games and VR, especially when attempting to maintain a high level of photorealism.

Product Visualization

People are often astonished when told that 75% of Ikea's catalog is completely CG. Why would Ikea choose the CG path? After all, they are not attempting to show a building that does not yet exist. They already have all those pieces of furniture in their warehouse, so wouldn't it be easier to just photograph them? The reality is that the costs associated with arranging, lighting, and photographing product sets like Ikea's can be much higher than creating them in CG. Add to that the advantages of working in a virtual environment – the ease of changing the lighting, colors, and textures, or moving the camera or objects around – and the CG option becomes very lucrative. Yet the success of such visualizations is completely dependent on photorealism. The product must look exactly as it does in real life, and the tiniest crack in believability shatters the illusion of photographic reality.

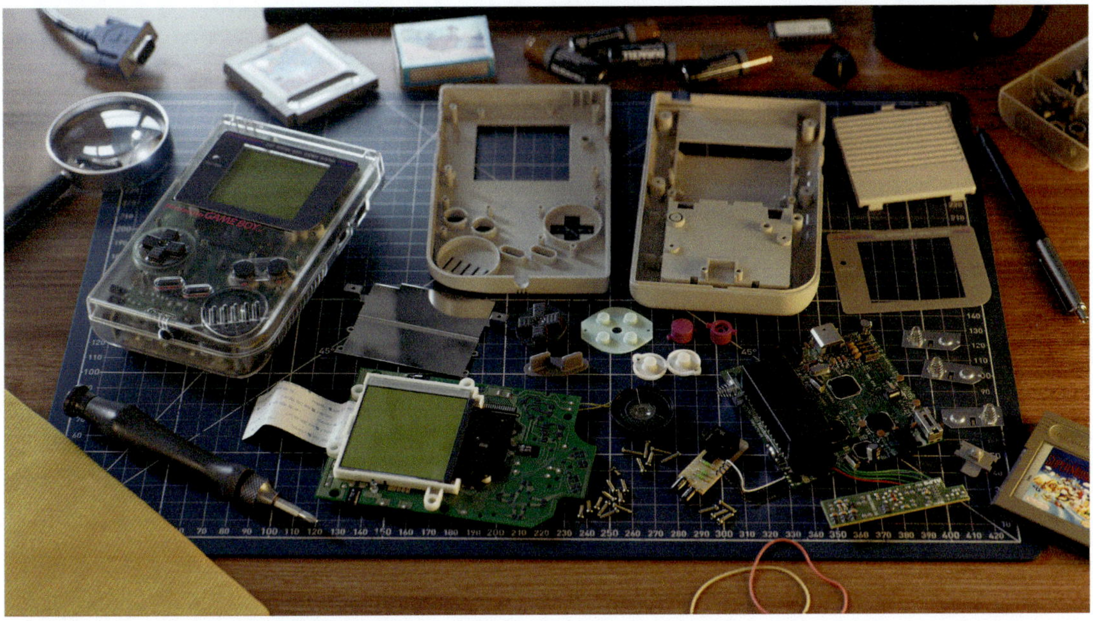

Game Boy by Raphael Rau. Cinema 4D, Octane Render.
© Raphael Rau.

Not all product visualizations rely on photorealism, but photoreal renders are now a vital part of product design and marketing in virtually every industry, from cars and furniture to phones and jewelry. They serve a wide range of media: printed and online catalogs, animated clips and interactive content, and immersive VR and AR presentations. In comparison to ArchViz, scale is generally smaller (not many products are as large as buildings or bridges), and renders contain less objects. On the other hand, the close-up presentation of smaller products necessitates a high level of accuracy. A cell phone may not have much exterior detail – a few buttons and sockets maybe, but this detail must be rendered with extreme precision to be convincing up close. So, while product visualizations may not require much of the kind of auxiliary detail (trees, plants, people) that is necessary to make ArchViz renders come to life, they present greater challenges in texturing and shading. Lighting can greatly affect the impact when focusing on a single small object, and lighting techniques in product visualizations often mimic studio photography.

Case Study: Nature in Architectural Renders

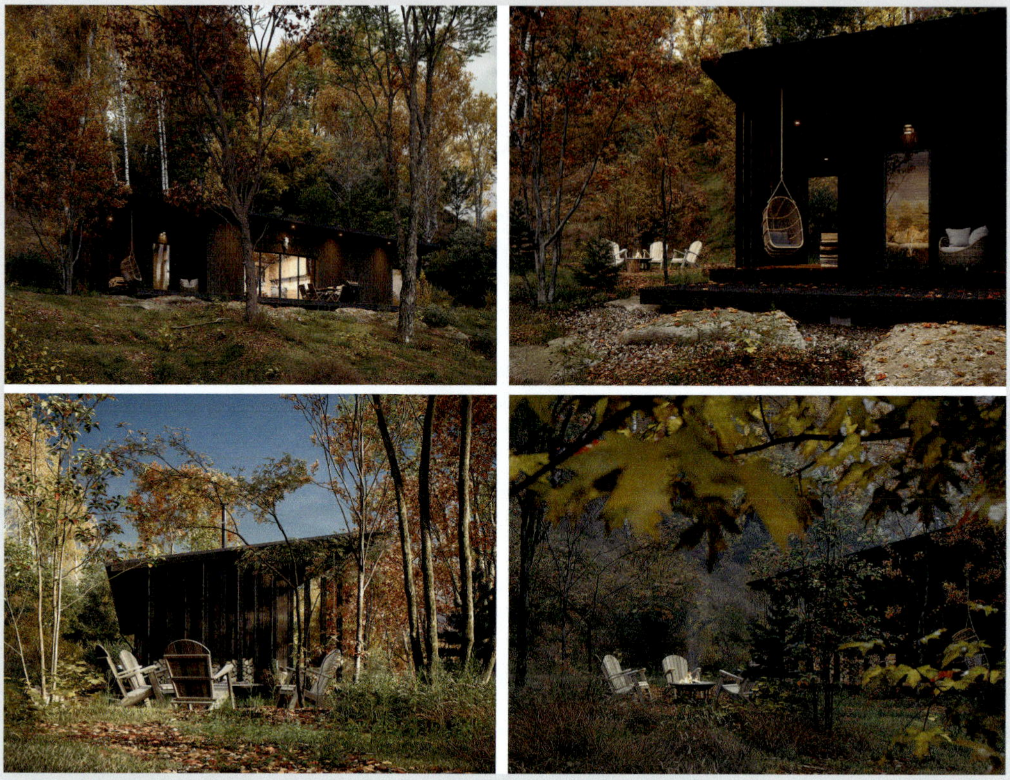

Wooden Black Forest Cabin by Mikhail Sizou.

Natural environments and elements like trees, shrubs, and rocks tend to play a secondary role in architectural visualization. As the modeling and surfacing effort is focused on the structure, natural elements are often treated as embellishments and portrayed with 2D cutouts or crude 3D models with minimal attention to photorealism.

Mikhail Sizou, a 3D artist with a background in traditional painting and design, goes the opposite way – he creates architectural visualizations where detailed, realistic natural environments play a leading role. "I love nature, plants, and landscapes" says Mikhail, "and my college art education in still life drawing and painting is certainly a contributing factor in my visual attraction to natural environments".

"The Wooden Black Forest Cabin project is my personal project, where I focused on the landscape, environment and natural mood", explains Mikahil.

> I have long dreamed of creating an autumn mood with a sense of tranquility, silence and solitude. I like to observe nature and natural lighting, and I often take pictures or draw sketches of moments that catch my eye. Observation and photography, to me, are the secret to realistically portraying nature. Technically, I use just a few ap-

plications in my process: 3D Studio Max and ItooSoft for modeling, Corona Render or FStorm for Rendering, and Photoshop for post-processing (if necessary).

Mikhail's lush and detailed environments show the importance of photoreal natural elements in boosting the impact of architectural renders. The surrounding environment forms a dialog with the structure, and plays an important role in conveying location, mood, and atmosphere.

Chapter 3

Color

Nearly every aspect of photorealism relies on color. Color defines not just lighting, but also surface properties, reflections, distance, depth, atmosphere, environment, and even camera and lens characteristics. Sometimes a simple color tweak is all that is needed to make things "look real". A little lift to the blacks, a slight reduction of red in the midtones, and suddenly everything falls into place. I cannot stress enough how beneficial it is for digital artists to master color, both in the inner, perceptual sense and in the ability to use digital tools to control it. But color is a vast subject. Navigating it can be daunting, the terminology is often confusing, and practices vary widely. Various software applications use different (and sometimes conflicting) names for color operations, and it is not always clear which tool should be used for a specific task. There are many books that cover color science, or the emotional and aesthetic attributes of color, and there are many ways to approach color theory and practices. In this chapter I chose to focus specifically on those aspects of color that I think are essential in the context of achieving photorealism. In particular, the way we perceive, observe, and interpret color, and how to think and work in additive color. I will simplify color terminology to the bare essentials, and then focus on the most relevant color operations.

Most humans can detect around ten million individual colors. We may not be champions of the animal kingdom (many reptilians, insects and birds can see more colors), but we still possess a highly developed color vision that allows us to separate what we see into very fine detail. Unlike comic and clip art, objects in the real world do not have a black outline that defines and separates them. Yet we can still effortlessly tell one object from another, or an object form the background, solely by color detection. Even if we take the hues out (for example, in a black and white photo), we can still make out the finest detail by sensing minute difference in gray shades. Our ability to visualize and interpret our surroundings based only on subtle color differences is truly astonishing – especially since we do it instinctively, without giving it a second thought, at every single moment of our life.

Yet there is a marked difference between our instinctive perception of color and how we think and talk about it. As toddlers, we learn to identify an object's dominant color: "What color is the ball? Red!". "What color is the truck? Green!". "What color is the cloud? White!". As we mature, we may talk about "that lavender tablecloth", or "this indigo shirt", or "a charcoal gray car", but despite our extended color vocabulary, we still by and large do the same as our toddler selves: we focus only on the dominant color of an object, and ignore the secondary colors. In our minds and in our discourse, we "paint back" what we see in uniform flat colors, like the simple illustrations in a children's book. Granted, this simplification is crucial to our everyday life. It is rather impractical to ask the car salesman about "this charcoal-gray car with orange-white highlights and slightly bluish shadowed areas and somewhat desaturated green and red reflections". But this is exactly

DOI: 10.4324/9781003590293-5

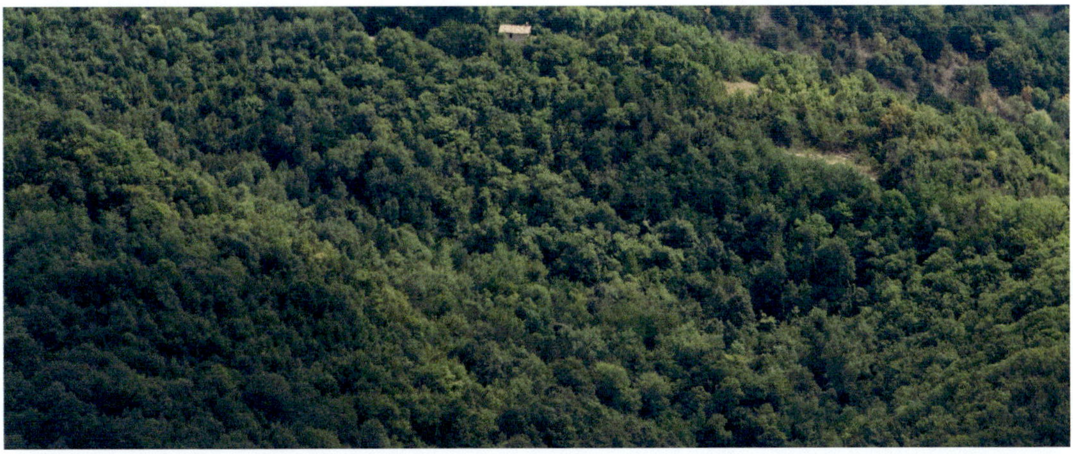

All the trees in this photo are green, but we can still make up the distinct shape of almost every tree just by detecting minute differences in brightness and hue.

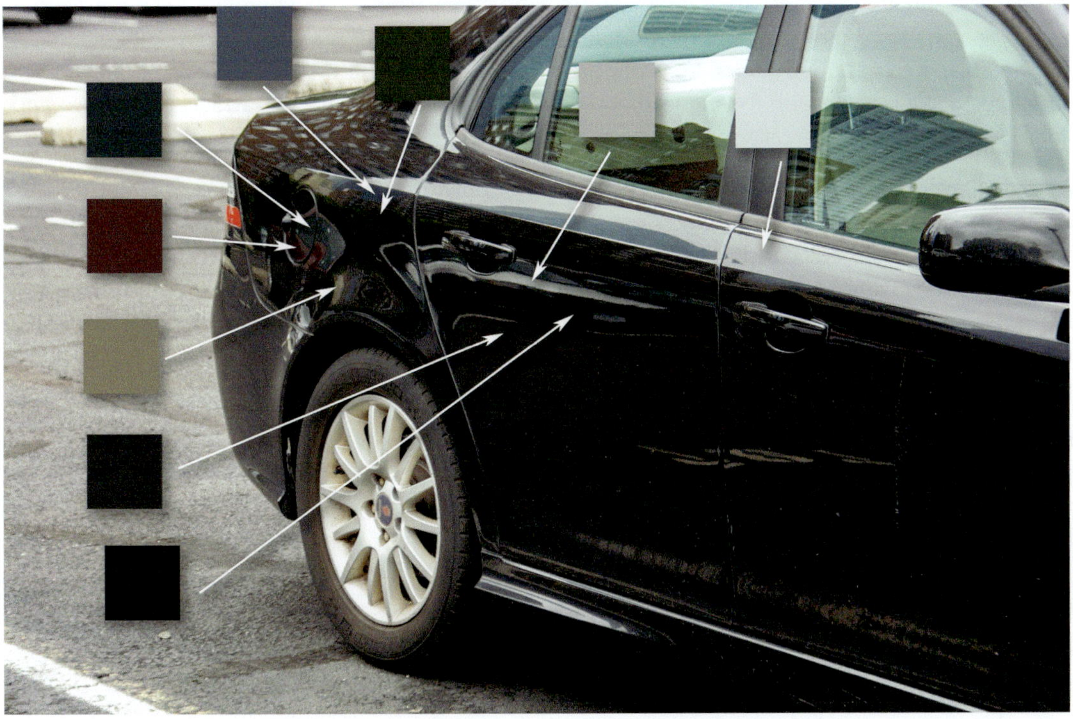

What color is the car? Black, we would normally say. But it's black only in a hypothetical sterile environment. In this particular photo, as captured by the camera, the colors of the car's black paint range from red and green to light blue and near-white.

how we must observe and describe color if we seek to recreate reality digitally. We need to rid ourselves from that lifelong conditioning of simplifying color and instead train ourselves to see colors objectively, in all their complexity. And we can surely do it – after all, we can see over ten million colors! Now, there is nothing new or groundbreaking in what I say – artists have been observing color this way for millennia. But I think that we, digital artists, are prone to becoming a little lazy in this respect. A painter has nothing but color to recreate reality on the canvas. We have all those tools at our disposal – we model highly detailed 3D objects and texture them and throw different lights in the scene and use physically plausible shaders and realistic simulations – and in the process, we sometimes tend to forget that, at the end, it all comes down to color. So, looking at the complexity of color in the world around us, is there a way to organize it so that we can make better sense of it?

The Six-Layer Approach

Let us revisit our charcoal-gray car and assume we photographed it on a sunny day. If we concentrate on a single surface, say, the dark gray painted body, we can observe a multitude of different colors: small imperfections like smudges and scratches shift the color to a lighter gray or reveal the underlying base paint layer. Areas that are directly lit by the sun appear as a brighter, warmer tone, while areas that are occluded from sunlight are darker with a slight bluish tint. Where the sky and sun are reflected, the highlights are almost fully white, and where light can hardly reach, the color shifts toward black. The environment surrounding the car is reflected in its glossy body, adding splashes of various colors, from the greens of the trees to the colors of nearby cars and structures. If we photograph the car from a distance, a subtle bluish haze is added to the mix, and finally, our camera's white balance and exposure settings shift the colors even further.

As is evident from this example, color represents not only the surface properties of a material or an object, but also the light that's hitting it, the environment that surrounds it, its distance from the camera, the atmospheric conditions, and even the camera itself. It is hard to process and analyze all these factors without a methodology, so allow me to suggest one here. I call it "the six-layer approach", and it is by no means scientific, nor is it set in stone. It is my own, basic method for splitting up a multi-factor process into manageable chunks, which I find very handy for organizing color interpretation into a simple hierarchy:

Layer 1 is the "base color" – essentially, the color of the surface if lit by a completely uniform, perfectly white light, in a complete void, with absolutely nothing to reflect, and is shot by a fully unbiased camera. These conditions are, of course, not achievable in the real world, on earth, or even in outer space. Ironically, this is just about the easiest condition to achieve in CG (add an object to an empty scene, assign a uniform color texture, add a uniform ambient light, and voila!). Under these imaginary conditions, a red ball, for example, will look like a perfectly uniform, flat red disk.

Layer 2 adds the surface detail and imperfections that vary the base color (subtly or drastically). The rule here is that these changes in color can be visible in our hypothetical void (i.e. they are not caused by any specific lighting, reflections, or atmosphere). The red ball still looks like a flat red disk, albeit now not uniformly red.

Lighting is added in **layer 3**, including any light sources in the scene, as well as the surrounding environment (which contributes by bouncing some light back on the surface). The red disk now takes the shape of a ball, and the colors shift depending on the intensity, direction and hues of the light source(s) (and to a lesser extent, the reflected light from the environment). The environment also affects the occlusion of light, which changes the color wherever less light can reach. For example, the ground causes a noticeable darkening around the lower part of the red ball.

Layer 4 is the reflection layer. Depending on the material, the reflections may be anything from isolated highlights where the brightest parts of the environment are reflected, to a fully detailed mirror view of the surrounding environment. This layer can therefore add a considerable amount of color variation to the surface.

Layer 5 is the atmosphere layer. The specific effect on the surface color (if any) depends on the distance of the surface from the camera, as well as the atmosphere itself (dry air has much less visible effect than fog, for example).

Layer 6 is the camera layer: factors like lens flares, white balance, or the exposure can subtly or drastically shift the colors of a given scene.

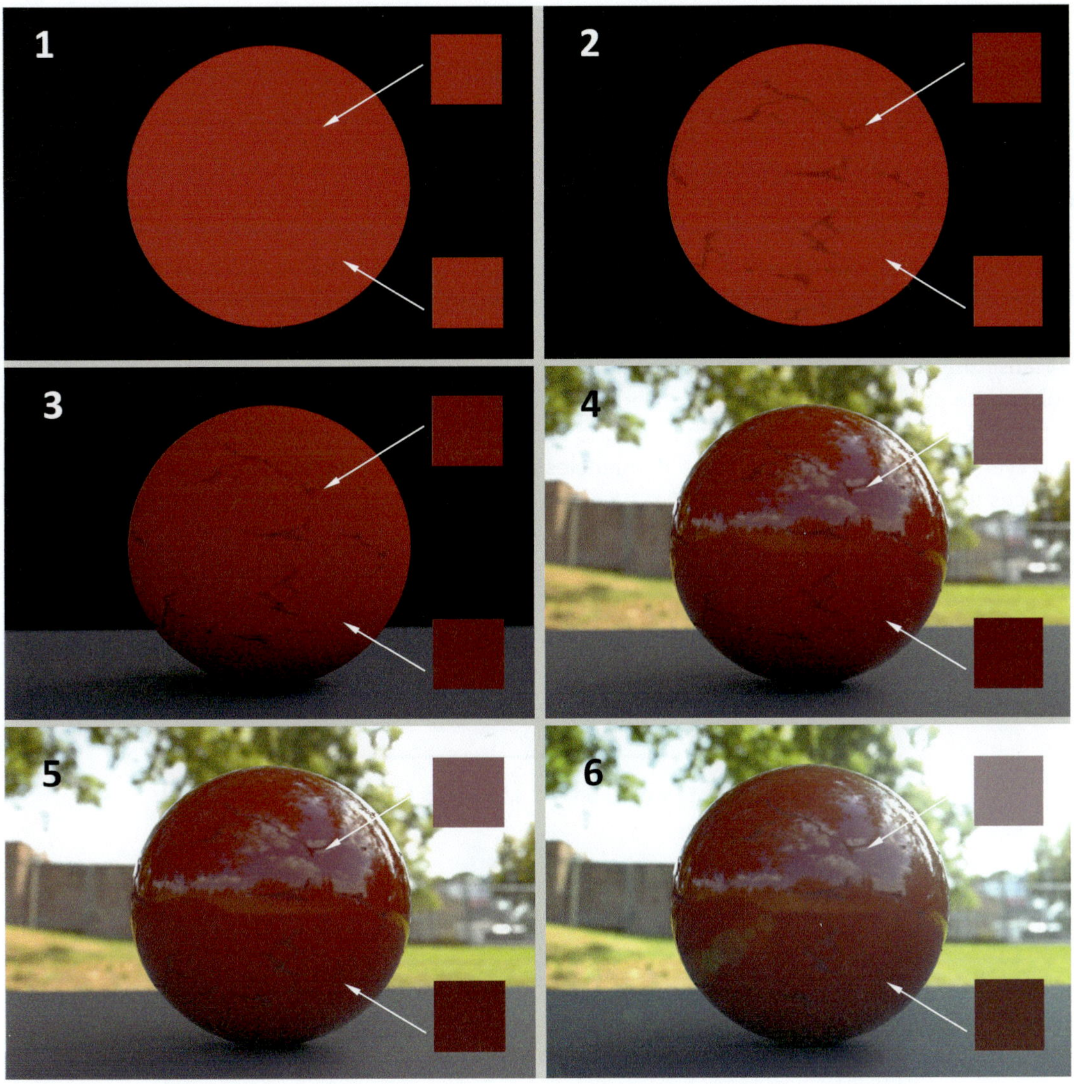

Since it's not possible to strip a photograph down through color layers, I used CG to illustrate the red ball example: 1: base color layer, 2: texture color detail, 3: lighting, 4: reflection, 5: atmospheric haze (very subtle here since we're close), 6: lens flares and camera color interpretation.

This layered approach comes in handy when analyzing a photograph or a shot. By stepping backward through the layers, we can "reverse-engineer" an image all the way back to the stripped-down basic colors of its components. By doing so, we can extract a great deal of information about the lighting, atmosphere and environment in the image. This kind of "color detective" work is an essential step when integrating elements into existing footage or emulating a certain lighting scenario in CG, and it can only be done successfully through a methodic interpretation of color. Incidentally, these six layers reflect the process of creating a CG scene and finishing it in comp: modeling (layer 1), texturing (layer 2), lighting (layer 3), shading (layer 4), adding depth/atmosphere effects (layer 5) and adding lens/camera characteristics and effects (layer 6).

Thinking Additive

Subtractive Color

Back in kindergarten, when we created our first artwork with finger paints or gouache, we discovered that the more colors we mix, the more likely we are to end up with a dark-brown mush. When we mix paint materials, we are in fact mixing reflected (not emitted) light. Each paint material contains pigments that define which light frequencies are absorbed and which frequencies are reflected. Green paint absorbs most of the red and blue light, while red paint absorbs green and blue light. When you mix red and green together you get a darker brown, because the combined pigments now absorb more of the light spectrum and reflect less. Subtractive color mixing is strongly governed by absorption – painters usually avoid starting out with primary colors (red, green, and blue) and instead use secondary colors (yellow, magenta, and cyan) because they reflect more frequencies than they absorb. Mixing two secondary colors reduces the reflected light to a single pure primary and increases light absorption (for example, cyan and yellow produce green). Since subtractive color mixes get progressively darker, painters rely on the use of white pigment (which reflects most wavelengths and absorbs little) to lighten up or desaturate the mix. Similarly, most ink printers use cyan, magenta and yellow as their base colors (usually with added black ink, the combination known as CMYK).

When mixing paint materials, two secondary colors produce a primary color, and mixing three secondary colors produces black.

Additive Color

Unlike paints and inks, display devices like computer monitors, televisions, tablets, phones, and digital projectors actively emit light. Since digital art is created with light-emitting devices, color is achieved and controlled as direct light – absorption and reflection play no role. Color mixing in computer graphics is akin to mixing light of various frequencies: with additive color, the result of mixing wavelengths will always be brighter, and the more colors are added, the more the mix approaches pure white. Unlike subtractive color, the additive color workflow is based only on the three color primaries (red, green, and blue), because every possible hue, intensity, or saturation value is achievable by additively mixing just those three primaries. The use of RGB primaries is also tightly connected to the mechanism of display devices: CRT monitors use three phosphor dots, red, green, and blue (called triads) for every pixel, and the specific color of the pixel is achieved by causing the triads to glow at varying intensities. Similarly, in LCD displays, red, green, and blue filters control the light as it passes through the liquid crystal. Moreover, our own color vision is achieved through three types of cone cells, each sensitive to either red, green, or blue light.

Understanding color through RGB values is therefore vital for the digital artist. If you are able to perceive any color as a specific combination of red green and blue or perform different color operations solely by varying the relationship between the three primaries, you are controlling color through its most fundamental elements. Achieving photoreal imagery often involves small, measured tweaks rather than broad stroke changes, and accuracy and control are paramount. Brightness, contrast, or saturation sliders are handy tools, but not always the best for precise fine-tuning or for reaching a very specific result. To get to the essence of color correction, digital artists need to learn how to control hue, saturation and brightness by balancing red, green, and blue light.

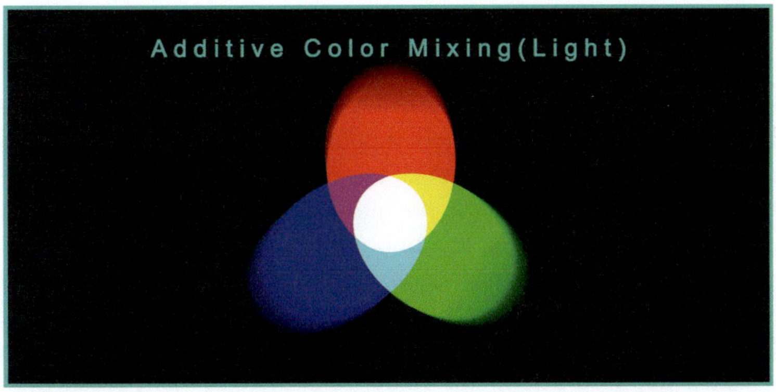

When mixing light, adding two primary colors produce a secondary color, and all three primaries produce white.

Hue, Saturation, and Brightness in RGB

Since we are able to distinguish between roughly ten million different colors, it is not feasible to assign a name for each possible color. There are hundreds of color names in English. For example, "ginger", "carrot", "tiger", "apricot", "clay", "sandstone", and "marmalade" are all names for different hues of orange. Designers or home decorators may be familiar with the subtleties of such naming conventions, but these names, lovely as they are, are far from a precise description of color. One person's "apricot" may be quite different from another's. There have been several attempts to come up with a standardized, accurate color catalogs. A good example is the Pantone color coding, where roughly 1000 different colors are specified based on measured mixing of 13 pigments and black. The Pantone color chart is widely used in print and manufacturing because it allows for very accurate color matching, but it is obviously not very useful for digital color work – it defines a rather limited number of colors and is based on subtractive mixing of pigments, not additive mixing of light.

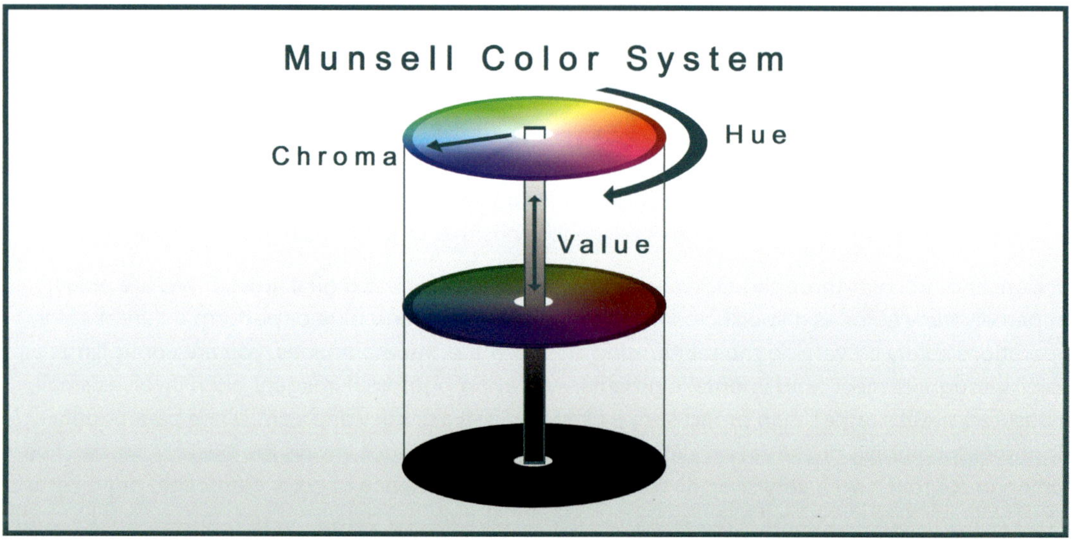

In Munsell's three-dimensional color cylinder, up-down represents value (brightness), rotation represents hue, and chroma (saturation) is changed by moving from the perimeter to the center.

In the early 20th century, Professor Albert Munsell came up with a color system based on hue, value (lightness, or brightness) and chroma (color purity, or saturation). He imagined this system as a three-dimensional cylindrical space in which the color values move independently on three axes: hue is changed by rotating around the perimeter of the cylinder, saturation by moving from the rim toward the center of the cylinder, and brightness is increased or decreased by moving up and down the cylinder's height. Munsell's goal was to find an accurate way to define specific colors, so he divided each dimension into small numbered steps. Munsell designed his color model long before computers existed, but his design foreshadowed the digital age, and is indeed the basis for the **HSV** (**Hue, Saturation, and Value**) color system. Many artists prefer to perform color corrections using HSV tools, as this is a somewhat more intuitive approach. But full mastery of color can be gained by working directly with RGB light mixing. To understand the correlation between HSV and RGB values, we will use three sliders for red, green, and blue, each going from zero (no intensity) to one (full intensity).

Brightness represents the intensity of light, not its frequency. In RGB light context, if we start up with all three sliders at zero (pure black, or absolute absence of light), and then raise just the red slider, we are effectively increasing the intensity of the red light while keeping both the hue and saturation fixed. When the red slider is all the way up at one, we reached the brightest possible shade of red (while increasing the value above one is possible, the perceived color is not affected). As we lower the slider back down, red becomes progressively darker as the intensity of the light decreases. Notice that the position on the color wheel is fixed throughout – there is no change to either hue or saturation, because we are moving only on the third, imaginary "up-down" axis of our 3D color cylinder (an axis which cannot be shown on a 2D color wheel). This is important to note, because it is a common mistake to refer to different intensities as "different hues". Using the word "shade" instead helps make the distinction between different hues and different intensities of the same hue ("a darker shade of red" versus "a more greenish red hue").

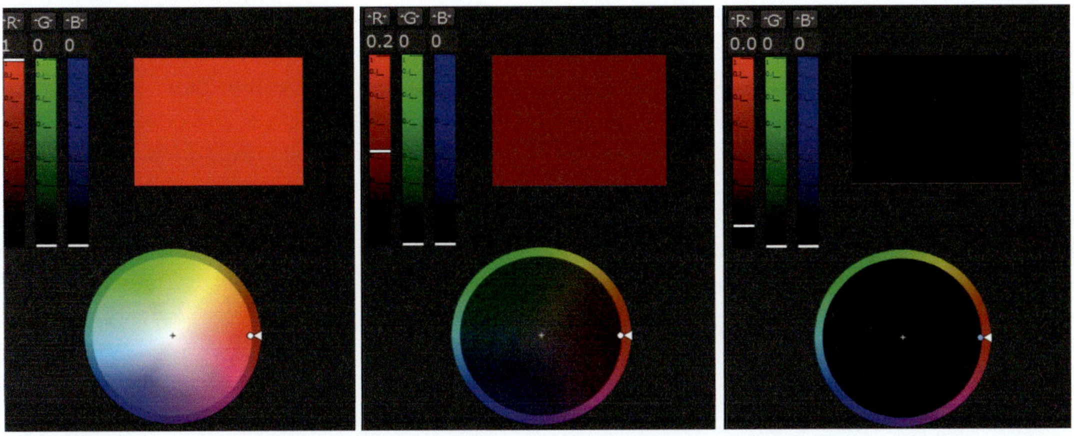

As the red slider is brought down, only the brightness (value) changes. The hue and saturation remain exactly the same on the color wheel, as we move along the third axis of the cylinder. This represents three shades of the same pure red hue.

Hue can only be changed by introducing another primary to the mix. If the red slider is all the way up at one, and we start raising the green slider, we are shifting the hue from pure red to orange and then to yellow. Notice that the color wheel indicator starts rotating along the perimeter of the wheel (hence the term "hue rotation"). The rotation from red to yellow is exactly one sixth of the perimeter. If we now keep the green slider at one and move the red slider down from one to zero, we continue rotating through the next sixth of the perimeter, from yellow to green. Continuing in the same fashion with the green and blue sliders, then the red and blue, we effectively rotate through the entire visible light spectrum, or all possible hues. When any two sliders are both at one, the result is a secondary color (yellow, cyan, or magenta).

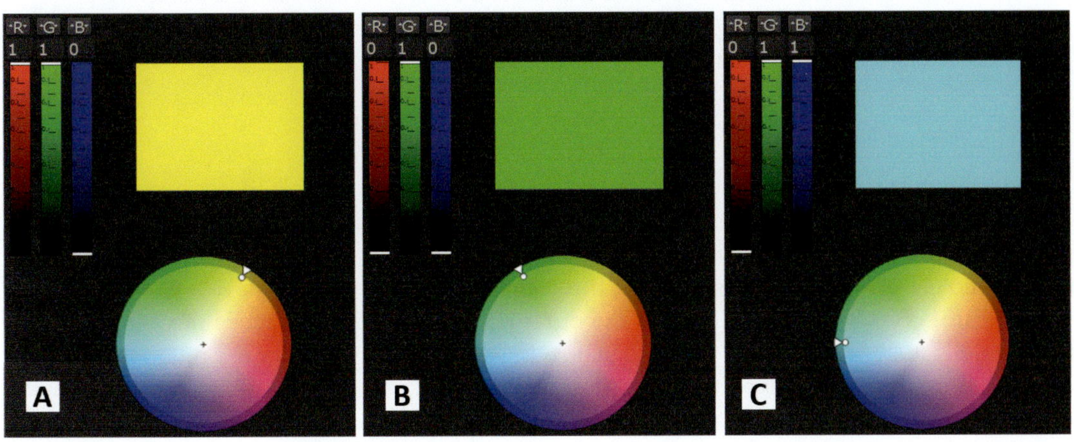

Hue rotation: using three combinations of primaries we rotate one third of the wheel, from yellow (A) through green (B) to cyan (C).

So far, we kept at least one slider up at one, so the brightness remains at 100% and only the hue changes. But once we move both sliders to positions lower than one, we start reducing brightness. If we keep both sliders ganged together, we get a darker or brighter shade of the same hue. For example, moving both the red and green sliders from one (yellow) to 0.4 gives us a much darker shade of yellow. Though this color may be referred to as "mustard" or "Dijon" in designers' jargon, it is still pure yellow, only darker (in other words, it is still pure yellow light, just less intense). But we can, of course vary the positions of both sliders independently, which shifts hue and brightness simultaneously. By using just two sliders (two primaries), and covering only one third of the color wheel, we can still achieve a staggering number of possible colors. Notice that up until now, the saturation remained 100% (the point on the wheel sticks to the outer perimeter). This is important to emphasize, because saturation is often confused with brightness. Darker shades, in general, feel more subdued and less vibrant than brighter shades, so we tend to think of them as less saturated. That mustard color, for example, is commonly considered less saturated than pure yellow, but technically it is still pure yellow. We should not confuse brightness (intensity) with saturation (purity of color, or how close it is to neutral gray).

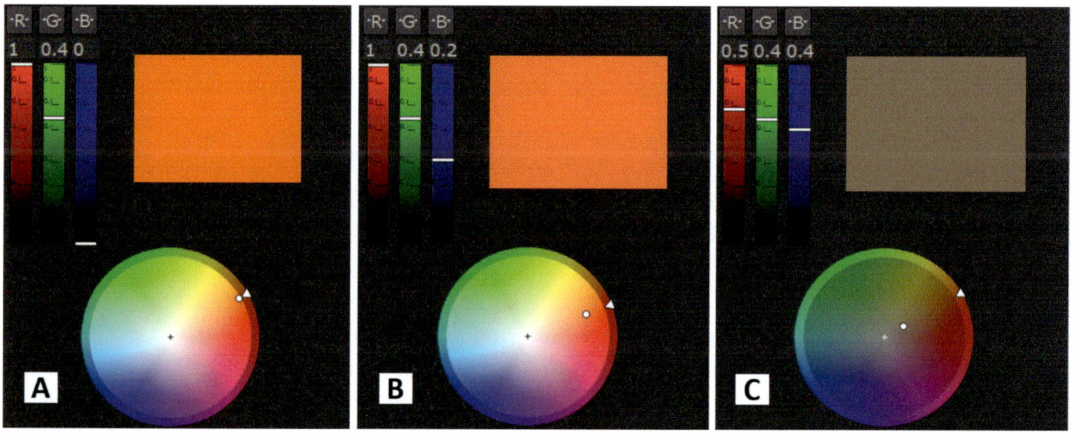

Pure orange (A) becomes less saturated when blue is added to the mix (B). Notice how the marker leaves the perimeter of the wheel and starts moving toward the center. When all three primaries are brought closer together, the saturation is further reduced and the color approaches pure gray.

To reduce **saturation**, we must introduce the third primary into the mix. If, for example, red and green are at one (pure yellow) and we start raising the blue slider, the color gradually gets closer to white as its purity decreases. If the red and green are both set to 0.4 (the darker mustard color), and we start raising blue, the color gradually becomes gray. In both instances, the indicator on the color wheel leaves the outer perimeter and approaches the center. Saturation is therefore affected by the distance in value between the three primaries: the closer those values are to each other, the more desaturated the color. One or two primaries always generates pure hues, and only a combination of all three primaries can reduce the color purity, or saturation.

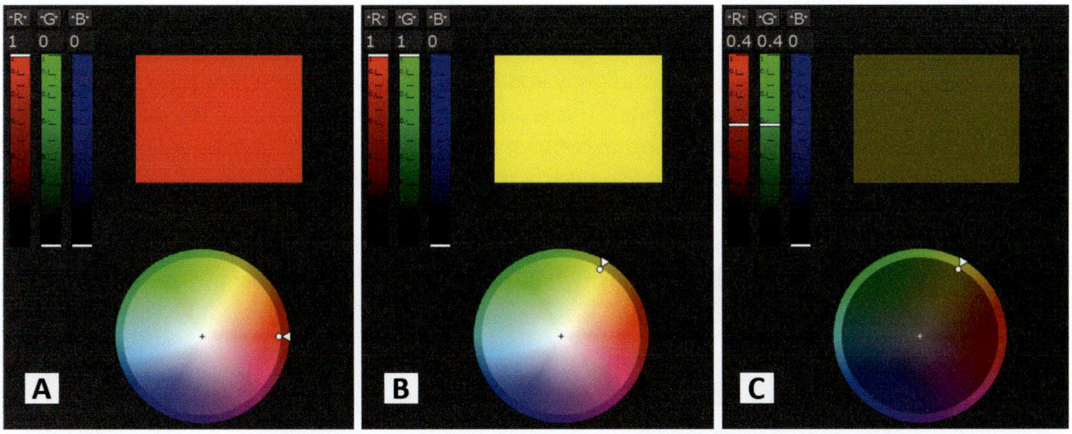

Red (A) with green fully at one produce pure bright yellow (B). With both at 0.4, the result is a darker shade of yellow (C). Notice that the marker is still at the edge of the wheel – the mustard color is still pure, fully saturated yellow.

Color Operations

Color operations terminology is a source of much confusion. There is no unified color vocabulary, and names and functions of color tools vary substantially between different software applications. Take a quick look at the basic color tools in three different applications, Photoshop, Maya, and Nuke. Photoshop offers Brightness/Contrast, Levels, Curves, Exposure, Vibrance, Hue/Saturation and Color Balance. Maya's texture map color controls include Exposure, Default Color, Color Gain, and Color Offset. Lastly, Nuke's grade node includes Blackpoint, Whitepoint, Lift, Gain, Multiply, Offset, and Gamma. This is a rather confusing list. Not only are similar operations named differently in different applications, some applications even offer multiple controls for the exact same color operation (for example, brightness, exposure, gain, and multiply are basically all the same operation). These redundancies and overlaps certainly don't make color handling any easier for the digital artist.

To simplify things, we can narrow down most color tools to just five basic operations: **gain**, **gamma**, **lift**, **offset**, and **saturation**. All these operations essentially control the brightness (intensity) of RGB light. However, they control brightness differently, which makes each operation suitable for a specific goal. It is helpful to look at the simple math behind each operation, and even more useful to understand the visual outcome of each operation and its functional place in the digital color toolset. You may have noticed that I did not include contrast as one of the previous five color operations. The reason is simple: each of these operations can be defined by an absolute value and applied to a single pixel. Contrast, on the other hand, is the relationship between dark and bright colors within an image. A "contrast value" is a relative term and cannot be assigned to a single pixel. Contrast can in fact be changed using a combination of two of the operations discussed here, as I will show later on.

Gain (Exposure)

The most basic color operation is a simple multiplier. Raise the gain to 2, and every pixel value is doubled. Set it to 0.5 the value is halved. A value of zero (absolute black) always remains zero. Gain raises or lowers brightness in a linear fashion throughout the entire range. Since it is a multiplier, it affects brighter tones much more than darker ones. For example, if you set the gain to 3, a black pixel with a value of 0.001 becomes 0.003 (just slightly less black), while a medium gray pixel with a value of 0.33 goes all the way to 0.99 (near-white). This is an important characteristic of the gain operation because it results in a relative preservation of contrast and tonal balance. Our eyes are much more susceptible to brightness changes in the low end, and much less sensitive to changes at the high end. This makes gain the best operation for increasing overall brightness, as it conforms to our vision – when gain is increased, the image does not simply wash out. Gain is also the most suitable operation for shifting the hue by changing the relative brightness of each of the RGB channels.

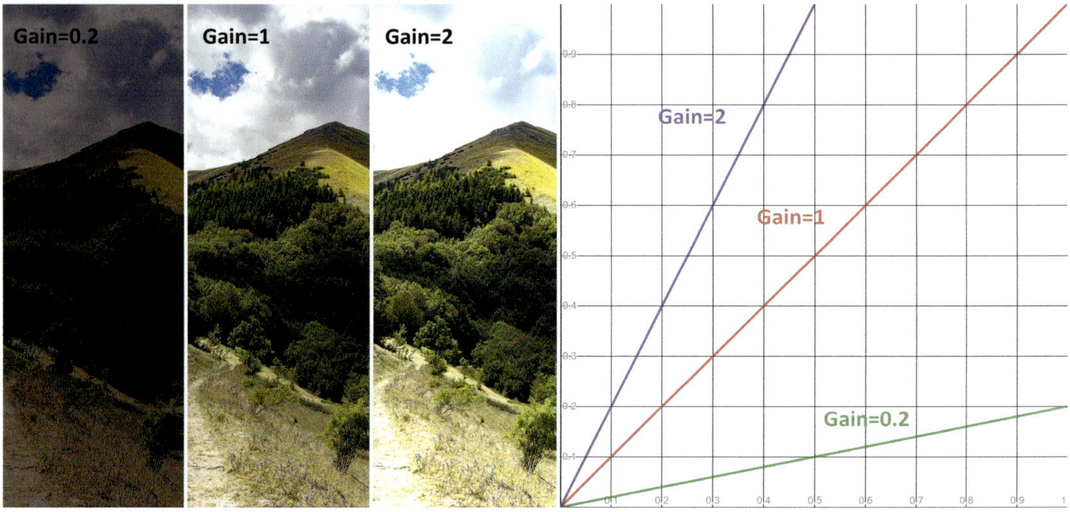

Because the gain curve pivots at the zero point, it affects brighter tones much more than dark ones. This makes it ideal for overall brightness control because we are much more sensitive to changes in the darks.

In some applications the gain operation is called brightness – a rather inaccurate name since all color operations manipulate brightness. Other applications offer gain as **exposure**, which is a correct terminology. Exposure is also a multiplier, but it uses stops as units. Stops are powers of two: raising the exposure by three stops is the same as setting the gain at eight, and lowering the exposure by four stops is the same as setting the gaining down to one-sixteenth (0.0625). Photographers and cinematographers may prefer to operate the gain in stops rather than multipliers, though the result is essentially the same.

Offset

Offset is an additive operation. An offset of 0.2 raises all the values across the color range by 0.2. Like gain, offset is a linear operation. However, the differences are obvious: offset affects every color value equally, shifting blacks and highlights by the same exact amount, and it also affects zero. Due to our biased perception of brightness, offset does not appear to brighten the color range equally: as the offset slider is raised, the blacks are lifted first, dragging the midtones and highlights along, until the entire image goes white. Clearly, offset does not work as well as gain for controlling overall brightness or hue, but it is useful as a tool for manually controlling saturation (as I will explain later).

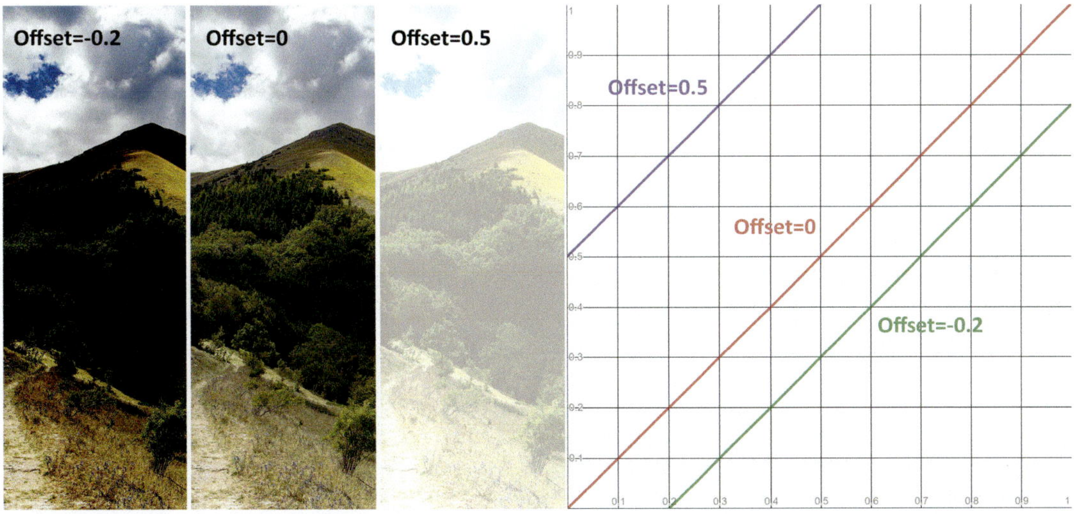

The offset curve affects bright and dark tones equally – notice how the blacks appear lifted when offset is raised and crushed when offset is lowered. Notice also that the cloud bright parts like the clouds are blown out when offset is raised.

Lift

The math behind the lift operation is slightly more complex than gain or offset: output = input*(1 – lift) + lift. Throw some numbers into this equation, and you will notice that the lower the input value, the bigger the change. For example, if lift is set to 0.2, a dark gray shade of 0.01 jumps to 0.208, while a near-white gray of 0.95 barely nudges to 0.96. Plotting this operation as a curve shows a sort of a mirror image of the gain curve: instead of pivoting on the zero point, the lift curve pivots at one. Lift is therefore the best choice for tweaking the low end of the spectrum with minimal impact to the high end. Accurate blacks are crucial for integration and photorealism, so this operation has an important role in color-correcting.

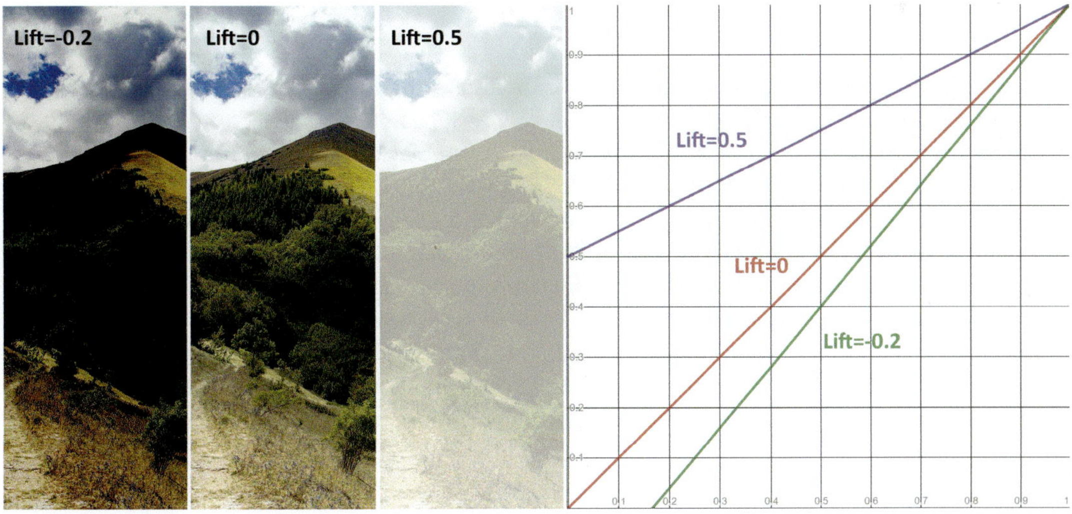

The lift curve pivots at one, so it affects the blacks much more than the whites. Notice that, unlike offset, lift does not blow out the clouds when raised.

Gamma

Unlike the previous three operations, gamma produces a non-linear curve: the pixel value is raised to the power of the inversed gamma value ($output = input^{1/gamma}$). For example, with a gamma of 2.2, a pixel with a value of 0.1 becomes 0.35 (the inverse of 2.2 is 1/2.2, or 0.454, so 0.1 to the power of 0.454 is 0.35), and a value of 0.9 becomes 0.95 (0.9 to the power of 0.454). It may seem that this is operation is similar to lift, where lower values are affected more than higher values, but the gamma curve is quite different: a logarithmic curve that has a "belly" around the low-to-mid end, then settles toward the top end. The curve also shows that unlike gain (which is anchored at zero) or lift (which is anchored at one), gamma is anchored at both ends. If you feed either zero or one as the input value into the equation, they will remain the same.

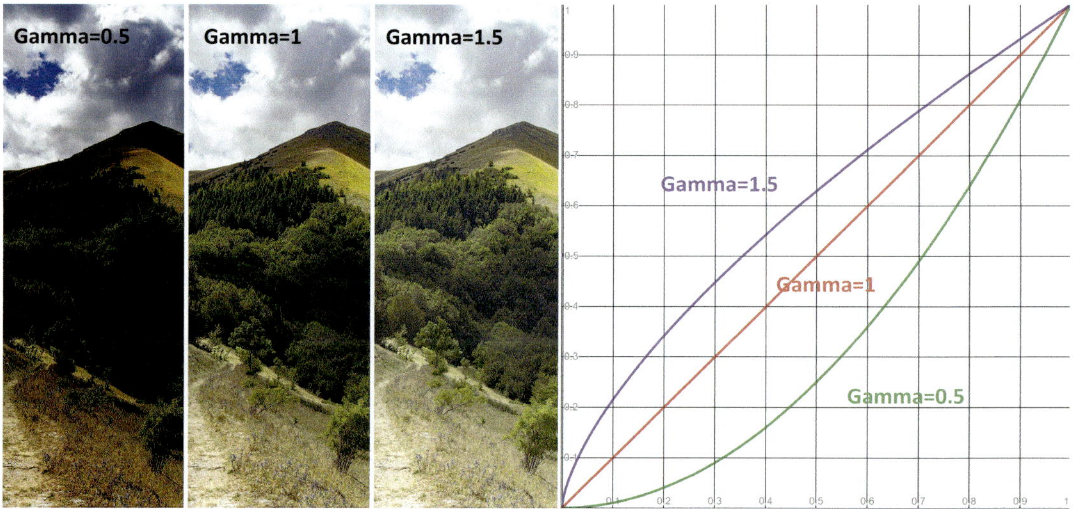

The gamma curve pivots on both the zero and one points, and affects mostly the low and low-mid range, which makes gamma ideal for controlling contrast.

The unique effect of gamma is that by raising or lowering the low and mid tones (while preserving the black and white points), it effectively reduces or increases contrast. Unlike lift or offset, when gamma is increased, it does not flash the blacks in an extreme way, but rather gently raises their brightness along with the lower midtones, bringing them closer to the higher midtones and highlights. Doing so, gamma reduces the difference between bright and dark areas, and thus reduces contrast. Lowering the gamma darkens the lower end without crushing the blacks, and preserves the higher end, effectively increasing contrast. The overall change in brightness can be counteracted with gain. For example, lowering gamma while raising gain increases contrast while preserving the general luminance of the image. Since gamma can be applied separately to each RGB channel, contrast can be controlled very accurately using the combination of gamma and gain. As previously mentioned, contrast is essentially a subjective term, and different contrast tools can vary greatly in their built-in math and the results they produce. Some contrast operations boost the highlights too much, while others muddy the blacks. In general, the perceived effect of contrast is highly dependent on the visual material and its context, and so is much better controlled using a combination of gamma and gain, and if necessary, also lift or offset.

Saturation

In the previous section I explained that all the shades of primary and secondary colors are 100% saturated, and saturation can only be reduced by mixing all three primaries and decreasing the relative difference between their values. Take for example a fully saturated yellow with the values red = 1, green = 0.9, and blue = 0. To reduce the saturation, we need to raise the blue toward the red and green. Gain or gamma are not the suitable operations for raising the blue in this case, because they do not affect zero values, but offset does. As we offset the blue value toward the red and green, we reduce the saturation, but we introduce a new problem – the desaturated yellow is now substantially brighter than the original, because we added a lot of blue light to the mix. To properly increase or reduce saturation, we need to maintain the original luminosity, as well as maintain the relative relationship between the primaries. It is very tricky to do so manually – and this is where saturation tools come quite handy, because they take care of this tricky balancing act for us.

Saturation should be used thoughtfully. As I mentioned before, we tend to view bright shades as more saturated than darker ones. A very bright orange, for example, may appear as overly saturated. But (and this depends on the context) desaturation may not be the best remedy. A simple gain reduction around the dominant color may be much more effective, as it affects only the intensity and not the hue. Often, it is not the overall saturation that needs to be increased or reduced, but rather one or two hues that are overly dominant, missing, or just don't sit well within the whole.

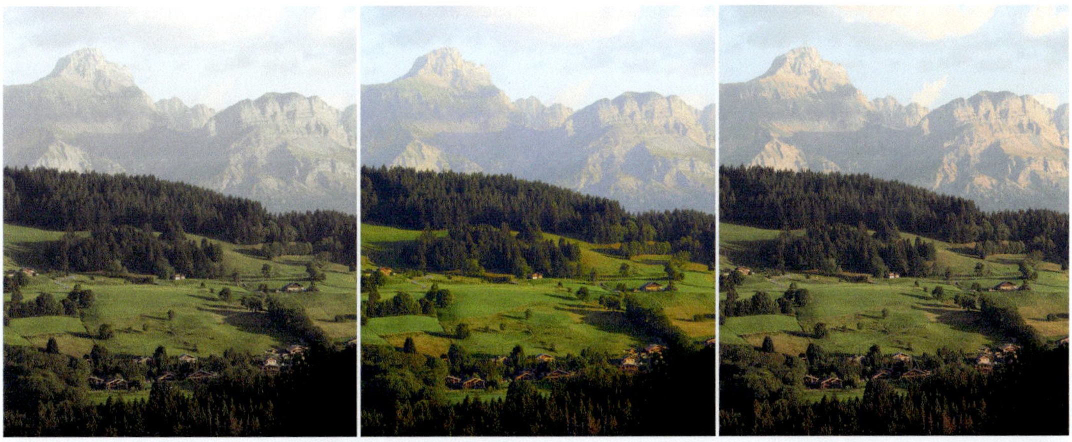

In this example, the goal is to increase the orange hue of the afternoon light and the blue of the sky (original photo at left). However, increasing the overall saturation (middle image) does not help, as it mainly increases green, the dominant color in the photo. A better approach is to use specific hue corrections (right).

Bit Depth and Dynamic Range

When you count in binary (only zeros and ones), digits accumulate fast. The number four in binary is 100, and 40 is 101000. An 8-bit color system can handle up to eight digits, so the highest number possible is 11111111 (or 255 in decimal). Including the zero, this allows for 256 possible shades for each of the three primaries, and a total of 16,777,216 (256x256x256) possible color permutations. Considering that we can detect up to about ten million colors, 16 million seems more than sufficient for representing color accurately. Indeed, a photograph saved as an 8-bit image preserves all the visible color detail. JPG is by far the most widely used 8-bit image format, and it holds more than adequate color detail for display and print. Now, a 16-bit color image has 65536 shades per primary, for a total of over 280 trillion (!) different colors. This number seems like a massive overkill, especially since there is no visible difference between a 16-bit and an 8-bit version of an image. But the difference is hidden within: 8- and 16-bit images may look identical, but they react quite differently to color manipulation. Once you start pushing and pulling the colors, an 8-bit image soon shows signs of distress, especially in areas of fine color gradation. Banding, stepped transitions, and artifacts are all possible results of manipulated 8-bit images. 256 steps per primary are simply not enough to withstand substantial color correction. Therefore 16-bit imagery is superior to 8-bit not because "it looks better", but because it is significantly more flexible.

Previously in this chapter we looked at RGB colors ranging from zero to one, where RGB 1,1,1 represents pure white. While this is perfectly fine for most digital display devices and for print, it is not enough to truly represent the range of light intensity in a given environment. Modern digital cameras can capture a wide dynamic range, and require a digital format that can preserve such a range. Eight-bit formats like JPG clamp all color information below zero and above one, but 16- and 32-bit color formats allow for a much extended range, especially with the help of floating point math (In a nutshell, floating point is a way of covering a very wide range of values by shifting the position of the decimal point between the digits). Extended dynamic range is invaluable for color manipulation because it maintains hidden information in the blacks and the whites – information that may not be visible at the photo's original exposure but is revealed once that exposure is modified. For example, reducing an overexposed sky's brightness in a JPG image will not show more detail, because all the values above one are clamped, and whatever

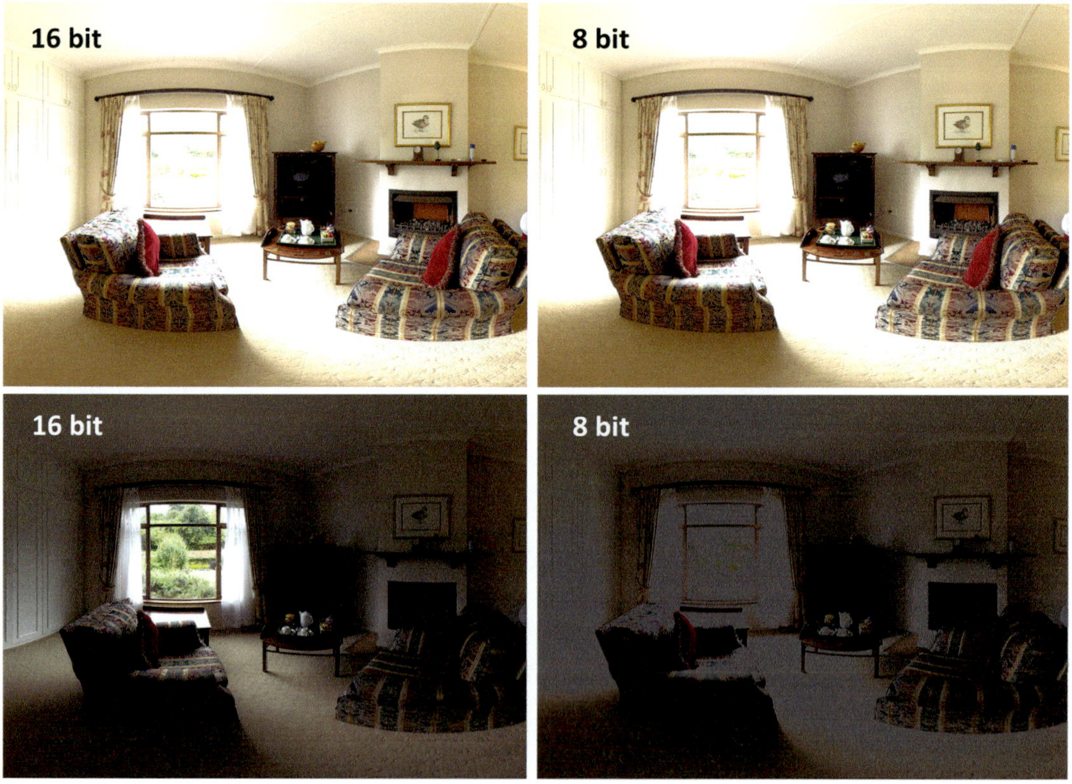

The 16-bit and 8-bit images (top) look exactly the same. But when the gain is pulled down, the "hidden" detail is revealed in the 16-bit image, but cannot be resurrected with the clamped values of 8-bit (bottom).

detail was there has been lost. The low range in 8-bit formats suffers a similar clamped fate. But 16-bit formats preserve hidden detail in the highlights or shadows, and thus provide an incomparably superior starting point for color editing and manipulation. The high dynamic range stored in 16- and 32-bit formats is also useful for image-based lighting (see Chapter 11), because photographed light sources preserve their true luminance.

The Low End

Blacks – in plural, not singular! In school we learn that "black is not a color, it's a shade", and that is very true. Although blacks are commonly referred to as "shadows" in photography and color grading jargon, this is a rather inaccurate term (not every shadow is black and not every black is a shadow). Blacks are simply the darkest shades of any hue. Hence the plural, as there is in fact an infinite number of blacks. Moreover, there is no precise definition of "blackness". There is no threshold that delineates when exactly a very dark blue becomes black, for example. Even for a single hue, there could be many levels of intensity around the range we would generally consider as black, even though such subtle differences may be nearly imperceptible. Black in digital art is therefore a relative term. One can argue that the only "singular", pure black is RGB 0, 0, 0. One can argue that such a color does not exist in nature, since it means an absolute lack of light (or a material that absorbs 100% of incident light). But to the digital artist, this argument is rather irrelevant. Photorealism is much about photographs as it is about nature, and areas of zero-black, or even negative values are common in photography and cinematography because of exposure settings, color correction, or the sensor's characteristics.

Blacks play a crucial role in integration and photorealism. Even minute tweaks in this low range can have a profound effect on the image, and how it conveys distance and depth. As mentioned previously, our eyes are much more sensitive to brightness changes in the low end than in the mid and high ranges. Some of the subjects explored in this book, like aerial perspective, lens flares, or light scattering, are most noticeable in the blacks. A mismatch of blacks between different elements can kill the believability of a composite or a matte painting, and inconsistent black levels in a texture map can ruin an otherwise photoreal render. There is a bit of a paradox here: we take, almost subconsciously, crucial cues of distance and depth from blacks, and we can easily sense that "something is wrong" when black levels are not consistent. Yet we normally pay little attention to blacks in everyday life – naturally, it's the midtones that attract our eye, because hues in this range are most vibrant and easy to discern. Untrained people may have a hard time distinguishing between different hues in the blacks,

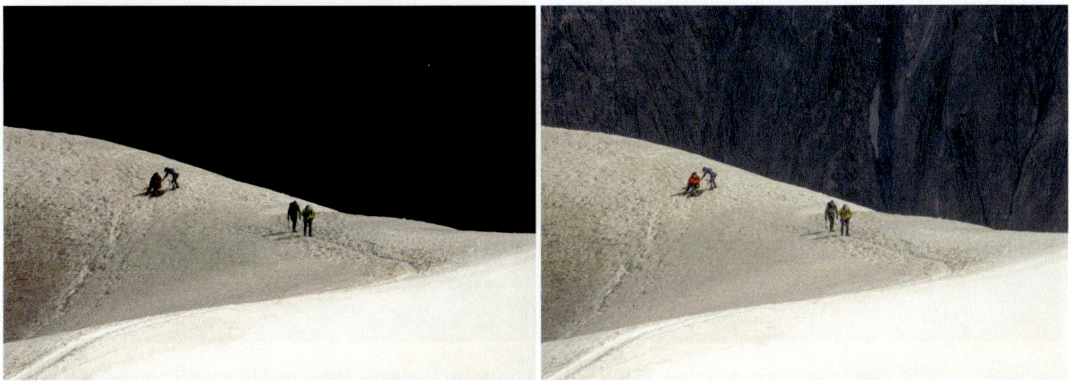

By lifting the low end, the "black" in the background (left) is revealed to be many colors (right). Compositors must be aware of the true values of blacks because a mismatch can be easily exposed when the final content is color-corrected.

but digital artists need to develop a keen sensitivity to the lowest range and an ability to identify the RGB makeup of the darkest of colors. Hardware plays a role here too. Most computer monitors (even those considered above average), fail at the very low end. For color-crucial crafts like texturing, lighting, matte painting, and compositing, a good monitor with accurate black representation is a must.

Blacks are usually controlled with lift or offset (Levels in Photoshop), or by range-limited tools such as Nuke's color correct node or Photoshop's curves. Naturally, changes are subtle and numerical variations are small. Since it is hard to "see" in the blacks, even with a good monitor, it is helpful to temporary raise the display gamma a bit. Some applications provide built-in viewer gamma and gain control. If no such controls exist, you can always add a temporary gamma color correction.

The High End

Whites, in plural, too: while most display devices peak at or near RGB 1,1,1, actual values maybe be higher (much higher, in fact) when working with full dynamic range. So, although white areas in an image may appear identical when viewed on a monitor, their numerical values can in fact be very different. A multitude of colors live in that "white zone", invisible when pushed above one, but visible at lower exposures. Noting the actual values at the top range of the spectrum is therefore important for rendering and integration, because it ensures coherent colors even at lower brightness settings or under extensive color manipulation.

Though the term "highlights" is generally used in art to describe the brightest parts of a painting or a photograph, it is often associated with specular reflections, and is not the best term for whites. The brightest parts in an image are usually light sources, as well as light-scattering elements like lampshades or the sky. Specular reflection on shiny surfaces preserves much of that brightness, so even if none of the light sources are visible in the frame, their reflections constitute the brightest whites in an image. Keeping whites consistent in digital imagery is as important for integration as matching the blacks, but the process can be somewhat more confusing with HDR material, because the true luminosity of white is often only revealed by inspecting the values. For example, a matte white plastic surface may be much less bright then a shiny black car paint, where that paint reflects high-intensity areas such as the sun or the sky. It is therefore important to sample the precise numerical values of whites in an image and match those contextually when integrating 2D or 3D elements.

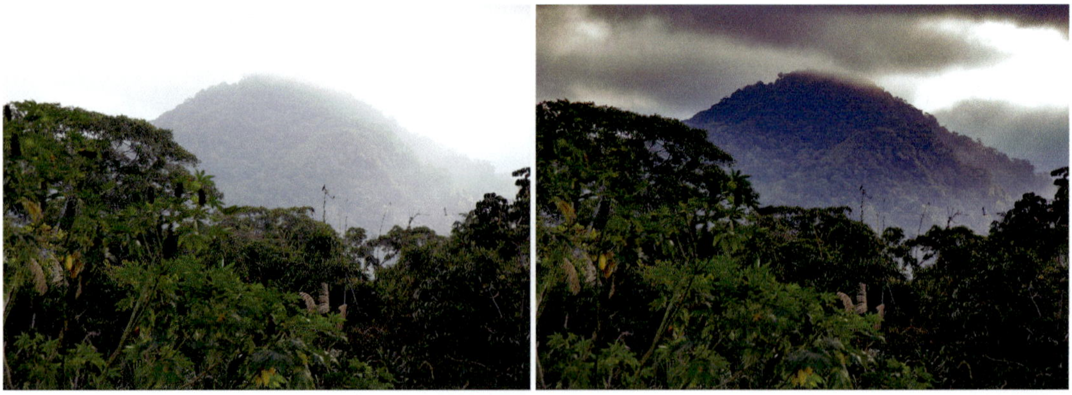

Just like as with blacks, what we perceive as "white" is often multiple colors. And just like with blacks, successful integration of elements relies on careful matching of white values.

PART 2

THE REAL WORLD

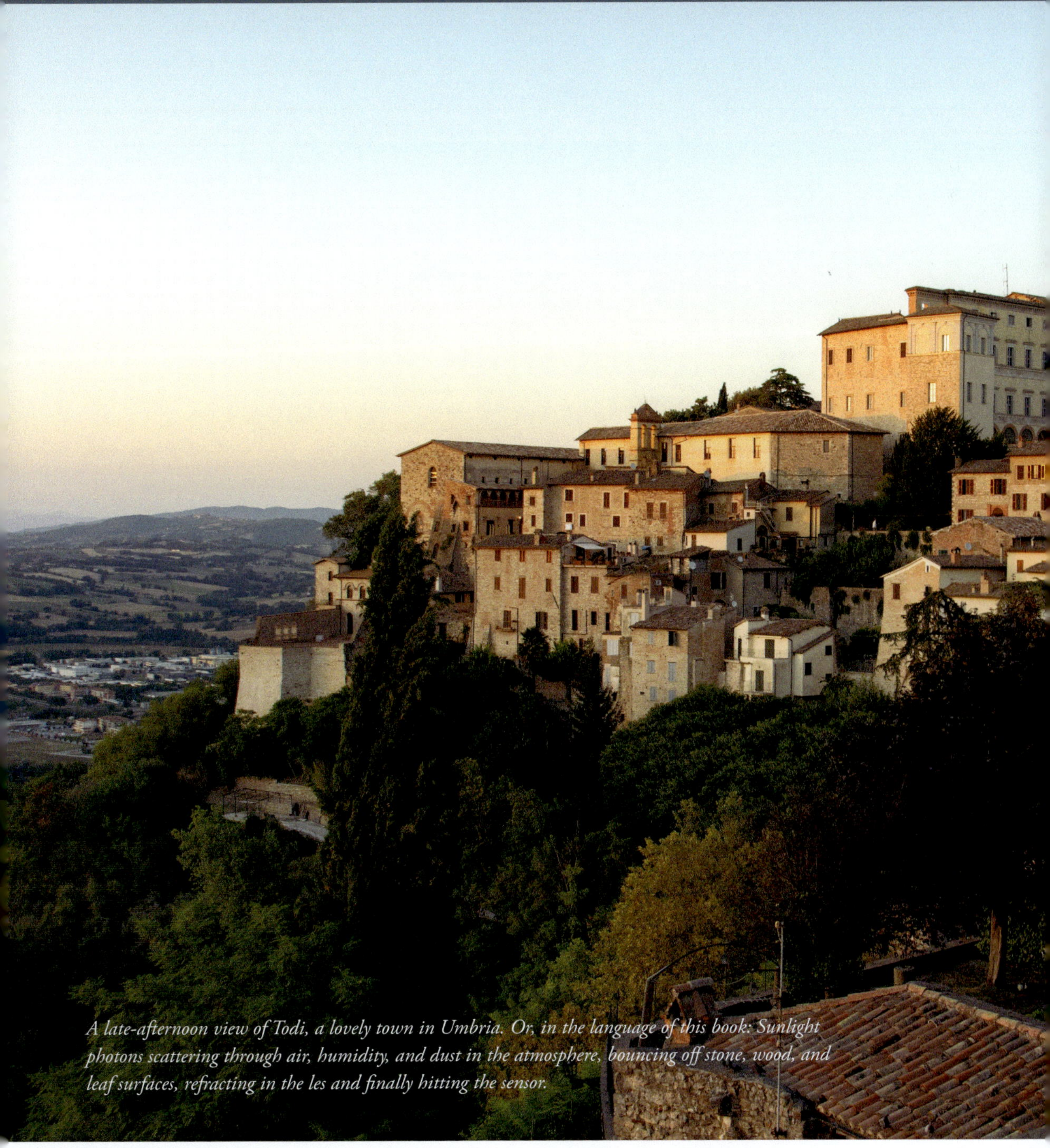

A late-afternoon view of Todi, a lovely town in Umbria. Or, in the language of this book: Sunlight photons scattering through air, humidity, and dust in the atmosphere, bouncing off stone, wood, and leaf surfaces, refracting in the les and finally hitting the sensor.

Chapter 4

Light Essentials

Is light particles or waves? It travels in a straight path and is reflected at a predictable angle – a particle behavior. But specific light phenomena like diffraction can only be explained through wave interference patterns. The waves/particles question was already a subject of debate back in the 17th century, when Huygens wave theory was pitted against Newton's corpuscular theory (which argued that light is particles). Newton's approach was widely accepted, and the wave theory went out of favor until the early 19th century, when fresh ideas from Augustin-Jean Fresnel and Thomas Young revived it. The understanding of electromagnetic waves continued to evolve throughout the 19th century and solidified the light wave theory. As a result, the particle approach was nearly neglected. Then came Albert Einstein, who, along with other revolutionary physicists like Max Planck and Niels Bohr, established the theories of quantum physics. The centuries-old debate was no longer relevant, as these theories proved that light can in fact be both particles and waves.

The wave-particle duality of light is certainly a baffling concept, as it seems to contradict basic notions of classical physics (Bohr called it the "duality paradox"). But since this book is not about quantum physics, the question to be asked is: does the concept of particle/wave duality bring any benefit to the digital artist? I believe it does! Looking at light from either a wave or a particle perspective (and sometimes both), makes it easier to understand different aspects of light behavior. For example, color is best approached from a wave perspective, while the difference between specular and diffuse reflections is easier to understand from a particle's perspective. Later in this book, when we explore CG rendering techniques, you will notice that some relate directly to either one of those perspectives. For example, thin-film diffraction calculates light as waves, while photon mapping calculates light as particles. Digital artists, therefore, can embrace the concept of light duality to advance their understanding of light behavior and optics, even without delving into quantum physics. Here is a quick description of light from each perspective:

Light as Waves

Visible light comprises a rather small portion of the entire electromagnetic spectrum. Waves on this spectrum are defined by their frequency (measured in hertz units) and their wavelength (measured in metric units). At the very bottom of this spectrum lie extreme low frequency radio waves (around 10 hertz), which can be as long as 100,000 kilometers. At the top end, gamma ray frequencies lie around 300 exahertz (300 x HZ^{18}), with wavelengths as small as one trillionth of a meter (1 picometer). In between (moving from long to short waveforms) are medium and

DOI: 10.4324/9781003590293-7

high-frequency radio waves, microwaves, infrared, visible light, ultraviolet, and X-rays. The visible light portion of the spectrum starts with red (frequency around 400 terahertz) and ends with violet (frequency around 780 terahertz).

There are a few evolutionary reasons why we (and most animals) developed eyes that can detect that range. First, this is the sun's peak emission range (higher frequency electromagnetic waves like X and gamma rays are mostly blocked by the atmosphere). Also, the energy of low-frequency waves like radio or micro waves is too weak to cause molecular changes, while the energy of high-frequency waves is so strong that it simply destroys biological tissue. Visible light frequencies sit right in the middle – just enough energy to be easily detected through molecular changes in our retinas without damaging tissue.

However, the visible light range is not identical for all living beings. Some animals can see only a limited part of it, others can detect frequencies below or above it (some types of snakes, for example, can see infrared, while many insects and birds can see ultraviolet). The colors along the visible spectrum (red, orange, yellow, green, cyan, blue, and violet) are called spectral colors. They are considered pure (fully saturated), because they represent a single light frequency. As I showed in Chapter 3, pure colors can be achieved by mixing just two of the three primary colors (red, green, or blue). However, most light sources radiate a mix of different frequencies, so most of the light we see is somewhat desaturated (sunlight, which encompasses nearly all visible light frequencies, is white, or fully desaturated when unfiltered by the atmosphere).

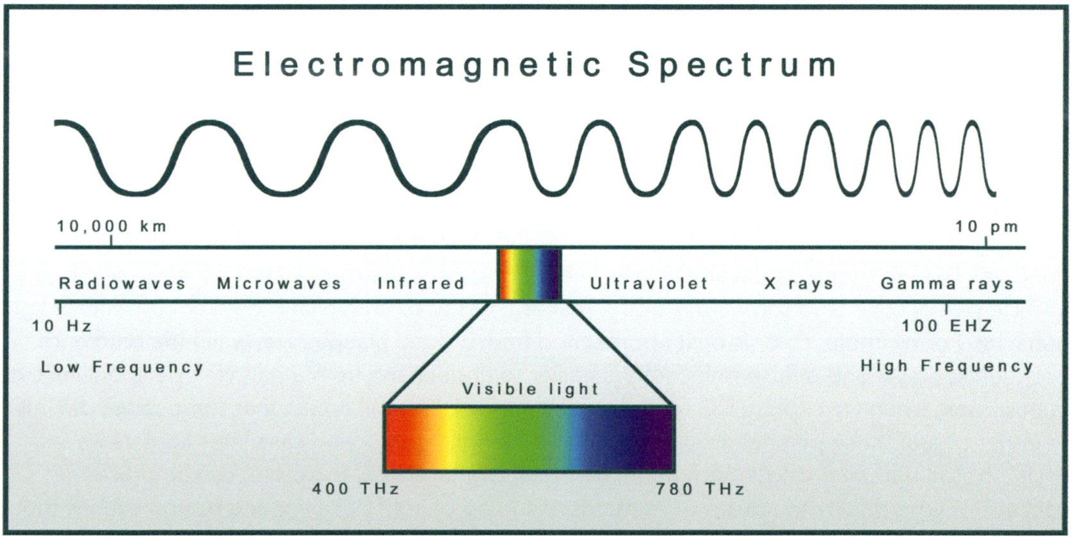

Light as Particles

Like all electromagnetic waves, the energy of visible light depends on its frequency. The smaller the frequency, the higher the energy. In quantum mechanics, this energy is said to be quantized (*quanta* = "how much" in Latin) into defined "energy packets" called photons. Blue light photons (higher frequency) have more energy than red light photons, however, all photons travel at the same speed, regardless of their frequency (in vacuum, the speed of light is roughly 300,000 km, or 186,000 miles, per second). We are surrounded by light photons traveling in all directions as they are emitted and then scattered/reflected by surfaces. But most of this activity is completely invisible to us because our eyes can only detect light when photons hit the retina directly. This means that we can only see light if we look directly at its source (the object that is emitting radiation in the visible spectrum), or if we look at a surface that reflects photons back toward our eyes. In other words, we cannot observe photons "from the side" like watching traffic on the highway. Photons must be on a straight collision path with our eyes (or cameras for that matter) for the light to be registered by the brain (or the camera's sensor). Therefore, we cannot see "light rays". Those shafts of light on a concert stage, or the beams of sunlight entering a room through cracks in the shades, are visible only because light is reflected off particles like dust and water droplets and redirected toward our eyes. Without some presence of a reflecting/scattering volume, even laser beams are invisible (indeed, and despite the sci-fi convention, you cannot see laser beams in space).

Light Decay

The intensity of light falls off with distance at a **quadratic rate**. The **inverse-square law** states that light intensity is inversely proportional to the square of the distance from the light source. For example, if we consider the intensity of light at a distance of a foot from the source as one, then that intensity will drop to a quarter at two feet (one divided by the square of two), and to a mere sixteenth at four feet (one divided by the square of four). Light decay has nothing to do with the energy or speed of the light, as both are not affected by distance (we can see stars that are hundreds of light years away). Light decay is only related to how thinly the photons are spread out. At any given moment, a certain number of photons are emitted from a light source. These are tightly bunched together close to the emitter, but as they travel out in all directions, the distance between them becomes progressively larger. Thus, the farther away an object is from the light source, the fewer photons are likely to hit it. When we look at a star in the sky, we look directly at the source, where the photons are closest together. Traffic lights, to use another example, are too weak to even illuminate the sidewalk just a few feet below them, yet can still be easily visible from hundreds of feet away, because we see the densely packed photons at the source.

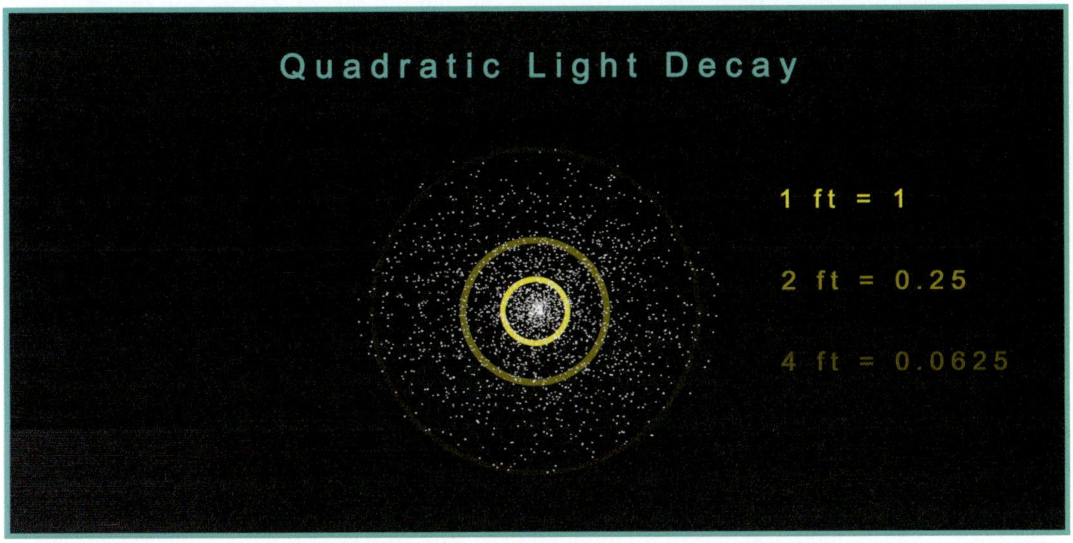

Photons retain their energy, but their density falls off with distance at a quadratic pace.

The quadratic decay of light means that there is a substantial increase in luminance in the vicinity of a light source. If you moved a light meter around a living room that is only illuminated by artificial lights, you would notice the luminance values spiking sharply when the meter is close to a lamp. If you would do the same during daytime, the values will spike even higher when you aim the meter at a window (the window is of course not a light source, but it is an opening to the considerably brighter environment outside). This is important particularly in the context of specular reflections. Typically, light sources are much more dominant in specular reflections then the rest of the environment, due to that pronounced spike in luminosity (more on this in Chapter 5).

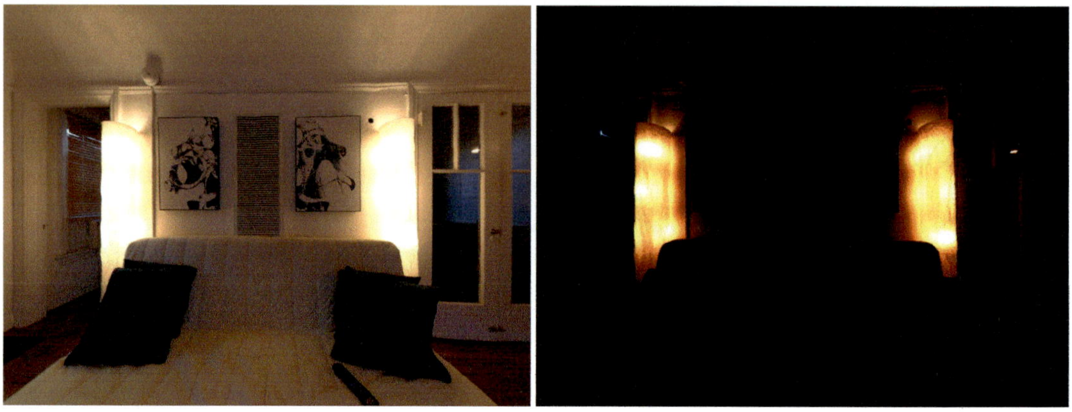

Pulling down the exposure shows the huge spike in luminance in the vicinity of the light sources.

Direct and Indirect Illumination

The only way our eyes (or the camera) can pick up direct light is by looking straight at the light source. When a camera is pointed at a light bulb, the photons emitted from the burning filament reach the sensor on a direct path. We can therefore say that the sensor is recording **direct illumination**. Everything else that is visible in the frame is the result of **indirect illumination** – photons that bounce off one or more surfaces before reaching the sensor. Simply put, most of the light that the camera captures is indirect (bounced light). This notion is, however, somewhat confusing for digital artists, because direct and indirect illumination is defined differently in CG jargon, where the distinction is based on the light-receiving surface, not the camera: if a certain surface is illuminated directly by a light source, it is said to be receiving direct light. If it is illuminated by light that is bouncing off surrounding surfaces, it is said to be receiving indirect light. This separation makes more sense from a rendering perspective: direct light is substantially faster to calculate and less noisy than indirect light (it is coming from a small, defined area rather than from multiple surfaces and directions). But in in real-world terms, even the separation between direct and indirect light is somewhat artificial. One can argue, for example, that a camera aimed at the sun is not recording purely direct light – by the time sunlight reaches the sensor it has passed through many miles of air, water droplets and dust, and has already been scattered, absorbed, and refracted to a certain extent. Categorizing light as direct or indirect is helpful when analyzing photographs and understanding light behavior, but it is important to remember that, unlike in CG lighting, the real-world interaction of light with different mediums is complex, irregular, and often difficult to categorize neatly.

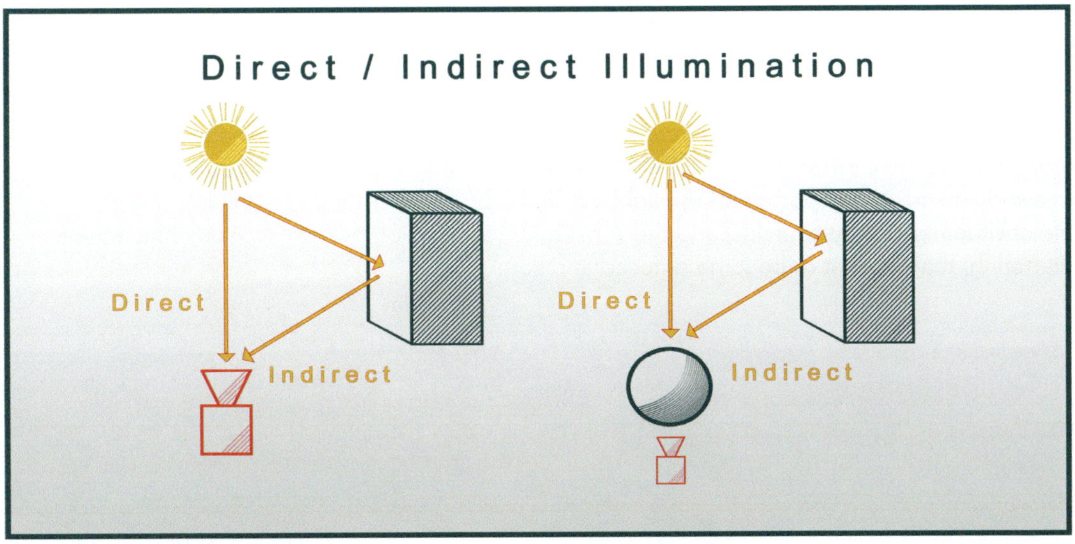

Left: in real-world/photographic terms, the camera is capturing direct illumination only if it is aimed at a light source – everything else is reflected/scattered light. Right: in CG, the distinction is made per surface: is it receiving direct or reflected light?

What is "Ambient Light"?

It is more of a notion than a physical definition. In other words, there really is no such thing as "ambient light". This may sound like an odd statement, considering how much we use this term in digital (as well as traditional) art. Let me explain: first, we need to acknowledge that the term ambient light has more than one meaning, depending on the context: when photographers talk about ambient light, they usually refer to the existing, available light in the scene (as opposed to additional lighting supplemented by the photographer). For an interior designer, the term ambient light distinguishes between the basic, general lighting layer in a room and more specific lighting layers such as task lighting and accent lighting. These usages of the term ambient light are quite different from (and not related to) the particular meaning I am referring to here: the common use of the term "ambient light" in visual art (and specifically in digital art) to describe the sort of complementary, soft, uniform light in the scene that does not seem to be coming from a distinct source or direction and is not casting distinct shadows.

This categorization can be misleading, because it infers that such a light exists as a discrete entity that is separate from other "direct" lights in the scene, or that some light is "directional" and some is "non-directional". In the real world, there is of course no separate "ambient light emitter", and all light is directional (every photon moves from point A to point B). All illumination starts out as directional light emitted from a distinct source, but some of this illumination becomes multi-directional, as it bounces off surfaces or is scattered by solid, liquid, or gaseous volumes. In CG lighting, for example, it is common to pair a directional "sunlight" with a separate spherical "sky light" to simulate natural daylight (see Chapter 11). This is an effective shortcut in CG, but it does not truly represent reality. The sky, of course, is not a light emitter, and the soft, multi-directional light that emanates form it originates exclusively from the sun and is tightly connected to it. The type of soft light in a room illuminated by a window on a cloudy day is commonly referred to as "ambient light", but all of it is coming from one source: the sun. It is thus more accurate to use terms such as **diffused light** or **scattered light**.

Does this distinction really matter? It does, especially nowadays, when CG rendering is getting closer and closer to a true simulation of real-world light behavior and relies heavily on accurate global illumination. When lighting a CG scene or integrating elements with footage, it is better to avoid looking at any portion of the light as "ambient", and instead see it in context of the environment's light source(s) and the subsequent interaction events (such as reflection or scattering) that cause it to become diffused and soft.

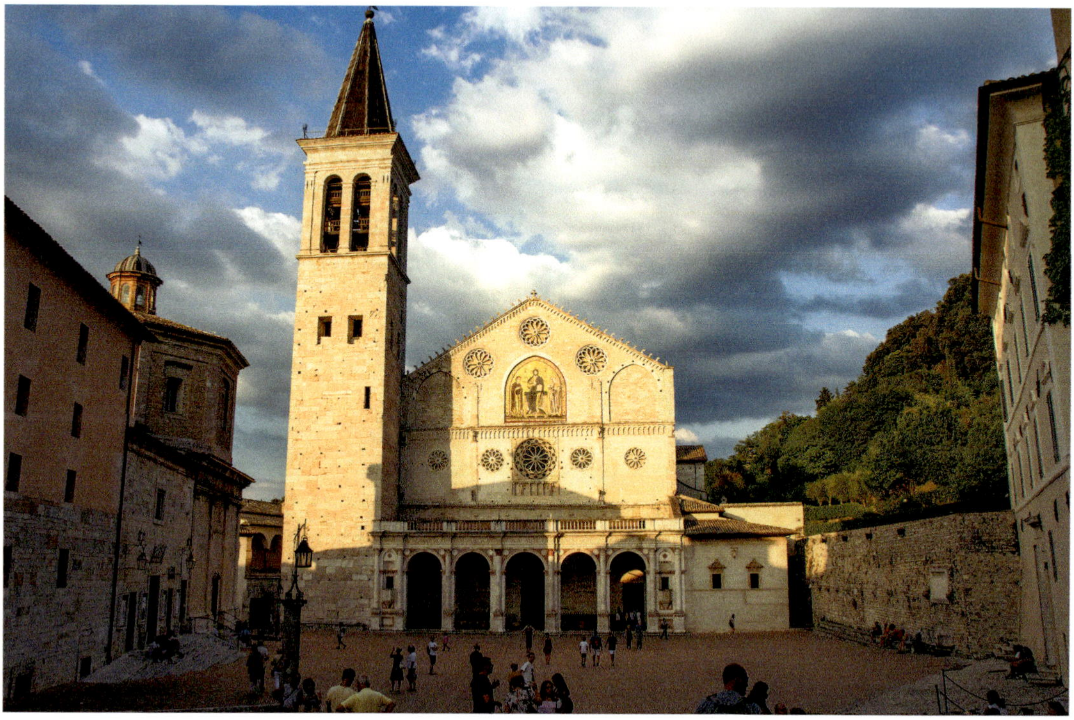

Direct sunlight is visibly hitting parts of the church. But the rest of the scene is also illuminated by sunlight that has been scattered by the atmosphere. All the light in this photo, therefore, originates from a single source, and there is no separate ambient light component.

Chapter 5

Light Interaction

As light travels through vacuum, nothing about it changes: the photons move at a constant speed and in a straight path, and carry the same energy (frequency). They may continue, unaffected, for an infinite distance and time – indeed, we can see through telescopes celestial bodies that are millions of light years away. But this perpetual motion is interrupted when light photons encounter an obstacle. That obstacle can be anything from a lone molecule in a thinly-spaced gaseous volume, to a wall of densely packed electrons in a solid metallic surface (for the sake of simplicity, I will use the term "volume" here to collectively refer to solids, liquids, and gases). When a photon interacts with the molecules in a volume, three main outcomes are possible:

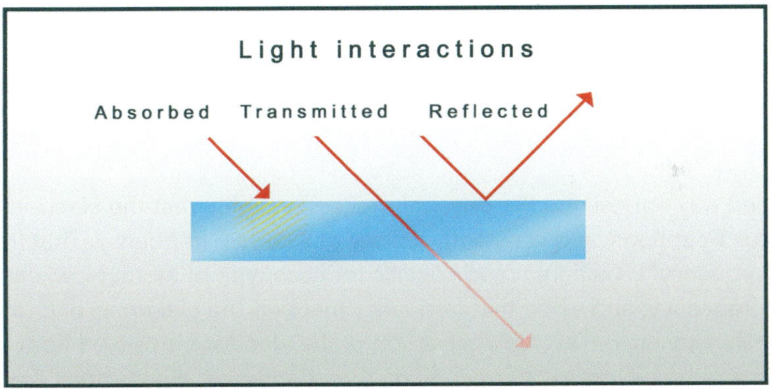

- **Absorption**: the photons' energy is converted into another type of energy (heat) which is no longer in the visible light spectrum.
- **Reflection and scattering**: the photons are reemitted from the volume's molecules at predictable angles or in random directions.
- **Transmission (transparency)**: the photons are transmitted through the volume and exit at the other end.

These basic interaction events are at the core of every aspect discussed in this part of the book, from the appearance of the sky or the visual characteristics of different materials to fog or lens flares. The following brief description of each of the three main interaction types should serve as a base for further understanding light interaction.

DOI: 10.4324/9781003590293-8

Absorption

Absorption has nothing to do with how soft, thick, or "spongy" a material is. Like all light interactions, it happens at the molecular level. A surface can be highly absorbent to certain light frequencies and still be hard as steel or thinner than paper. Think of different car paints – each paint's color is determined by the light frequencies it absorbs, yet white, blue, or black car paints all have the same material consistency. The term absorption is thus somewhat misleading – photons are not "swallowed" by the volume, rather, their interaction with the electrons in the material's atoms transforms their energy.

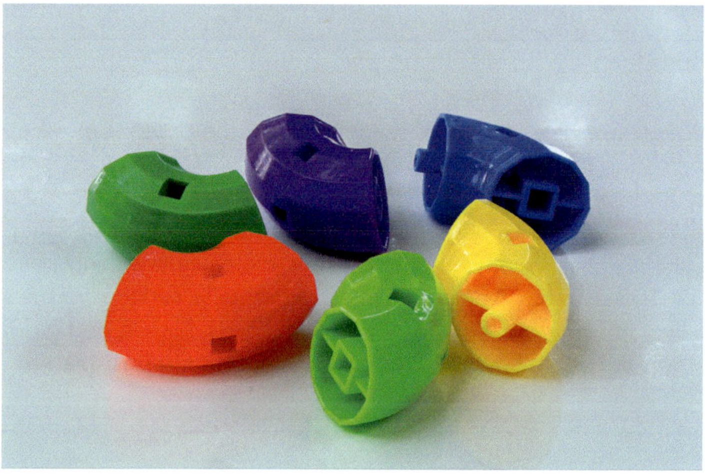

These toy pieces are all made from the same plastic material, but different paint pigments on each surface absorb different wavelengths.

Here is a simplified way to describe the physical process: imagine that the electrons are attached to the atom's nucleus by springs, and constantly vibrate at a certain frequency. That frequency (known as the "natural frequency") is different and specific for each type of atom. Now, each photon also carries a certain frequency, and when that frequency matches the electrons' natural frequency, the electrons vibrate in resonance. This high excitation of the electrons is passed from atom to atom and transforms the photons' energy to heat. If, on the other hand, the photons' frequency does not match the electrons' natural frequency, a brief period of electron excitation re-emits the photon (the photon is reflected/scattered). Absorption therefore affects the color of a surface by way of negation: the perceived color of a certain surface is the incident light color minus any absorbed frequencies. Absorption does not only affect the reflected light's frequency balance (i.e., its hue), it also affects its intensity. Black surfaces are much darker than white surfaces because most of the light energy is absorbed and very little is reflected. A similar loss of energy happens when light is transmitted through a semi-transparent material. Tinted glass absorbs some of the light, which results in darker transparency, as only a portion of the original light energy is transmitted.

Reflection and Scattering

Any surface that is not emitting light is visible only because it is reflecting light. Every photograph and every frame of a movie is essentially a capture of (mostly) reflected light, and every CG render is first and foremost a representation of reflected light. Reflection is therefore the most important aspect of the three main light interactions, and understanding reflection is key for any digital artist who strives to create photoreal imagery.

First, let us clarify the definition of reflection. In common discourse, we use the word reflection to describe the mirror-like appearance of the environment on smooth, shiny surfaces, like reflections in a lake, in a window, or on the surface of a chrome ball. But dull, rough, matte surfaces like wood, rubber, fabric, or paper reflect light too. The difference between what we commonly call "reflective" and "non-reflective" (or "matte") depends on whether the photons are reflected at a unified, predictable angle, or bounce off in random directions. An "orderly", predictable reflection is called a **specular reflection**, while a random, chaotic reflection is called a **diffuse reflection**. While most CG workflows separate the two types of reflections (diffuse controls the color of the surfaces, and specular control its shininess), in reality they are not so neatly divided, and there is no precise boundary or distinct separation between specular and diffuse reflections (one could say that a diffuse reflection is just a "messed-up specular reflection"). Still, it is important to examine each of the two reflection types separately to get e better understanding of the differences and how they affect the characteristics of various materials.

Specular Reflection

Squash players can anticipate the ball's trajectory even before it bounces off the wall, because they know that the bounce angle will be the same as the incoming angle. A ball that travels straight toward the wall will bounce straight back, while a ball that travels at an angle will bounce off at the same angle but mirrored along the wall's facing direction. The law of reflection states that: "the angle of reflection is equal to the angle of incidence". If you draw a line that represents the facing direction of a flat surface (the **surface normal**), then a photon hitting the surface at a certain angle relative to the surface normal will reflect at exactly the same angle in the opposite direction. Now replace the squash wall with a mirror, and the single squash ball with billions of photons: in a perfect specular reflection, every single photon bounces off the mirror at exactly the same angle. This means that every detail in the reflected image is preserved – no blurring, no displacement, no distortion, or degradation.

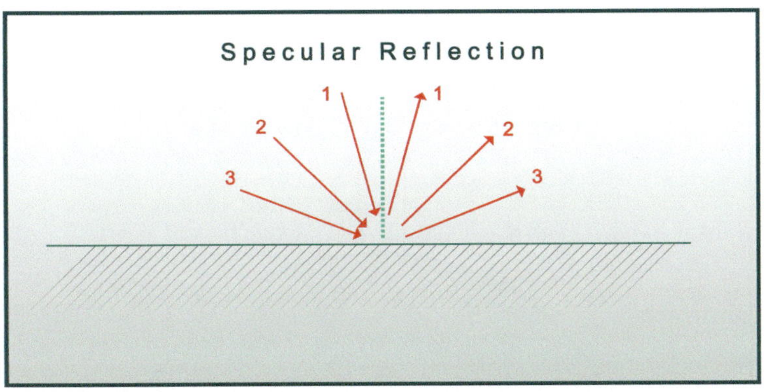

The angle of reflection is equal to the angle of incidence, mirrored along the normal (cyan line).

Mirrors are an excellent example for specular reflectivity because they are manufactured to produce the clearest reflection possible. A mirror must allow light to follow the law of reflection with minimal deviations. This requires a material that is consistently smooth even on a microscopic level. Any irregularities in the material must be smaller than the wavelength of the incident light, to ensure that photons are uniformly reflected at a predictable angle and do not scatter. The amount and scale of a material's micro-roughness is the principal deciding factor between specular and diffuse reflections. Glass is a non-crystalline material and has no grain boundaries, which makes it an ideal material for mirrors (especially when coated with silver or aluminum to boost its reflectivity). But on top of its microscopic smoothness, the mirror glass must also be perfectly polished and even. A curved mirror is still highly specular, but it produces a distorted reflection. This does not happen because the photons are being scattered in different directions – they still bounce at a predictable angle, but the facing normal of the mirror change along its curvature, which affects the angle of the reflected photons and distorts the reflected image.

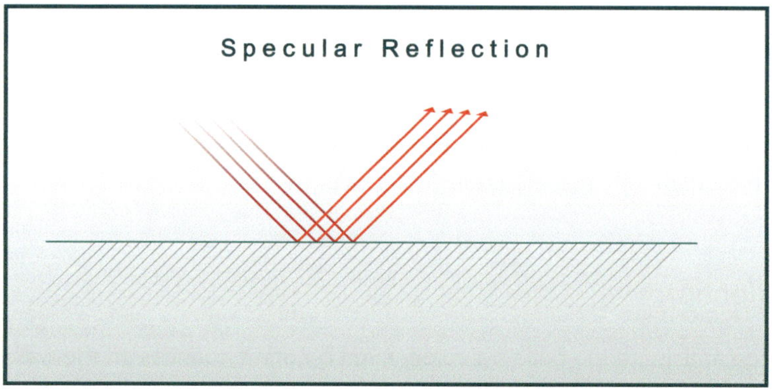

A perfect specular reflection transforms the reflected image without any distortion or degradation.

Most specular surfaces do not reflect as perfectly as mirrors. Imperfect or partial specular reflections blur the reflected image to some degree, and when that happens, the brightest areas in the reflection are naturally the last ones to remain distinct. Since light sources are substantially brighter than their surroundings, their appearance in a soft specular reflection can greatly affect its visibility and intensity. Where a light source (including the sky or even a window) is reflected, the specular reflection is substantially brighter and more noticeable. This dominance of light sources or very bright areas of the environment in specular reflections is commonly called **highlights.** If we want to catch a highlight on a flat surface, we need to aim that surface directly toward the light source. But what happens when the surface is curved?

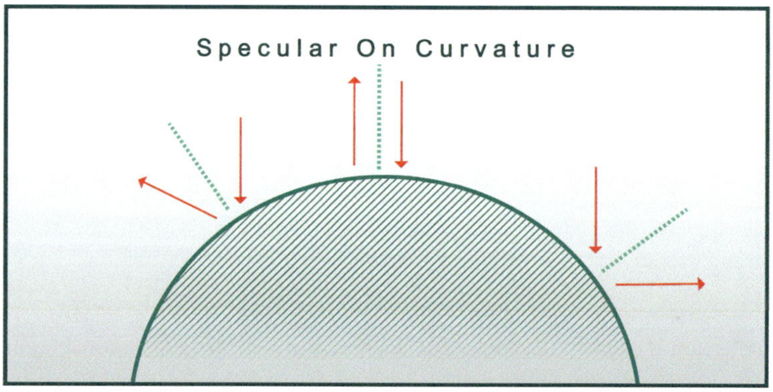

The direction of the normal varies along a curved surface, which changes the angle of reflection and distorts the image – just like a fish-eye lens.

Replace a flat mirror with a chrome ball. The spherical surface acts very much like a fish-eye lens – it covers a much wider angle, reflecting a much larger portion of the surrounding environment. Such a curved surface has a much higher chance of reflecting a light source than a flat surface. If you photograph a reflective cube and a reflective sphere on a sunny day, you will need to carefully rotate the cube to catch the direct reflection of the sun. The sphere, on the other hand, stands a 50% chance of reflecting the sun – if the sun is anywhere in in the hemisphere behind the camera, it will be visible in the specular reflection. This is the reason why specular reflections are much more noticeable on curved surfaces. What we generally call **rim highlights**, the appearance of bright, distinct specular reflections along beveled or curved corners, contours, and surface bumps, is the result of each of these curvatures acting as a "mini-fisheye lens", catching and reflecting bright areas like light sources, windows, or the sky within a very narrow area on the surface. This is one of the most important visual characteristics of specular reflections, because rim highlights play an important role in visually defining the object's contours and surface curvature.

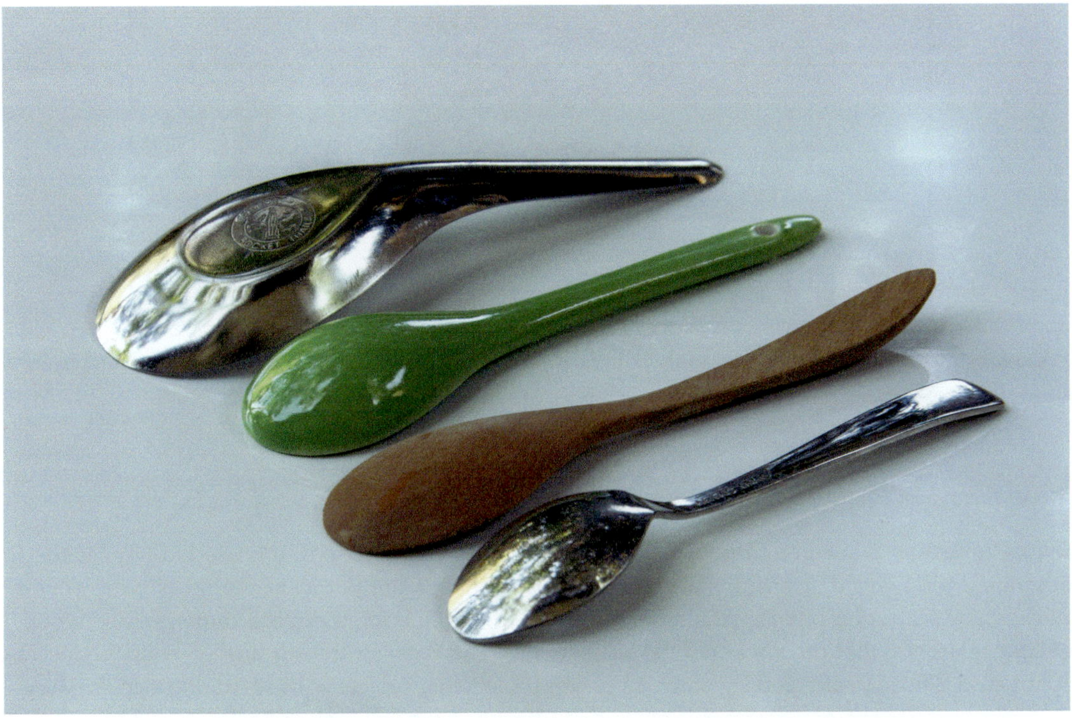

These spoons display different specular characteristics. Notice the mirror-like, detailed reflections on the metal spoons. The top spoon's surface is just slightly rougher, showing a softer reflected image. In contrast, the reflection on the non-metallic green spoon is much more subdued, showing mainly the brightest parts. Notice the rim highlight along the side. Finally, the wooden spoon may have some specularity, but the roughness is so strong that reflections are all but visible.

Diffuse Reflection

The micro-structure of highly specular materials has very few irregularities at the scale of the light wavelengths – it is flat and smooth, just like a squash wall. Now imagine playing squash against a rocky cliff full of outcrops and cavities: it would be virtually impossible to anticipate the trajectory of the bouncing ball. Many materials are composed of microscopic "pieces" (called **grain**) bound together. These pieces can be crystals (polycrystalline materials like rock and ceramics) or fibers (materials like wood, paper, and fabric). The irregularity of grain and grain boundaries causes the photons to scatter in random directions as they hit the surface. Light is reflected, but the arbitrary direction of the photons means that detail of the reflected image is not preserved. Perfectly diffused reflection appears as a uniform color rather than an image of the environment, as the outcome of photons reflecting in random directions is an average of the light intensities, not unlike applying a very heavy blur to a photo. That color represents those frequencies that have not been absorbed, so the balance between absorption and diffuse reflection is what gives a surface its color (hence the term "**diffuse color**").

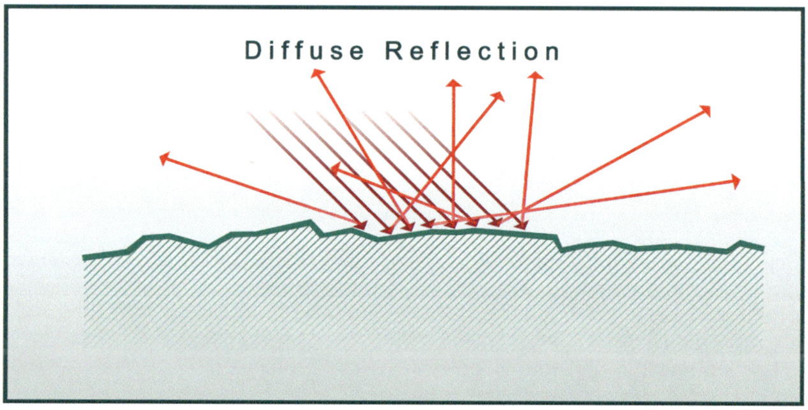

Micro-roughness causes reflected light to shoot in random directions. The reflected image becomes completely diffused.

Scattering

Diffuse reflection is in fact a form of scattering, a much broader term that is used in physics to describe a wide range of interactions and disturbances between particles, atoms, and molecules. While many scattering phenomena are not related to visible light, there are some specific types of light scattering that greatly affect the visual world. The term scattering is often used to describe light being diffusely reflected by particles or gas molecules. Typical light scattering occurs when light hits dust, smoke or water particles, and in Chapter 6 we will look at Rayleigh and Mie scattering and their important contribution to daylight and the appearance of the atmosphere.

Subsurface Scattering

Going back to the example of playing squash against a cliff full of outcrops and cavities, the ball may sometimes enter one of the cavities, bounce inside a few times and then exit at a different location from where it first hit the cliff. In most non-metallic materials, diffuse reflection also involves at least some penetration of photons below the surface. This internal scattering, and the slight displacement of photons off their original course contributes to the characteristic "soft" feel of many diffuse surfaces. When the photon penetration is deeper and more pronounced (as in marble), materials take on a sort of milky appearance, as the surface features become slightly blurred and less defined. When the penetration is even deeper and the scattering radius larger (as in wax or human skin) the material appears to be almost translucent, although light is not really transmitted through it (see next section).

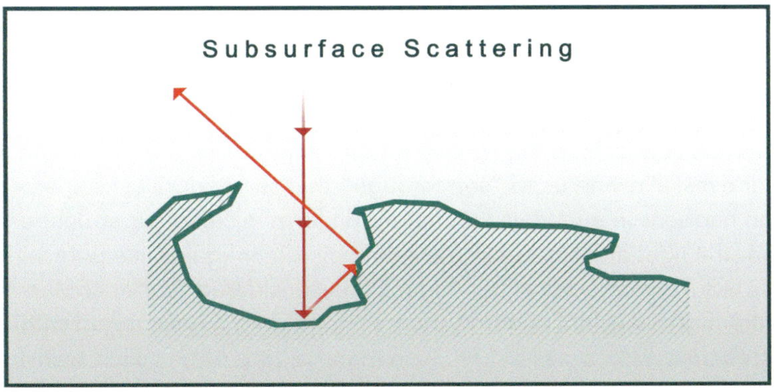

Subsurface scattering is in fact a CG term, and it is usually an additional shader parameter that is separate from diffuse reflection (see Chapter 12). Although this makes sense on a practical level (subsurface scattering is computationally intensive, and many diffuse materials can be emulated well enough without it), it is important to note here that subsurface scattering is not

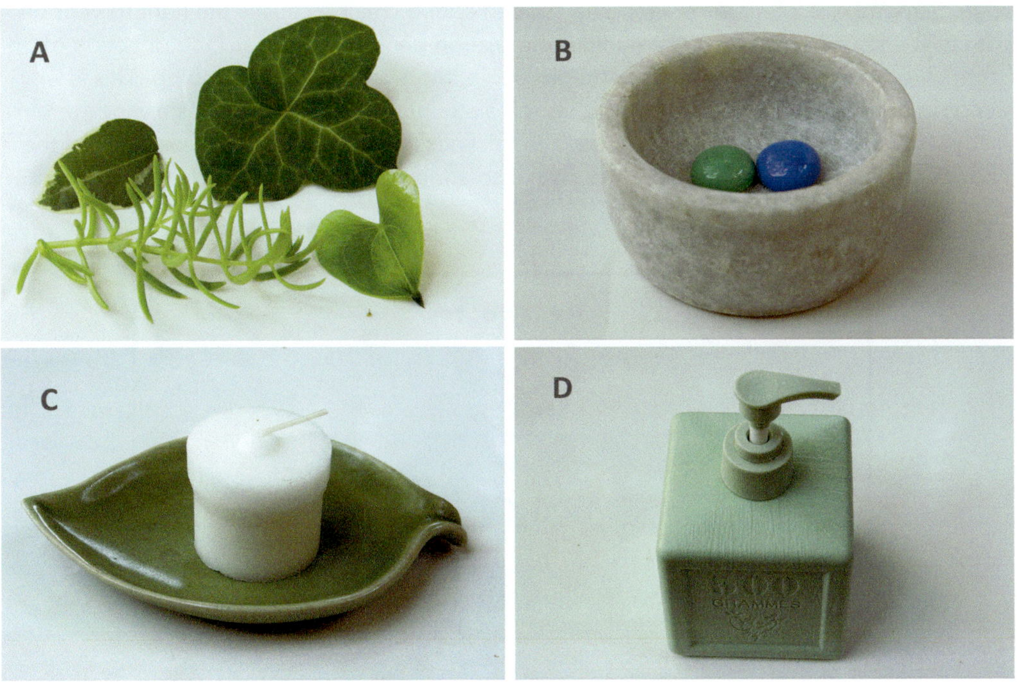

Each of these objects shows some amount of subsurface scattering, giving them a subtle “soft” look: various leaves (A), a marble bowl and polished marbles (B), a wax candle and lacquered clay candle holder (C), a plastic soap dispenser (D).

exclusive to only a few special materials, neither is it a separate process from diffuse reflection. In fact, it is an integral part of the diffuse characteristic of almost all dielectric (non-metallic) materials.

Transmission and Refraction

Whenever light is not completely absorbed or reflected by a volume, the photons are transmitted through the volume and exit on the opposite side. We perceive a surface as transparent when reflected light from objects behind the surface (or direct light from a source) is transmitted through it and reaches our eyes, allowing us to "see through" that surface. Just like specular reflections, clear transmission happens in materials that have a uniform microscopic structure with little or no grain to scatter the light and cloud the transparency. Clear liquids like pure water are a good example. Glass is not a typical solid either (In fact, it is considered "super-cooled liquid"). When the structure of the material is less uniform, the transmitted light is scattered within the volume, and the transmitted image becomes blurry. Such materials are often called **translucent**, but one can say that the difference between translucency and transparency is similar to the difference between diffuse and specular reflections – translucency is a "messed up transparency", as light is transmitted and scattered at the same time.

The visual characteristics of transparent materials cannot be properly simulated in digital art without factoring in the all-important side effect of transmission: **refraction**. The transmission process slows the photons down. In gases such as air, the slowdown is very small, just a mere fraction. But when light passes through glass it is roughly 1.5 times slower, and when it passes through diamond it is 2.5 times slower. The sudden speed change causes the light to veer off its path as it enters the transmitting medium. Refraction bends light toward the direction of the normal, so when the normal is at a straight angle, there is little or no bending. This is why refraction is not noticeable when looking through a straight glass panel such as a window, but is clearly visible when the surface is curved. The thickness of the transparent medium and the number of different mediums in the light's path also affect refraction. A glass full of water shows stronger refraction than an empty one.

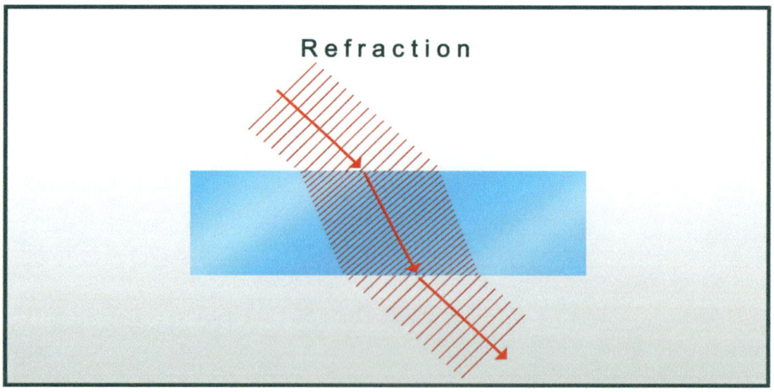

The ratio between the speed of light in vacuum and the speed of light within a certain volume is called the **refractive index** (also **index of refraction** or **IOR**). The higher the ratio, the stronger the deviation, and subsequently, the distortion of the transmitted image. The IOR of water is

1.33, and most liquids like alcohol or oil have an IOR between 1.33 and 1.5; the human cornea has an IOR of around 1.38; plate (window) glass has an IOR of 1.5, and most transparent plastics hover around this value; some gemstones and minerals have values higher than 2, with diamond at 2.42 and silicone at 3.45. However, as you can tell, the IOR of most transparent materials lies within a rather limited range of 1.33 to 2.5. So, in most cases, incorrect IOR is not as detrimental to photorealism as some may think, especially since we are not inherently sensitive to this aspect (most people will not notice anything wrong in a render of a wine glass that has an IOR of 2.5). That said, refraction index plays other roles in physically-based rendering, such as determining the amount of Fresnel effect (see Chapters 9 and 12), so adherence to correct IOR in CG rendering is still advisable.

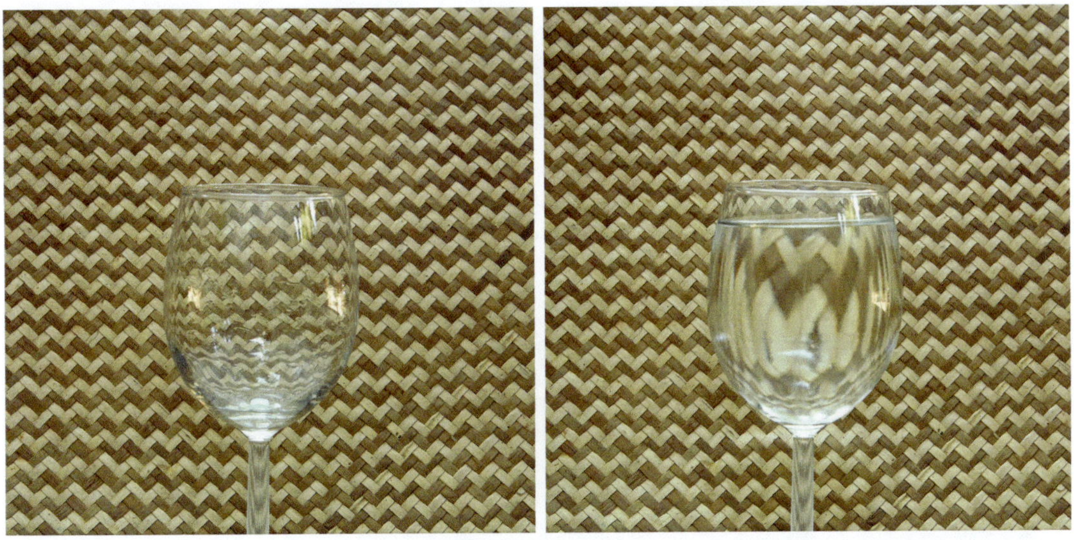

An empty wine glass shows only minimal refraction, but when filled with water, the much thicker volume, and the transmission through two different materials, increase the effect of refraction substantially.

Albedo

According to the **law of conservation of energy**, a surface cannot reflect more light than it receives. The combined luminance of both the diffuse and specular reflections can never exceed the luminance of the incident light. Albedo is a simple measurement of a surface's reflective luminosity on a scale of zero to one. A surface that reflects none of the incident light has an albedo value of zero, and a surface that reflects all the incident light has an albedo value of one. Naturally, the albedo of most materials lies somewhere in between those extremities, as no material is 100% absorbent, and even the whitest or shiniest material still absorbs at least a fraction of the incident light. For example, black velvet has an albedo value as low as 0.01, while fresh snow reaches 0.9. Since real-world albedo includes both diffuse and specular reflections, it can be used as an effective measure for the overall brightness of a material, regardless of whether it is dull or shiny, metallic or non-metallic. This measure is especially important in the context of CG (where the law of conservation of energy can be easily broken) and serves as a check for photoreal consistency. In CG, separating the diffuse and specular albedos can help gauge the balance within a specific material, and how it is affected by absorption and transmission. For example, the diffuse albedo of clear glass is very low (nearly black), even though there is almost no absorption. Instead, most of the light is either transmitted or reflected.

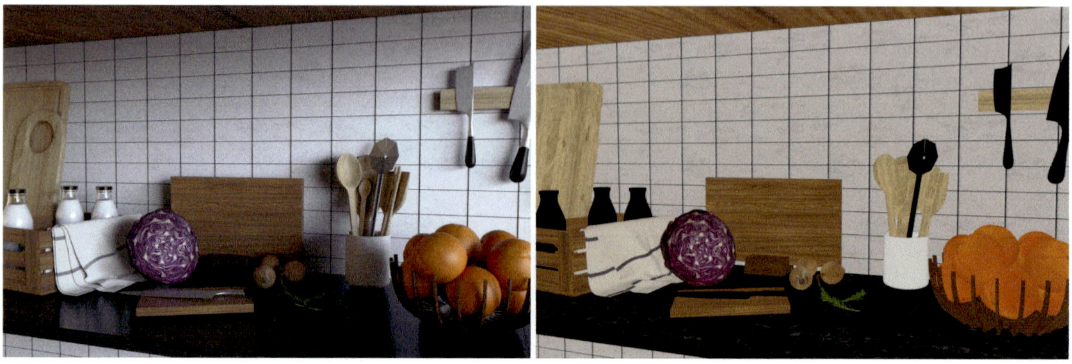

It is, of course, not possible to extract an albedo image from a real photograph, but a CG render engine can easily do the separation. Here. The image on the right is a diffuse albedo pass of the full render (left). Notice that metallic (fully specular) and transparent surfaces are black (zero diffuse albedo).

Case Study: Side by Side Comparison

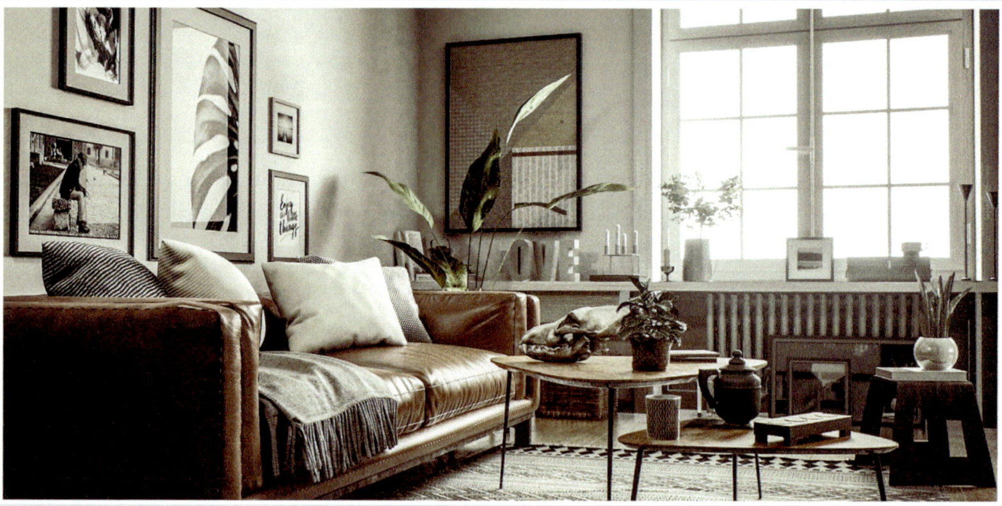

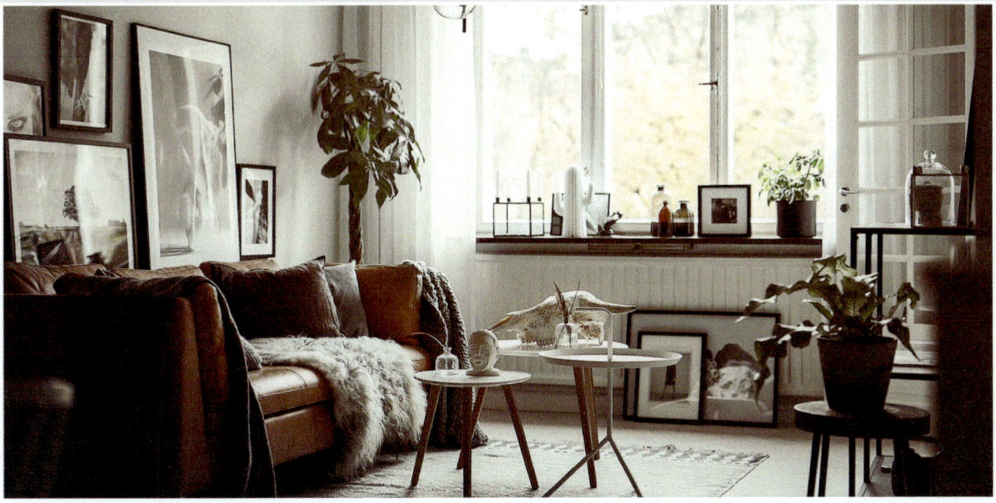

© Pasquale Scionti.

Pasquale Scionti is an architectural visualization artist with an impressive talent for producing photoreal interior renders (several examples of his artwork are featured in this book). In this particular work, Pasquale offers us a rare opportunity to compare, side by side, his CG render and the photograph that he used as reference. This is unique, because Pasqual's render (top image) is strikingly similar to its real-world inspiration (bottom image). When I first saw these images, I thought that the bottom image was simply an improved version of the render. The fact that I did not immediately realize that one is a CG render and the other a photo speaks volume about the quality of this artist's work. Conversely, the fact that I preferred the photograph over its CG emulation shows how elusive the goal of photorealism is, even when we get as close as Pasquale did here.

What makes this render so successful, and where can it still be improved? The area of the sofa and the wall behind it, in my opinion, is outstandingly photoreal, and cannot be distinguished from the photo. The leather texture, the blanket and pillows, the plants, the carpet – the entire left half of the image is incredibly well done. Where the CG may be lacking compared to the photo is mostly in the window area: the trees outside, though nothing more than blurry yellowish smudges in the photo, add the kind of depth and detail that is missing from the render. A subtle bloom effect is a good addition around areas of bright light (see Chapter 10), but here it seems a tad too strong, almost sucking the color out of the objects on and around the window sill. Adding some out of focus exterior detail, reducing the light bloom and bringing back some color and definition around the window could lift this fantastic render even higher. But, to be perfectly honest, would I even point out these specific factors without being able to compare the render to a photograph? Most likely not. Photorealism is tricky – it is easier to achieve when we reference real-world footage, yet this process also exposes the tiniest facets that can otherwise go completely unnoticed.

Chapter 6

Daylight

Earth is illuminated by a single source of light: a G2V type yellow dwarf star, roughly 93 million miles (150 million km) away. But the light generated from this lone star would have felt very different without the other, inseparable component of daylight: earth's atmosphere. Even Though it does not produce any light of its own, this thin layer of gases acts as a multi-purpose diffuser/reflector/filter that affects the constant, unchangeable sunlight in a myriad of ways and generates the richly varied qualities of daytime light. Understanding the complex interaction between sunlight and the atmosphere is the key to simulating realistic outdoor daytime lighting and plausible environments.

The Sun

The sun is a sphere of hot plasma roughly 860,000 miles in diameter. Light radiates from it in all directions; in the same way light radiates from a light bulb. However, what makes the sun such a unique light source on earth is the fact that it is so far away. It is impossible to illustrate the actual distance between earth and the sun on a single page, but this great online true-scale illustration by Josh Worth, "If the Moon Were Only 1 Pixel" (http://joshworth.com/dev/pixelspace/pixelspace_solarsystem.html), provides a frighteningly true sense of how incredibly far the sun is. That distance, as well as the minuscule size of earth relative to that distance, means that we are receiving only an extremely narrow slice of sunlight rays. If you draw many lines emanating from a circle (the sun) and then place the earth at the correct relative scale and distance, you'd see that the earth sits on a single line, with the nearest lines virtually parallel (here is another useful animation: https://javalab.org/en/parallel_rays_of_sun_en/). So, is sunlight on earth parallel? Not exactly: even when factoring our distance from it, the fact that the sun is such an enormous light source means that some of the rays that reach us do so at slightly converging (or diverging) angles. But the difference is minuscule, no more than half a degree, and while mathematically sunlight rays on earth are not absolutely parallel, for all practical (and visual) purposes they are. Direct sunlight hits every object in our view at the same angle, and its shadows are always parallel. This is distinctly different from any other light source we experience on earth, natural or man-made. The inverse-square rule applies to sunlight as it applies to any light. Indeed, Mercury's sun-facing side receives substantially more radiation than, say, Mars, and even on our planet, there's a significant difference between the equator and the poles (which are farther away from the sun). But for all visual purposes, sunlight has no perceivable distance-related decay. On a clear, dry day, there is no perceivable difference in sunlight intensity between an object right next to the camera and one

DOI: 10.4324/9781003590293-9

five miles away (the atmosphere, of course does have an effect over distance, as will be discussed later in this chapter).

To experience raw, unfiltered sunlight, we need to go up above the atmosphere. Nowadays we are used to seeing space photography, so it's easy to forget that up until October 24, 1946 (when the first grainy photograph from space was taken by a camera mounted on a German V2 rocket), people had no way of experiencing such visual oddities as bright sunlight in a pitch-black sky, the stark contrast between light and shadow in space, or the eerily sharp and depthless horizon on the moon. But thanks to more than 70 years of space photography, we have substantial visual reference to how raw sunlight looks like. First, the color: the sun emits radiation across much of the electromagnetic spectrum, including X-rays, ultraviolet, infrared, and the full range of visible light. So, without atmospheric filtering, sunlight is white – not yellowish or orange as we experience it on earth.

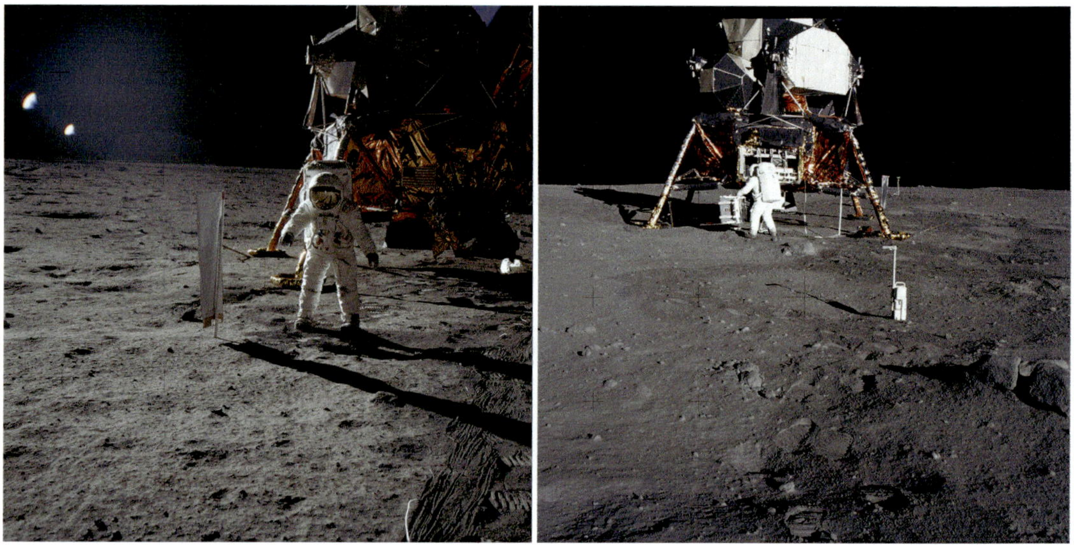

Two photographs from the Apollo 11 mission. Notice the bright unfiltered light, harsh, pitch-black shadows, and complete lack of atmospheric depth. The shadowed parts on the spacesuit are much less dark than the shadows on the ground because of reflected light from the ground.

In the absence of any scattered sunlight, fully occluded areas are very dark. Shadows are only illuminated by reflected light from nearby surfaces. If no such bounce light exists, shadows are truly black. For example, in many of the iconic Apollo moonwalk photographs, the astronaut's shadow on the lunar ground is pitch black, yet the fully occluded areas of his spacesuit are still relatively bright. This is because the spacesuit is illuminated by light bouncing off the ground, yet there is nothing above the lunar surface to bounce light back onto the shadows on the ground. Imagine daylight on earth without atmosphere: sunlight is white, harsh, and consistent day-in day-out. There is extreme contrast between lit and occluded areas, and shadows are always sharp and well-defined. Sunsets are a rather dull affair, a simple transition from full illumination to complete darkness as the sun unceremoniously disappears behind the horizon. The sky is always pitch black and the horizon always clear and well-defined, without the familiar visual cues for distance and depth. But thanks to our atmosphere, this harsh white light is transformed into the constantly shifting visual magic we call daytime light.

The Atmosphere

The layer of gases that envelopes earth is comprised mostly of nitrogen and oxygen, about one percent of argon, and small traces of other gases like carbon dioxide, helium, and methane. It also contains some water vapor, with amounts varying from mere traces in dry areas to about 5% in humid tropical areas. Most of the atmosphere's mass (about three quarters) is concentrated in the first 6.8 miles (11 kilometers) above earth's surface, and the rest gradually thins out over a distance of several hundred miles (thus there is no precise definition of the atmosphere's height). Closer to the surface, additional elements like dust, pollen and other particles (mainly from human pollution) may also be present in the atmosphere.

Like other gases, air molecules are spaced far apart. However, the thickness of the atmosphere layer means that collision events between air molecules and photons are much more likely to happen. With the additional influence of other elements in the atmosphere such as water and dust, the atmosphere becomes a considerable obstacle that substantially alters sunlight. When a sunlight photon enters the atmosphere, three main outcomes are possible:

- It can be absorbed or reflected by atmospheric elements (in which case it never reaches earth's surface).
- It can pass through the atmosphere and reach earth's surface as direct sunlight.
- It can be scattered by air molecules or larger particles and reach earth's surface as multi-directional scattered light.

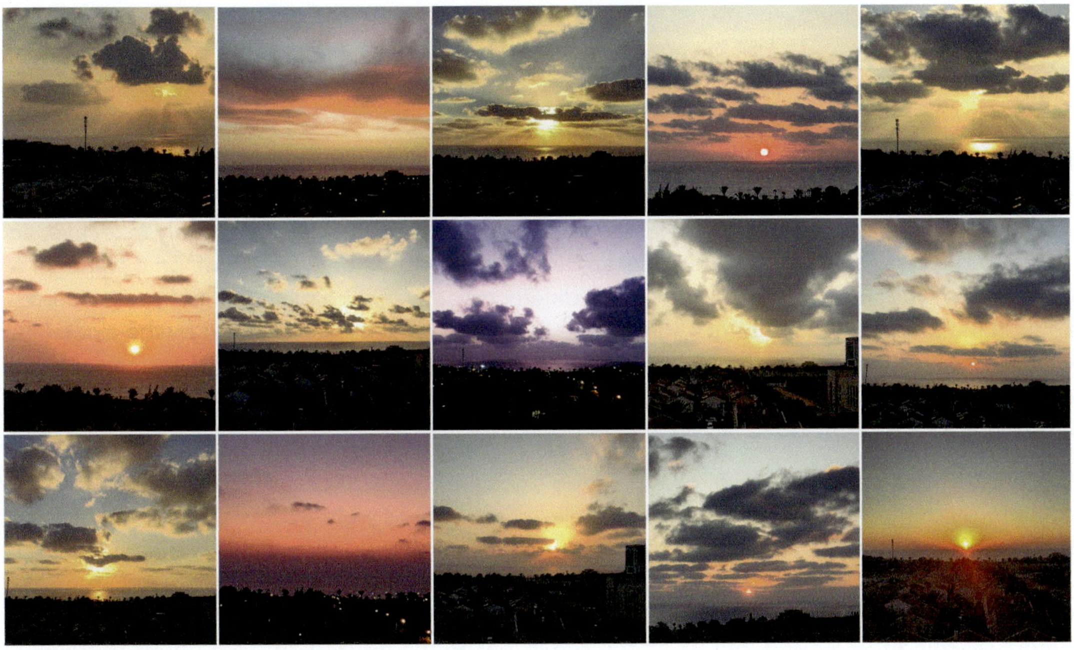

Every evening for several years, my father would take a photo of the sunset from the balcony with his phone. This small selection shows how differences in the balance of atmospheric elements can produce a stunning variety of moods and colors at one single location.

© Ouri Dinour.

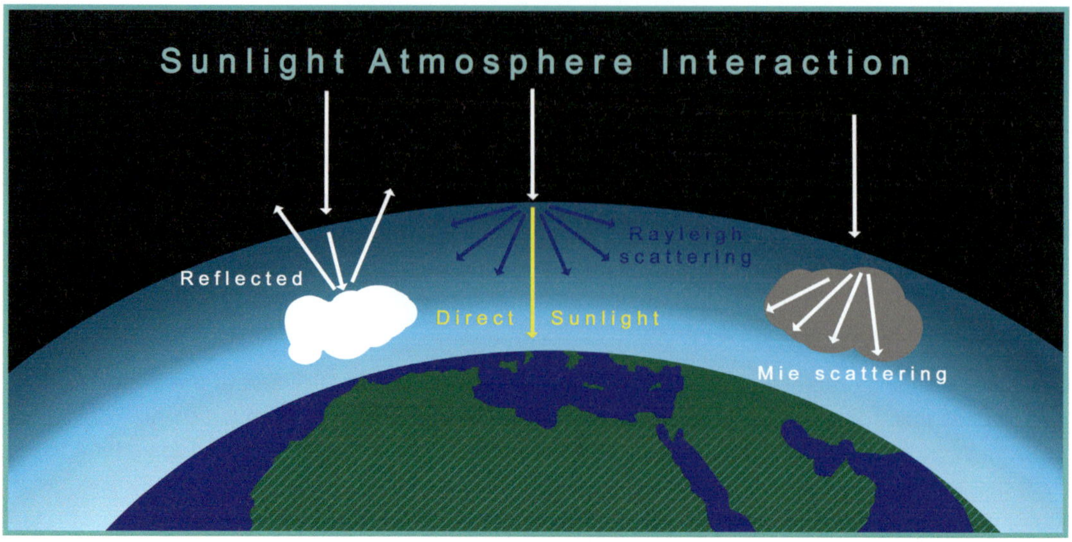

The three main outcomes of sunlight interaction with the atmosphere: reflection/absorption, scattering, and direct sunlight.

Daylight is a combination of all three outcomes, and to better understand their collective effect, we will examine each of them separately.

Reflection and Absorption

Just a little less than half of the sun's radiation that reaches earth never makes it through to the ground. About a quarter is reflected into space, and another fifth or so is absorbed by the atmosphere (most of the absorption happens in the ozone layer and affects mostly ultraviolet radiation, which is outside the visible spectrum). Water droplets in clouds are highly reflective, and thick cloud layers can reflect a substantial amount of sunlight. When you take off on a plane under a cloudy sky, you may experience the strong difference between the relatively dark, attenuated light on the ground, and the bright light above the clouds, which is augmented by the cloud layer which acts as a formidable light reflector. Seen from above, even the most massive storm clouds are white, but they appear particularly dark from the ground because most of the sunlight has already been scattered and reflected in their upper reaches.

While the amount and density of clouds in the sky significantly affects the attenuation of sunlight on the ground, other types of particles contribute as well. Absorption and reflection by dust, smoke, or pollution further reduces sunlight brightness. Like all atmospheric light interaction, the amount of attenuation strongly depends on the position of the sun in the sky. When the sun is at the zenith, sunlight travels through the least amount of atmosphere before reaching the ground, but when the sun is low on the horizon, its light needs to travel a considerably longer path through the atmosphere (and atmospheric elements like clouds or dust) before it reaches our eyes. Sunlight attenuation is thus substantially stronger during sunrise and sunset, so much so that when the sun is low enough, we can stare at it directly without any harm to the eyes.

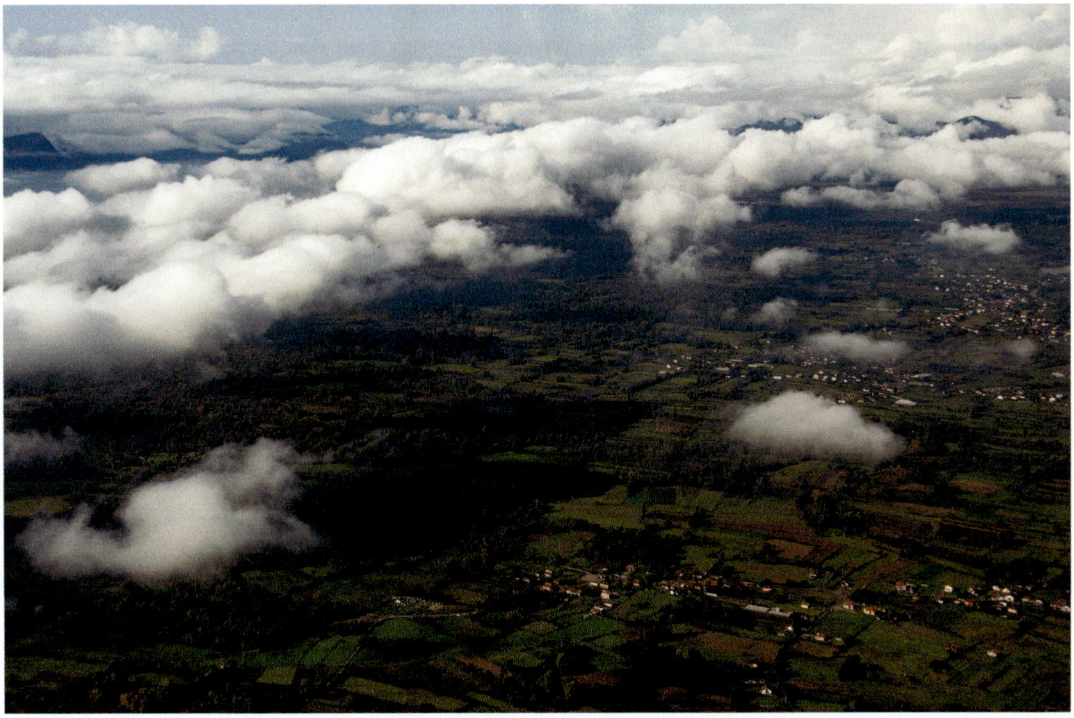

Clouds act as a formidable reflector, bouncing sunlight back into space and substantially attenuating daylight on the ground.

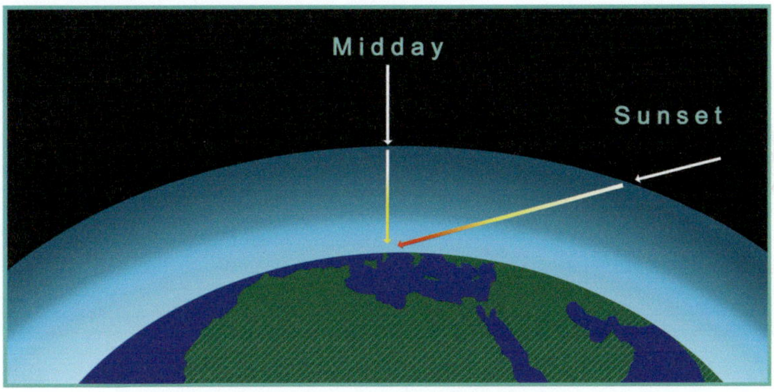

The effects of atmosphere on sunlight substantially increase during sunrise and sunset, as sunlight travels longer through the atmosphere.

Atmospheric Scattering

Scattering happens when a light photon collides with a molecule or a particle and is redirected while retaining most of its energy. Scattered light is widely propagated through the atmosphere with little attenuation, and hits the ground from multiple directions, from the zenith down to the horizon. Scattering effectively converts the sky into a giant diffuser (or softbox) which transforms direct sunlight into multi-directional diffused light. This effect is called **diffuse sky radiation,** or sometime just **skylight**. The latter term is both accurate and inaccurate – the light is indeed coming from the sky, but the sky is not a light source, it merely propagates the light of the sun. Two principal types of scattering, **Rayleigh scattering** and **Mie scattering**, happen in the atmosphere, and the balance between them, as well as the balance between direct and scattered sunlight, all contribute to the infinitely varied color and character of daytime light and the appearance of the sky.

Rayleigh Scattering

The most transformative effect of the atmosphere on sunlight, Rayleigh scattering occurs when sunlight photons hit air molecules (mostly nitrogen and oxygen) which are smaller than the light's wavelengths. Rayleigh scattering affects shorter wavelengths more than longer ones, so blue and violet light is scattered considerably more than red and yellow light. This gives the clear sky its distinct blue color (and diffused sky radiation a bluish tint). It also strips some violet and blue from the white sunlight, causing direct light to tint toward yellow. At sunrise and sunset, when the sun is low on the horizon and its light passes through substantially more atmosphere before reaching the surface, the increased scattering removes even more of the short wavelengths, tinting sunlight color further toward orange and red.

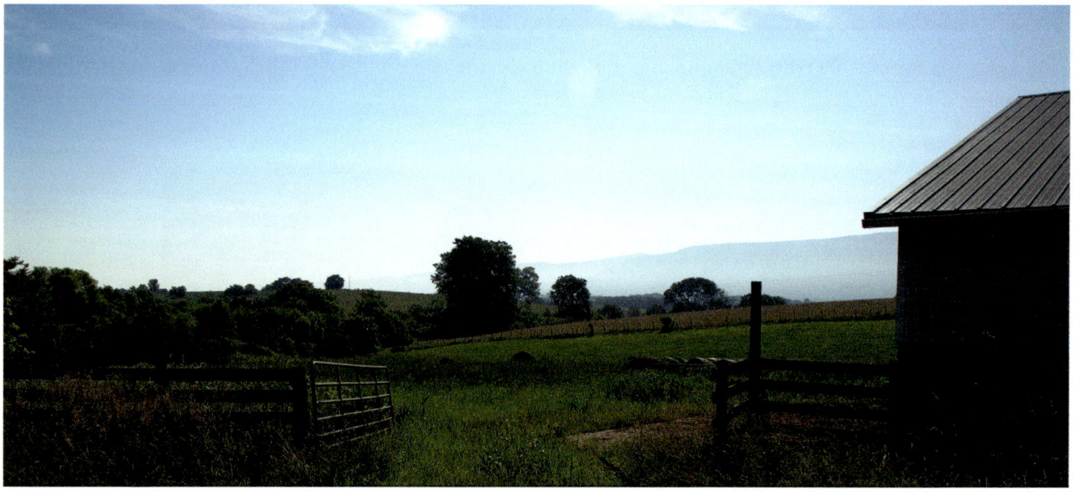

Rayleigh scattering caused by air molecules accounts for the blue color of the sky, but Mie scattering, caused by low-altitude water droplets and dust particles shifts the horizon color toward white.

Mie Scattering

Mie scattering happens when photons hit particles that are larger than the radiation wavelength. These particles are commonly water droplets (in humid air and clouds) or dust, but they can also be smoke and pollution particles. As with Rayleigh scattering, sunlight photons are redirected in all directions, but Mie scattering is not wavelength-dependent, and since all frequencies are scattered equally, sunlight color remains unchanged in the process. The perceived color of Mie scattering depends on how sunlight color is shifted by Rayleigh scattering. When the sun is it at the zenith and sunlight is closest to white, clouds are white too (or gray, where attenuation happens). As the sun moves toward the horizon, Rayleigh scattering shifts sunlight color toward orange and red, which in turn affects the color of Mie scattering. During sunrise and sunset, clouds, humidity, dust and pollution take an orange/red tint, especially toward the direction of the sun.

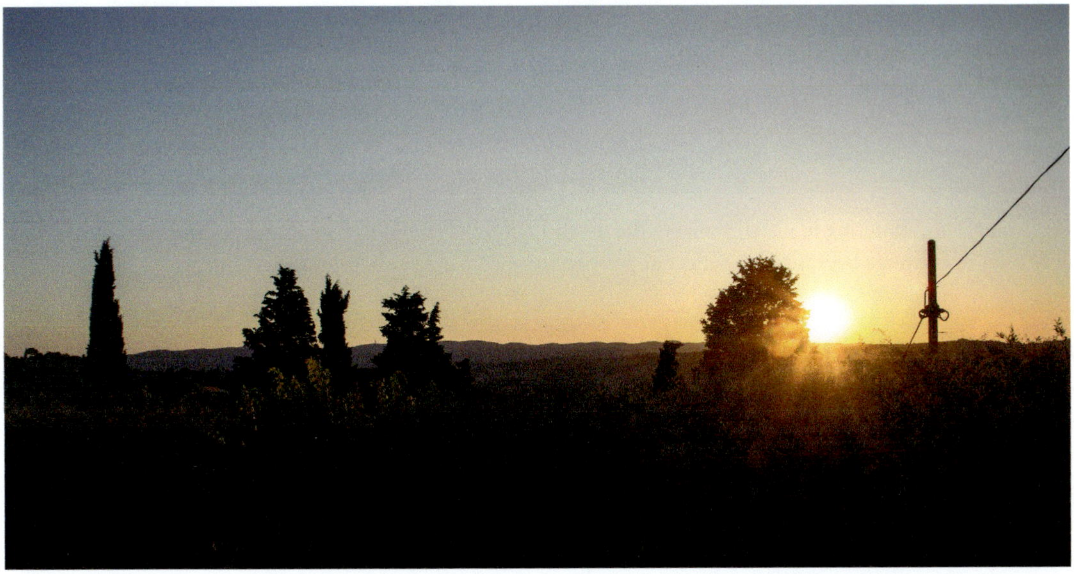

When sunlight travels through more atmosphere to reach the camera, more of the blue and violet has been removed from it, and it takes a reddish hue. Mie scattering scatters that red light at lower altitudes where larger particles lie, but the blue color of Rayleigh scattering still dominates the higher altitudes.

Since all these elements lie mostly in the lowest layers of the atmosphere, Mie scattering is considerably more visible at the horizon than at the zenith. In general, the horizon appears brighter and more desaturated, but the color shift and the rate of gradation from zenith to horizon can change drastically depending on the time of day and the atmospheric conditions – from pale blue or white or gray at midday, to yellow/orange/red during sunset and sunrise. It is important to note that the air in the atmosphere remains blue during sunset and sunrise. It is the additional atmospheric elements (clouds, humidity, dust and pollution) that turn red and create the familiar effect of "flaming sky" (indeed, some of the most spectacular sunsets occur when there is a strong presence of particles in the atmosphere).

Volumetric Light

The term volumetric light comes from the CG world and is used for a wide range of scenarios where light beams are distinctly visible due to molecular or particle scattering. These scenarios include atmospheric phenomena like **crepuscular rays** (aka sun beams or god rays), shafts of sunlight entering a room, or spotlight beams on a stage. The two ingredients that are needed for such light beams to appear are a dense enough accumulation of particles like humidity, fog, smoke, or dust, and some kind of obstacle that creates distinct separation between light and shadow. In the case of sunlight rays in a room, the obstacle could be narrow slits in the blinds, an

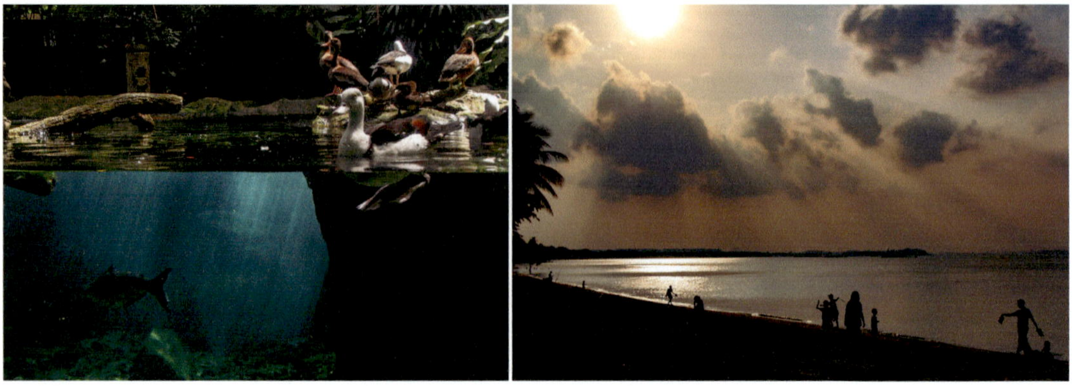

Left: volumetric scattering in water, showing the parallel nature of sunlight. Right: god rays are parallel too, and only seem to converge because of the effect of perspective over distance.

opaque/transparent pattern in the curtains, or a crack in a wooden wall. For crepuscular rays, it is usually a cloud formation with relatively small and well-defined gaps. In all these cases, it is the sharp contrast between light and shadow that makes scattered light appear as distinct rays rather than a general diffused glow. The direction of volumetric rays usually spreads out from the light source (though stage spotlights limit the spread to a narrow, focused area). The two exceptions are laser beams (essentially, an extremely focused light) and sunlight. God rays are in fact parallel, and only appear to be fanning out from a single point because of perspective (just like railroad tracks appear to be converging in the distance).

Aerial Perspective

The moon landing conspiracy theories must have been at least somewhat motivated by the Apollo moon photographs. After all, everything in those images, from nearby rocks to far away mountains appears to be at the same distance from the camera. The complete lack of atmospheric perspective does make the lunar photos look as if they were shot in a studio with a black screen as a background. The accumulative influence of the atmosphere on the way objects appear at a distance is strongly ingrained in our spatial perception, and it constitutes a fundamental aspect in visual art, traditional or digital. **Aerial perspective** is the commonly used term for the atmospheric depth, but contrary to what the name implies, air is not the only factor in atmospheric depth.

The atmosphere is not just above and around us – it engulfs us. All the atmospheric light interactions described so far (absorption, reflection, attenuation, Rayleigh and Mie scattering) affect not only sunlight entering the atmosphere, but also the reflected light within the environment. Any object we photograph is visible because of light reflecting from its surface, and this reflected light needs to travel through a certain amount of atmosphere to reach the sensor. The farther the reflected light photons travel through the atmosphere, the more likely they are to be affected by it. On a clear, dry day, light reflected from an object 50 feet (15 meters) away does not pass through enough atmosphere to be affected in any perceivable way, and the air between the camera and the object appears to be completely transparent. However, light reflected from an object 5 miles (8 km) away passes through 500 times more atmospheric volume, and the effect of the air affects the look of the object: its features become less pronounced, contrast and saturation are reduced, and the blacks are lifted.

Aerial perspective is affected by the same atmospheric elements that shape daylight: air, water droplets, dust, smoke, and pollutants. We commonly use the term "haze" to describe the effect of any one of these elements, but technically, haze only refers to the effect of dust and pollution, not air or humidity. A common phrase like "the bluish hills on a hazy summer afternoon" is not quite accurate – the bluishness of the hills is mostly due to of air molecules (Rayleigh scattering), not haze (Mie scattering). Let's examine each atmospheric depth component separately:

Air

Rayleigh scattering is the dominant factor in the way air visually affects distant objects. Light reflected from that object is partially scattered by air molecules, and the farther away the object is, the more scattering happens. In terms of color values, the object does not get brighter nor darker. Rather, the blacks are lifted toward the midtones, which reduces contrast and "flattens" the object, making it harder to see distinct features. On a clear, dry day with minimal presence of clouds, fog, or aerosols, distant objects take on the classic bluish tint (aka "blue haze"). It is important to remember that brightness is not squashed by clean air. A bright specular reflection on a lake or a metallic surface can be as bright even when viewed from miles away. It is the low end that is the most heavily affected with distance, starting with the darkest blacks and gradually onto the lower midtones.

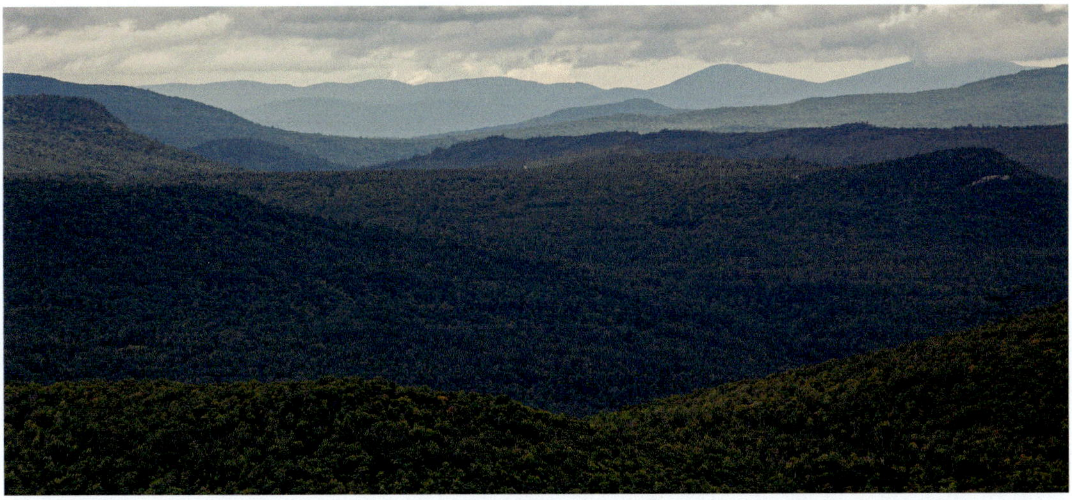

Rayleigh scattering of the light reflected from the hills increases with distance. For each layer in depth, contrast is reduced, blacks are lifted, and RGB values unify around desaturated blue. Notice that brightness is not reduced with distance.

Water Droplets

Humidity in the atmosphere has a similar effect to air, but Mie scattering does not tint the light, so distant objects on an overcast or foggy day retain their hues while the blacks are lifted toward neutral gray. In most environments (except in the driest deserts or high-altitude settings), there is usually some humidity hanging in the lower atmosphere.

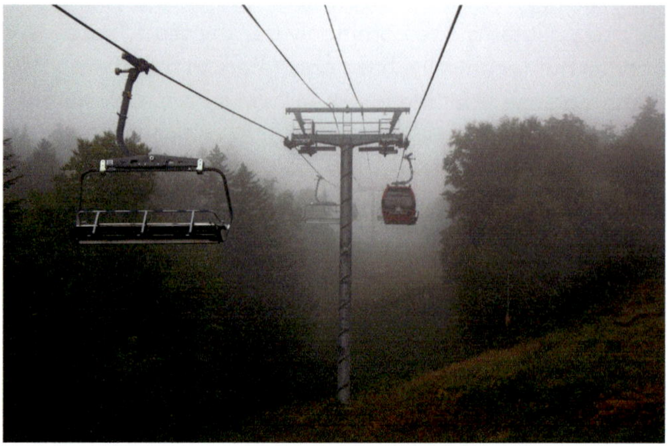

This causes some desaturation at the horizon, which shifts the deep blue hues of distant objects toward a more desaturated cyan or white (colors of distant objects may change to orange or red during sunset and sunrise). While the effect of air is always subtle, dense accumulation of water droplets in clouds and fog can dramatically change the scale of aerial perspective. Reflected light that passes through heavy fog, for example, is strongly scattered, reducing visibility from several miles to mere feet.

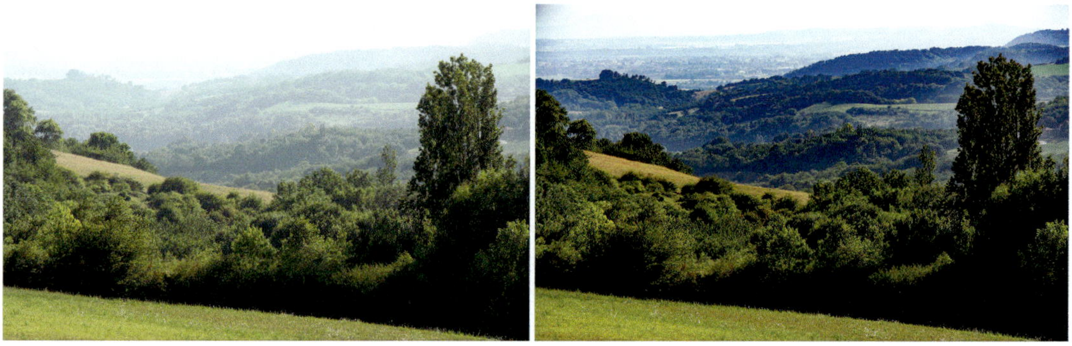

Rayleigh and Mie scattering create a mixed aerial perspective in this photo (left). When the photo is artificially "de-hazed", bands of white fog are revealed along the valleys and lower altitudes.

Haze

Haze is formed by dry particles of dust, smoke, and pollutants. Like fog, heavy haze can dramatically reduce the scale of aerial perspective, but it also has a greater light attenuation effect. Haze does not only lift the blacks and decreases contrast – it can also reduce brightness, as light is not only scattered through the particles but also heavily absorbed. Haze colors vary depending on the type of particles. Dust usually appears as brownish-orange haze while smoke and pollution cause gray or bluish haze. Lighter amounts of particles in the lower atmosphere often form a yellow-gray band at the horizon, and attenuate sunlight even further when the sun is low.

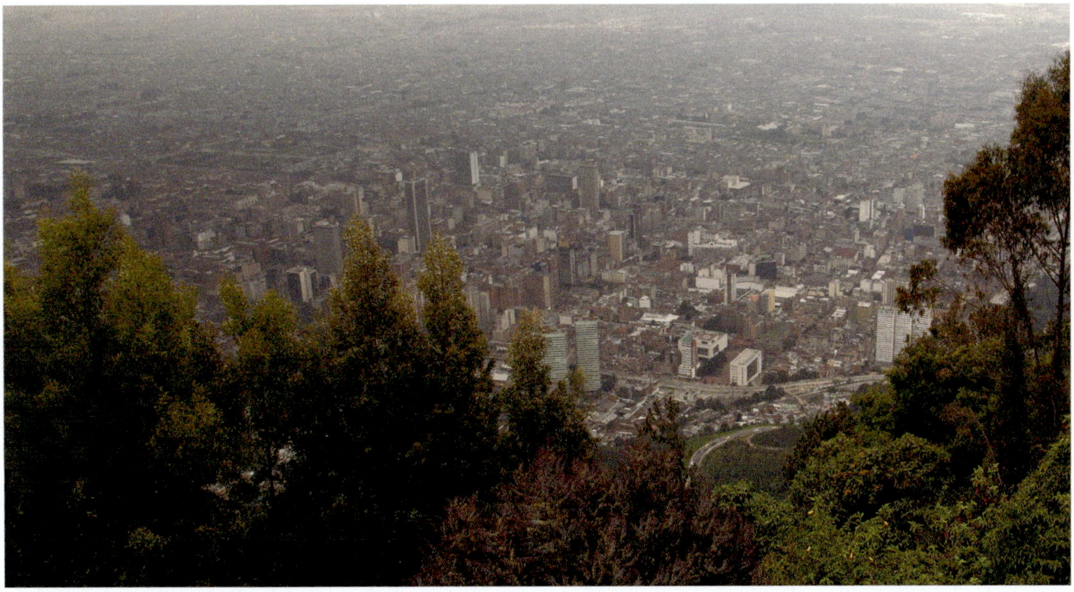

A combination of pollution aerosols and water droplets lies over Bogota, Colombia. Notice the difference in black levels between the foreground trees in and the city.

Chapter 7

Nighttime and Artificial Lighting

Natural Nighttime Light

Nighttime light comes from only two sources: the moon and the stars. The moon is just a reflector of sunlight, and even when full, its luminance is but a tiny fraction of the sun's (based on the scale of apparent magnitude, which is used to measure the brightness of stars, the sun is about 400,000 times brighter than the moon). The luminance of the stars is much lower than the moon, and since the sky is rarely clear of clouds, humidity, or pollution, starlight contribution to nighttime illumination is negligible. Put simply, nighttime is simply too dark to produce visually meaningful photography without the help of a long exposure and/or high ISO. Long exposures can produce some beautiful nighttime sky photos, but as far as subjects on the ground, the results are usually flat, desaturated, and overall, rather uninspiring.

It is therefore not surprising that almost all nighttime movie scenes take advantage of man-made lights, either by using light sources that are integral to the scene (street lamps, lit windows and doorways, car headlights, a flashlight, an actor's lit cigarette), or by using movie lights to create an illusion of natural night light (the ubiquitous "Hollywood moonlight"). A similar approach is taken in architectural visualizations or video games – nighttime renders and environments are illuminated by artificial lights, whether the purpose is to show the beauty of a building or create drama. This section is therefore brief. By and large, nighttime photorealism largely falls within the subject of artificial lighting, which will be discussed next.

The Purkinje Effect

One question about natural nighttime light deserves special attention here: what color is moonlight? The previously mentioned "Hollywood moonlight" is usually distinctly blue – is this a photoreal depiction or movie fantasy? The moon is merely reflecting sunlight, which is white, so one would expect moonlight to be white, not blue. The truth is somewhere in the middle. Light reflected from the moon is slightly warmer than pure sunlight (the moon reflects long wavelengths more strongly than short wavelengths). However, our eyes tend to shift colors toward blue in low-light situations, a phenomenon known as the **Purkinje effect**. Therefore, moonlight does not appear warm to us, but rather neutral or slightly bluish. The blue color of artificial moonlight in movies is thus not completely incorrect, though cinematographers often like to exaggerate it for visual drama.

DOI: 10.4324/9781003590293-10

Artificial Lighting

The incredible diversity of man-made lighting is a combination of many lamp types and an endless assortment of lamp shades, diffusers, reflectors, filters, and lenses. Different **light fixtures** (also called **luminaires**) can produce an almost infinite range of light intensity, color, shape, and focus, and these variations affect the feel, mood, and atmosphere of artificially illuminated environments in countless ways. Approaches to real-world lighting design are as varied as the different fields they serve (architectural and interior lighting, stage lighting, film and TV lighting) and each of these fields embodies different goals, aesthetics, and techniques. While the art and craft of real-world lighting is well beyond the scope of this book, there are some key aspects that are relevant to digital artists, particularly in the context of emulating man-made lighting in CG and achieving seamless integration in 2D.

Color Temperature

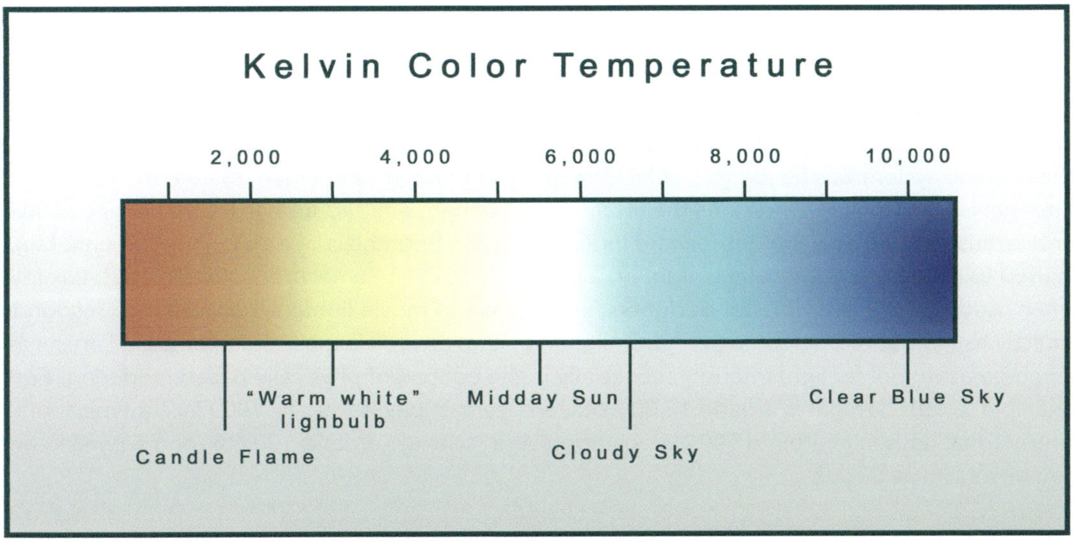

The color of emitted light is measured in **Kelvin** (K) units. The saturated orange color of sodium-vapor lamps is around 1700K, while a typical CRT monitor produces a blueish light at around 7000K. The common range for household lamps lies roughly between 2400K (standard incandescent bulbs) and 6000K (cool white fluorescent lamps). The Kelvin scale may be somewhat confusing for artists, since low-temperature values indicate what we consider "warm" colors, and high-temperature indicates "cool" colors. Moreover, a high-temperature blue light is not necessarily physically hotter, because many lamp types do not produce light through a burning filament. For example, fluorescent lights produce much less heat than incandescent lights, yet their Kelvin temperature is higher. It is important to note that Kelvin temperature does not indicate saturation. Highly saturated light is rarely used (except for special effects or stage lighting) because it produces poor color definition, and most bulbs, either warm or cool, produce a rather desaturated color.

Light Intensity

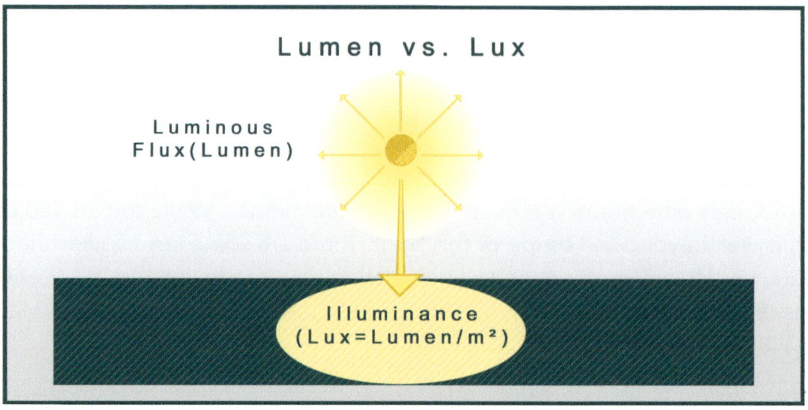

There are several ways to measure light intensity, but no single universally accepted measurement unit. In the past, household light bulbs were commonly differentiated by their wattage. Watt is a power consumption unit, not a light intensity measure, but it still worked well enough when most light bulbs were incandescent (the higher the wattage, the brighter the light). However, as new types of bulbs that produce brighter light at significantly lower wattage became more common, the wattage system was no longer useful. The most common measurement system for light intensity is **Lumen,** which indicates the total amount of light emitted from a source (luminous flux), and is thus suitable as a straight-forward indicator for the brightness of a bulb. Another unit, **Lux,** is used to measure the intensity of light on a surface (1 Lux = 1 Lumen per square meter). Lux is often used by architects, interior designers, and stage and movie lighters, because it corresponds directly to the desired outcome at a given lighting situation. Lux is thus useful for digital artists as a comparative tool for light intensity, especially in the context of physically-based rendering. For example, bright daylight is around 12,000 Lux, an overcast day is roughly 1000 Lux, a typical office or supermarket is lit at around 500 Lux, corridors and stairways around 100 Lux, and a parking area can be as dim as 50 Lux.

Common Lamp Types

- Good old incandescent lamps have been our lighting workhorses since the 19th century, and despite their relative inefficiency are still the most common household lamps. They use a burning filament in a vacuum as the light-producing element. Incandescent lights are typically in the warm color range (2000–3000 K).
- **Halogen lamps** use a Tungsten filament inside a small transparent envelope filled with a halogen gas (Iodine or Bromine). Halogen lamps have similar characteristics to incandescent, but can operate at much higher temperatures, and thus produce much brighter light. They are used both at home and for many professional photography, film, and stage lights.
- **Fluorescent lamps** do not use a burning filament, and instead produce light by electrically exciting mercury vapor within a tube, which in turn causes the tube's phosphor coating to glow. They are substantially more efficient than incandescent and halogen lamps (and

gradually replacing them). Originally, fluorescent lamps where large, elongated tubes that produced a rather cold bluish light (around 5000 K), and were used mainly for impersonal spaces like hospitals, supermarkets, factories, and offices. But nowadays, **compact fluorescent lamps** produce a much wider range of color temperature and are small enough to effectively replace incandescent lamps in most fixtures.

- **Neon lights** preceded fluorescent lamps by about 25 years. They use electricity to ionize one or more gases in a tube, which causes them to emit colored light. The specific color depends on the gas used: neon for orange, hydrogen for red, helium for yellow, etc. The energy output is relatively weak and not very suitable for illuminating spaces, but neon tubes have been used for decades for colorful signs and decorations. Despite their historical and artistic appeal, neon lights are ceding their place to the much more flexible LED's.

Sodium-vapor lamps are extremely efficient but produce a rather ugly orange light and poor color definition, so are primarily used to light streets, roads, parking areas, and industrial spaces.

LED lights (Light Emitting Diodes) have been around since the early 1970s, but their limited color production and low intensity restricted their use to (mostly) little red signal lights in electronic appliances. However, development of blue and white LEDs and a substantial increase in output power completely transformed LEDs into today's most versatile light type. LEDs can be combined to create RGB lighting units that can produce a wide range of light color and intensity, and today's usage of LEDs runs the gamut from small flashlights through efficient household lighting to giant LED advertisement screens and revolutionary LED lighting domes (see epilogue).

Light Modifiers

Just as the atmosphere transforms sunlight, light modifiers such as lamps shades, diffusers, reflectors, filters, and lenses transform the light of man-made emitters in a myriad of ways. While bare fluorescent tubes are a familiar sight in industrial spaces, homes are rarely lit by naked bulbs,

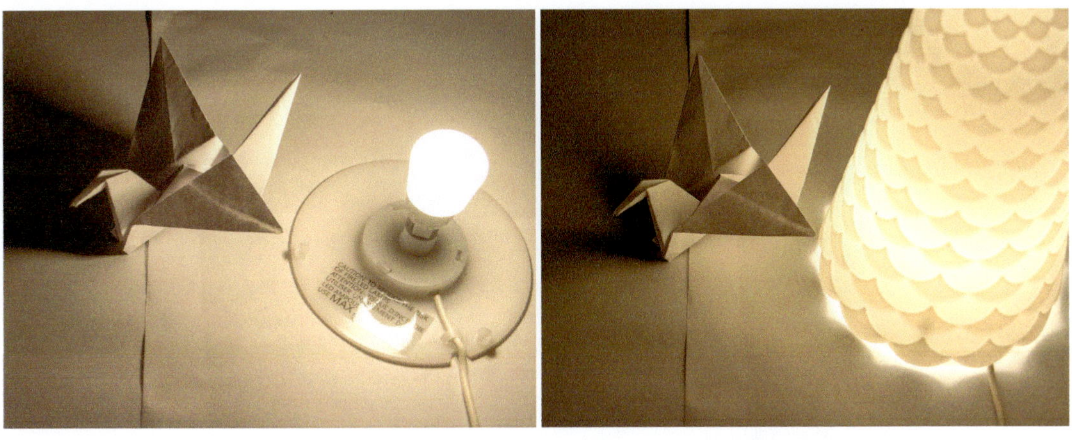

The same lamp, with and without its plastic shade. Notice the difference in luminance, contrast, hue, shadows, and light diffusion.

which are uncomfortable to stare at and throw a harsh, unpleasant light. Lamp shades have been used to diffuse and soften electric light from the very start, and today they exist in infinite shapes and made form a wide variety of materials such as plastic, glass, linen, burlap, metal, and paper. Opaque shades are coupled with reflectors to direct and intensify the light, while lenses are used to create focused beams in some home lighting fixtures and many stage lights. Soft boxes and other types of diffusers are used in professional photography and cinematography to create larger, softer light emitters.

From the perspective of digital art and photorealism, lamp modifiers are in fact more important than the actual bulbs, because it is those modifiers that define the final visual characteristic of the light fixture. A given lamp generates a specific light intensity and color temperature, but the eventual shape, pattern, color, intensity, and softness of the fixture depend on the light modifiers. For example, an incandescent bulb inside a square-shaped diffuser generates a square-shaped soft light, even though the bulb is spherical. When emulating man-made lighting, therefore, it is important to look at the characteristics of the entire fixtures and its final output.

Chapter 8

Shadows

Shadows are the glue that holds elements together in a CG environment, 2D matte painting, or composite. The photoreal coherence of an image easily falls apart when shadows are missing, mismatched, or incorrect. Shadows play an important artistic role in visual art, and of course in photography and cinematography, but do not always receive enough attention in CG lighting or 2D compositing, as they are often seen as a mere by-product of light. Understanding shadows is crucial for achieving photorealism. What happens when two or more shadows overlap? What affects the softness of a shadow? Could there be shadows within shadows? To address such questions, we need to move away from common categorizations of different shadow "types" and instead adopt a more holistic approach.

To most people, the word shadow brings to mind the clearly defined dark area that an object casts on the ground on a sunny day, such as our own shadow on the sidewalk, or the shadow of a tree on the lawn. Such shadows are commonly referred to as "cast shadows". But what about shadows on an overcast day? Photographers like to call overcast days "shadowless days". But there are plenty of visible shadows around on such a day. You may not see your own shadow on the sidewalk when the sky is overcast, but look under any parked car, and you will notice a very distinct, dark shadow. One can also spot similar shadows inside bushes, in the canopy of trees, at the bottom of rocks, in between bricks in the wall or cracks in the pavement – generally inside any cavity, large or small, and in every nook and cranny. These shadows are often called "ambient" or "occlusion" shadows, but are they really any different from "cast shadows"? On a clear day, our own shadow on the sidewalk represents the area where most of the sunlight is most of the diffuse sky radiation is blocked. In both cases, shadows form in areas where less light can reach, because the light source is partially occluded. We can thus state that **shadows are simply the relative lack of light**. It does not matter if the occluded light comes directly from a source or from a diffused version of it. The separation between "cast shadows" and "ambient shadows" is rather artificial – there is no clear line of separation, and in most lighting situations (especially man-made lighting) there can be a variety of shadows in the same environment that range from sharp to diffused, dark to light.

Taking this notion even further, the term "cast shadows" forces the assumption that one object casts a shadow onto another. But shadows appear wherever some of the light cannot fully reach, and this can happen between two separate objects or within the same object. The shadow inside our nostrils is not cast by a separate object, just like the shadow inside a half-opened bag or in a rock crevice. In fact, one can argue that, from the "light's perspective", everything – the ground, the trees, the cars, the people – everything is just one single giant "object". And within that

DOI: 10.4324/9781003590293-11

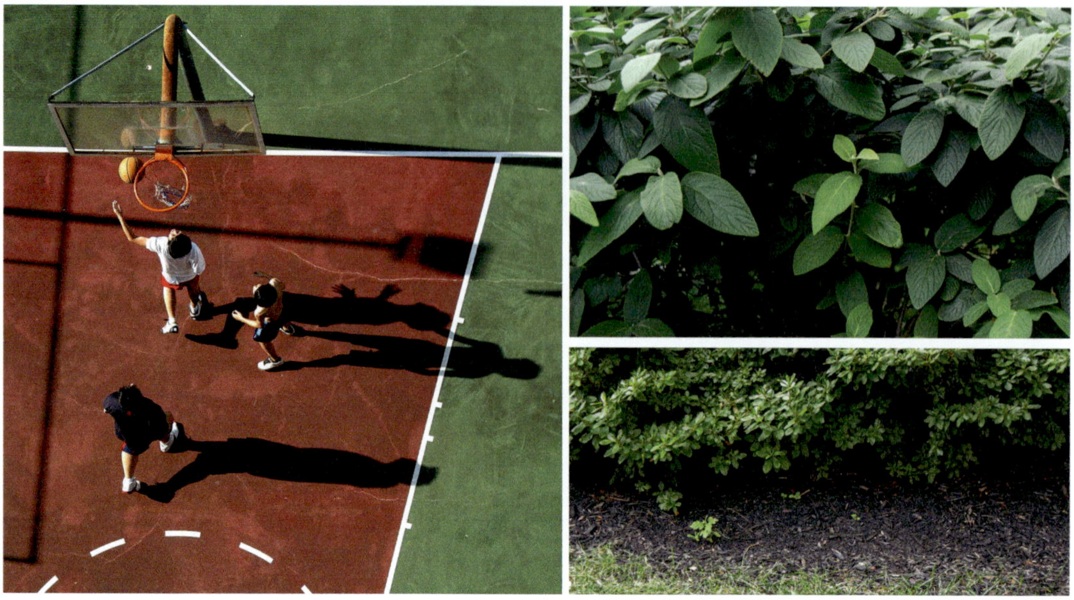

The word shadow brings to mind the defined dark area that an object casts on the ground on a sunny day (left). But "shadowless", overcast days also produce shadows, albeit not as sharp and well-defined (left, top and bottom).

object, any area that is partially or fully occluded from one or more light sources is shadowed. Rather than thinking of an environment as a collection of objects that cast shadows on each other, we can approach the scene as a unified entity where the intensity, color and softness of shadows is determined by the characteristics of the light sources and their interactions with the various surfaces.

Shadow Softness

Imagine standing behind a wall, in its shadow. If you cannot see the light source at all, you are standing in the umbra, the darkest part of the shadow where all the light form that source is occluded. As you slowly move sideways toward the edge of the wall, part of the light source becomes visible – you are now in the penumbra, the lighter part of the shadow where only some of the source's light is occluded. There is a clear correlation between the size of the light source and the transition between umbra and penumbra. If the source is an infinitely small point, the light source is either fully occluded or fully revealed. There is no transition between the two stages, so there is no penumbra, and the shadow is 100% sharp. But point lights exist only in the virtual world of CG, and every real-world emitter has a size and a shape, so all real-world shadows have at least a minimal amount of softness. With larger light sources, the changeover from full occlusion to full reveal is more gradual, and there is a noticeably softer transition between umbra and penumbra.

This however does not explain why shadows also fall off with distance. In fact, the scale of a light source is only one of several factors that affect the softness and falloff of the shadow. The following illustrations show how each factor affects the size of the penumbra and its falloff direction:

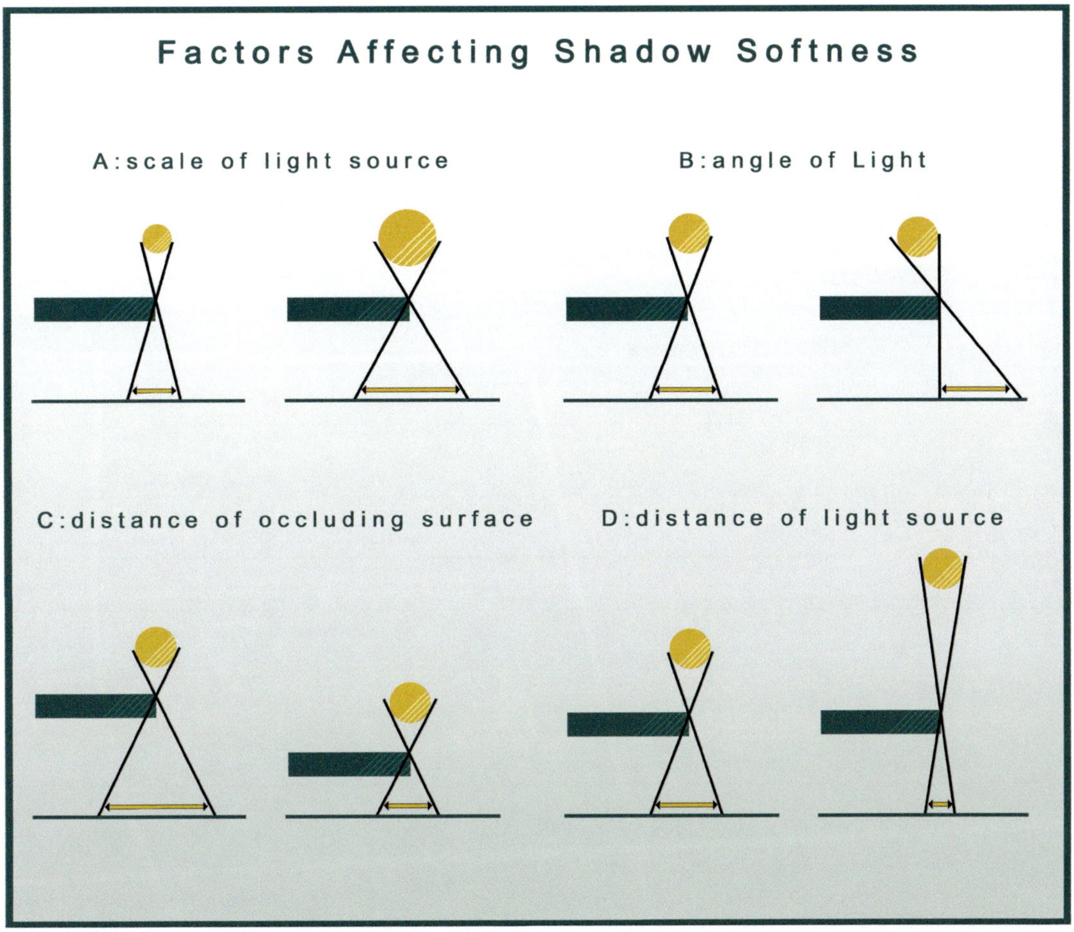

A: The scale of the light source affects shadow softness (larger light = larger penumbra).

B: The angle of the light source relative to the occluded and shadowed surfaces affects the direction of the penumbra (when the angle is straight, the penumbra surrounds the shadow equally, but as the angle becomes shallower, the penumbra stretches in the opposite direction to the light).

C: The distance between the occluding surface and the shadowed surface directly affects shadow softness (larger distance = larger penumbra).

D: The distance between the light source and the occluding surface inversely affects shadow softness (larger distance = smaller penumbra).

Factor B and C explain why the shadow of a tree on the ground, for example, is strongest and sharpest at the base of the tree and why this softness is accentuated at sunset. The base of the tree is closest to the ground while the tree top is farthest away (factor C), and at sunset the angle of the light stretches the penumbra in the opposite direction of the light (factor B).

However, factor D is somewhat confusing: it is usually hard to see the shadow getting sharper as the light is move farther away, because at the same time, the light's intensity also falls off with distance. So, if you move a lamp away from an object, the increased sharpness of the shadow is somewhat negated by the reduced light intensity (which reduces the intensity of the shadow).

The one exception here is the sun, which the only light source that is extremely far yet still strong enough to produce visible shadows. Indeed, sunlight shadows are sharp, but that sharpness is of course affected by factors B and C, and, most importantly, by the atmosphere.

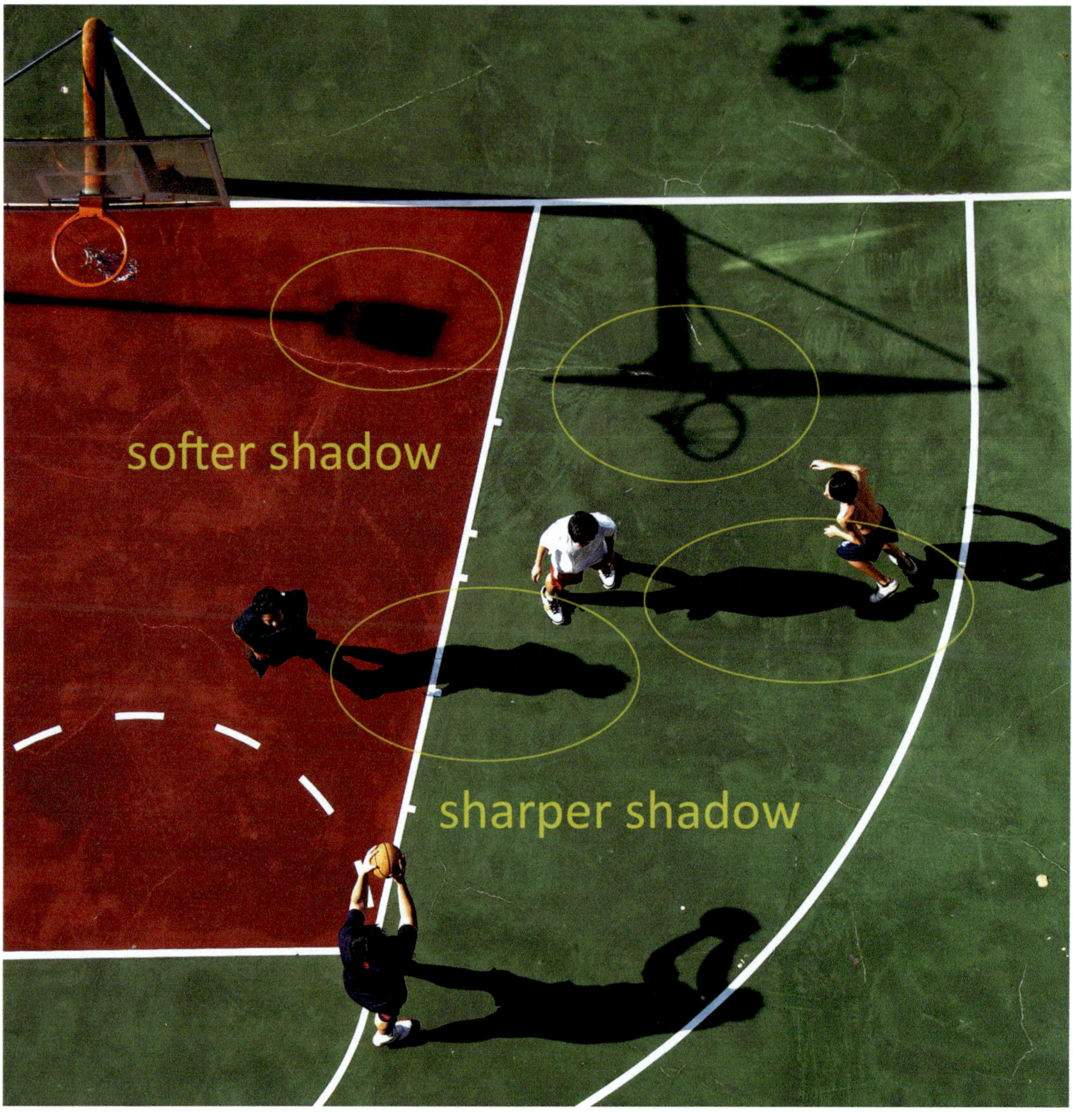

The shadows of the basket and the light pole are softer than the shadows of the players, because their top is farther away from the ground (factor C, the distance between the occluding and shadowed surfaces).

Shadow Color

Since shadows represent the relative lack of light, their color is determined by the balance between the occluded and non-occluded lights in the environment. In other words, it is the sum of all lights in the scene, minus the occluded light/s. On a clear sunny day, our own shadow on the sidewalk is slightly blue, because the warm sunlight is occluded, and the shadowed area remains lit mostly by diffuse sky radiation. Shadow color in a daylight scenario is simple – there is only direct sunlight and diffuse sky radiation to consider. But in an environment with multiple lights of different colors (for example, an urban night scene), the color of the shadows represents a varied and complex balance between occluded and non-occluded light colors.

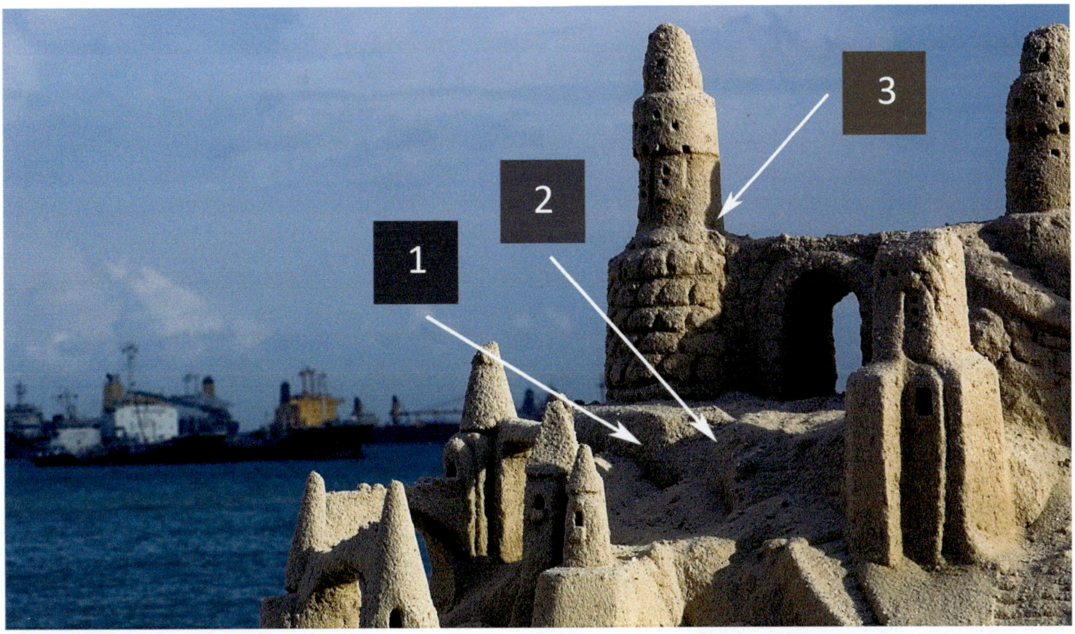

On a sunny afternoon, lit areas are warm-colored while shadowed areas are bluish-gray. But the exact color of the shadow depends on the surface texture and amount of occlusion: darker and cooler in the most occluded parts (1), brighter in more exposed areas (2), and warmer where the shadow blends with reflected light (3).

It is important to note that while the occluded area is darker than the fully lit area, it is not necessarily blacker. In any given environment, the darkest shades usually represent the areas with the highest amount of occlusion. On a sunny day, cracks in the sidewalk may already be quite dark even in the fully lit areas, because they occlude both the direct sunlight and some of the sky radiation. So, while our shadow on the sidewalk is darker wherever the direct light is removed, those cracks still maintain the same black value, because no additional light is removed from them. In other words, while a shadow is darker than its surrounding, it also has less contrast, as the midtones are reduced while the blacks are preserved (pushing the blacks too low while darkening to create a shadow is a common compositing and matte painting mistake, as discussed in Chapter 17).

Overlapping Shadows

What happens when shadows from two or more objects overlap? Do they add up to a darker shadow, or merge seamlessly? This depends on which lights are occluded. If both objects are occluding the same light source, then the shadows merge seamlessly. A group of trees on a sunny day all block the same light source (the sun), so their occlusion forms one continual shadow. However, if each object is blocking a different light source (for example, in a room with two lamps), then the overlap area represents the relative lack of both lights and is therefore darker.

Nested Shadows

Although the combined shadows of elements that occlude the same light source blend together seamlessly, additional occlusion can still happen within a shadow. A car that is parked in the shadow of a tree is not occluding any additional direct sunlight (it's already in the shadow of the tree), but its floorboard, which is much closer to the ground than the tree's canopy, occludes more of the diffuse sky radiation, and thus creates a darker shadow within the tree's shadow. If we now place a small rock underneath the car, the proximity area between the rock and the ground will be even darker, creating a third nested shadow, or more accurately, a third level of light occlusion. In a typical evening interior scene of a living room, the combination of several light sources with different characteristics and various pieces of furniture and other objects generates an intricate mosaic of multiple overlapping and nested shadows with different intensities, sharpness levels, and hues.

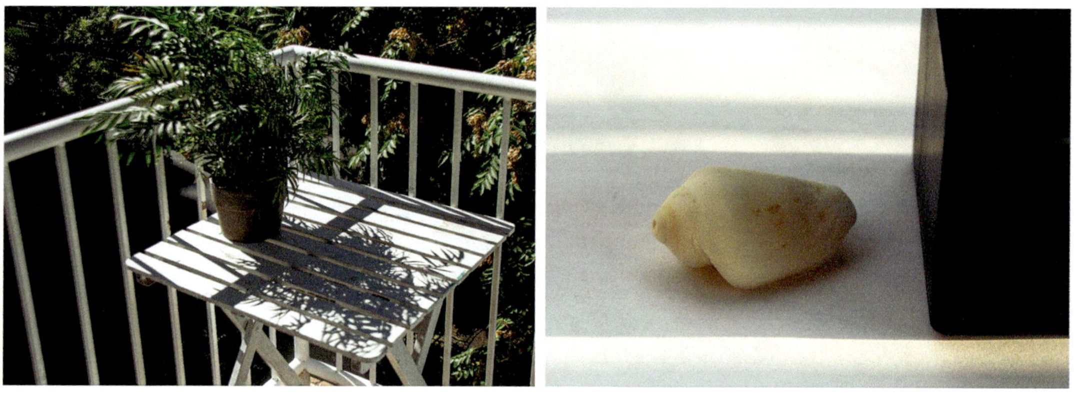

The potted plant and the railing (left) occlude the same source (sunlight), so their shadows merge seamlessly. The conch sits in the shadow of the box (right), but generates a nested shadow by occluding more of the diffuse sky radiation.

Contact Shadows

At the beginning of this chapter, I suggested that there are no separate "types" of shadows. So why this separate discussion of contact shadows? Indeed, contact shadows are not different from any other shadows. However, when simulating photoreal imagery, there is a unique importance to proximity and contact shadows, especially when blending CG elements with 2D footage (or blending 2D elements from different sources in compositing and matte painting). When objects are at close proximity, the area between them is occluded from most of the light and is therefore strongly shadowed. It does not matter if the scene is dominated by a directional light or softly lit by diffused or scattered lights – when surfaces touch (or nearly touch) each other, the contact area is distinctly dark. We are particularly sensitive to such contact shadows, because we instinctively use them for spatial orientation (for example when we reach out to touch or grab something). Our eyes and mind are therefore quite unforgiving toward missing or mismatched contact shadows. When a CG table is composited onto an image without generating the appropriate contact shadows where the table's feet are touching the floor, the table appear to be floating. The innate importance of contact shadows to our spatial orientation makes them somewhat on an "Achilles heel" in digital art. So, although technically no different than any other shadow, they do require special attention, and later in the book we will examine 3D and 2D procedure for generating and implementing contact/proximity shadows.

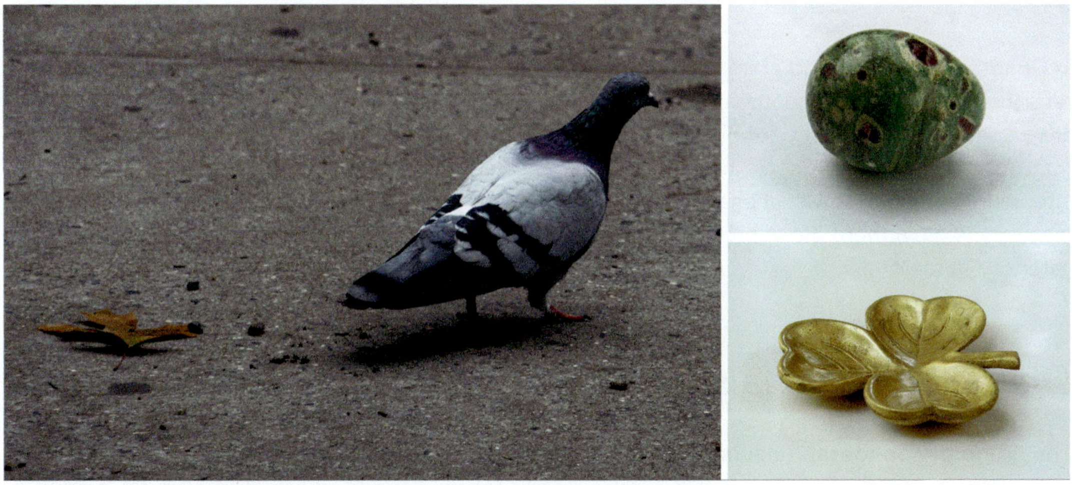

Examples of contact and proximity shadows. The photo on the left was taken during a fully overcast day. I could not see my own shadow on the sidewalk but the pigeon and leaf box generated a clear shadow, because the bulk of their occluding body is much closer to the surface (proximity shadow).

Chapter 9

Basic Material Properties

In Chapter 5 we discussed absorption, specular reflection, diffuse reflection, and transmission/refraction as separate outcomes of light interacting with surfaces. The visible attributes of materials are a combination of some (or all) of these interactions. Tinted glass, for example, absorbs, transmits, and reflects light both diffusely and specularly. A specific combination and balance of interactions define the visual characteristics of each and every material. While these characteristics may vary widely, there are some common aspects that apply to all materials, or to a specific category of materials.

Dielectric Materials

The group of **non-metals** (**insulators**, **dielectrics**) encompasses an enormous range of materials: rocks, leaves, wood, skin, clay, water, plastic, paper, fabric, cement, glass, diamond, flesh, milk, to name but a few. While these materials greatly differ in their characteristics and appearance, they all share some key properties that make them distinctly different from metals:

Dielectric Diffuse/Specular Balance

All dielectric materials display a combination of diffuse and specular reflections. The material's base color is defined by the diffuse reflection, and in addition (and depending on the glossiness level of the material), specular reflections add the colors of the environment, independent of the diffuse color. Clear dielectrics like water or glass do not tint specular reflections, and the reflected environment retains its original colors. Likewise, most shiny dielectrics derive their glossiness from a thin, transparent coating (like the resin shell in some plastics, the lacquer finish of polished wood, the wax layer on an apple or the thin oily layer on human skin). In all these cases, the specularity happens at the outer transparent layer and is thus not tinted.

In CG shading, diffuse and specular reflections are usually separate parameters, but from a real-world perspective we can consider specular and diffuse reflections as two facets of the same thing. On a highly irregular surface, almost all of the incident light is scattered and reflected at random angles, while a more organized microscopic structure reflects at least some of the light back at

DOI: 10.4324/9781003590293-12

predictable angles. This mixed diffuse-specular balance characterizes all dielectric materials, from the very dull to the super-glossy. It is also distinctly view-dependent, as I will explain in the next section.

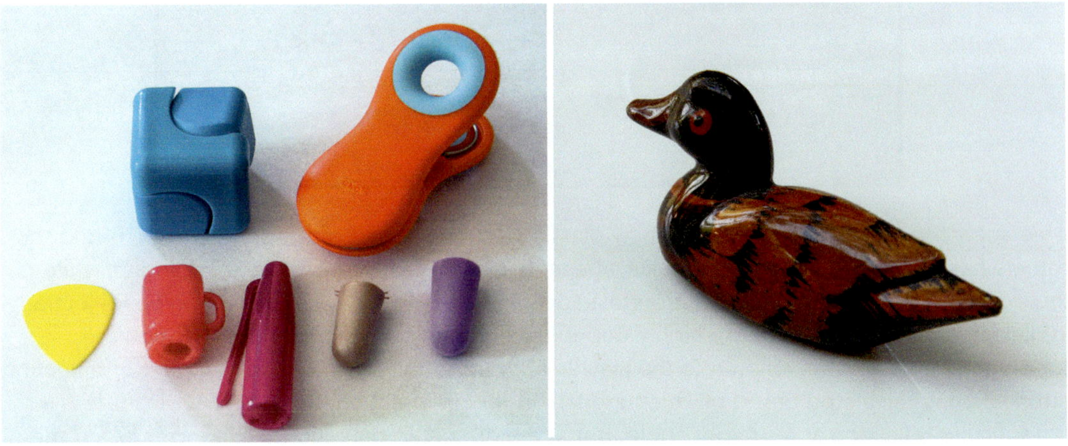

Left: different plastic materials displaying varying ratios of specular/diffuse balance. Right: wood specularity is fairly rough, but the lacquer finish on this wooden duck adds a layer of clear specular reflection.

Fresnel Effect

Augustin-Jean Fresnel (the French physicist who was largely responsible for the widespread adoption of the wave theory of light), derived a series of equations (known as the **Fresnel equations**, or **Fresnel coefficients**) that established the ratios and relations between reflection, transmission, and refraction. While it is not necessary for the digital artist to delve deep into the math, the basic principle is crucial for the correct simulation of specular reflections in dielectric materials. This principle is often dubbed the "Fresnel effect" by CG artists.

At its essence, the Fresnel principle ties the prominence of specular reflection to the incident angle of the viewer's line of sight. At the shallowest angle (near parallel to the surface) reflection is strongest, and at the steepest angle (perpendicular to the surface) reflection is weakest. Here's the classic example: if you stand on the shore of a calm lake and look down toward the water at your feet, you notice that it's transparent, and you can see right through to the bottom. But as you shift your graze straight toward the opposite shore, you notice that the water gradually becomes opaque and reflective, with the reflection strongest further away. Your viewing angle is much steeper looking down at the water near you and becomes progressively shallower as you shift your gaze toward the horizon. This example demonstrates not only the effect of viewing angle on the intensity of specular reflection, but also the kind of inverse correlation between specular reflection and transmission: as the angle of incidence becomes shallower, the surface becomes less transparent and more reflective. Another typical example: the transparent glass cover of your cellphone screen. Look at it from the side, and it becomes more reflective and less transparent.

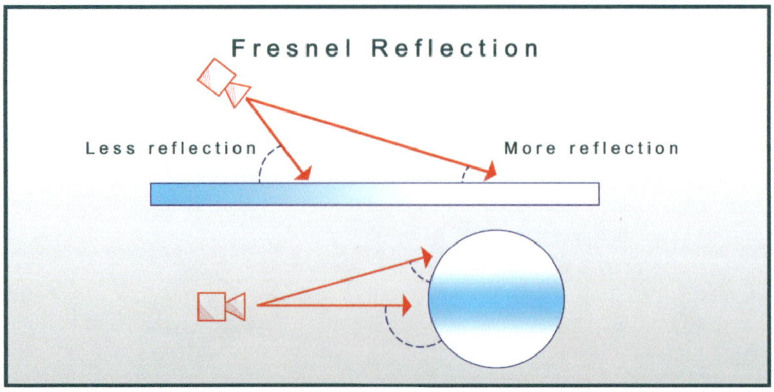

If we apply the Fresnel principle to a shiny plastic sphere, the part of the sphere that's facing the camera has the least amount of specular reflection (viewed from a perpendicular angle), while the sides of the sphere exhibit the strongest specularity (near parallel, grazing angle). Reflection in dielectric materials is therefore highly dependent on the viewing angle, and thus on the topology of the object in relation to the viewer. In the Fresnel equations, the amount of reflection attenuation is calculated based on the material's index of refraction. A lower IOR means less (or no) specular reflection at a straight angle, but at a higher IOR, reflections "expand" toward the straight angle zone, diminishing the Fresnel effect. There is a stronger influence of the viewing angle on reflection in water (IOR = 1.33), than there is in diamond (IOR = 2.4).

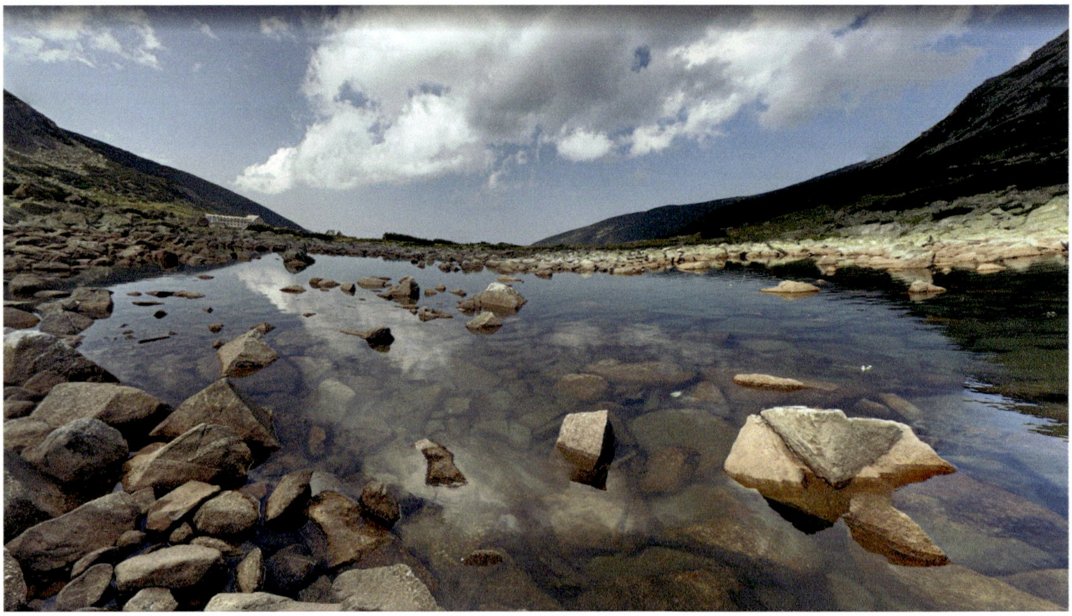

The water surface gets progressively more opaque and reflective as the incident angle becomes shallower.

© Google Street View.

Metals

The distinctly shiny and lustrous quality of metals is related to the looseness of the electrons in their molecular structure. The force of the atoms' nuclei is not strong enough to hold the outer shell of electrons, which causes them to float and interact freely. This "sea of electrons" acts like an imaginary impenetrable force shield around the molecule that either bounces photons back or "kills" them right at the surface. Because photons never penetrate the surface and never scatter, metals reflect all the light specularly, with no diffuse component. While some of the light energy is never reemitted, a strong, entirely specular albedo characterizes the shiny look of polished, freshly prepared metals. Many metals (like silver, chrome, aluminum, or platinum) reflect across a wide range of wavelengths, which gives them their neutral, silvery color. But some metals like gold, bronze or copper do not reemit all wavelengths equally, and thus reflect with a specific tint, giving such metals their distinct color. Note that the diffuse color of all metals remains black (zero diffuse albedo), and their visible color is only affected by their specular tinting.

The Fresnel effect in metals is fundamentally different from dielectrics. Because of their unique molecular characteristics, the simple refractive index calculation described previously does not apply to metals. A **complex index of refraction** describes the relation between viewing angle and the attenuation of specular reflections in metals. Unlike dielectrics, metals do not show much

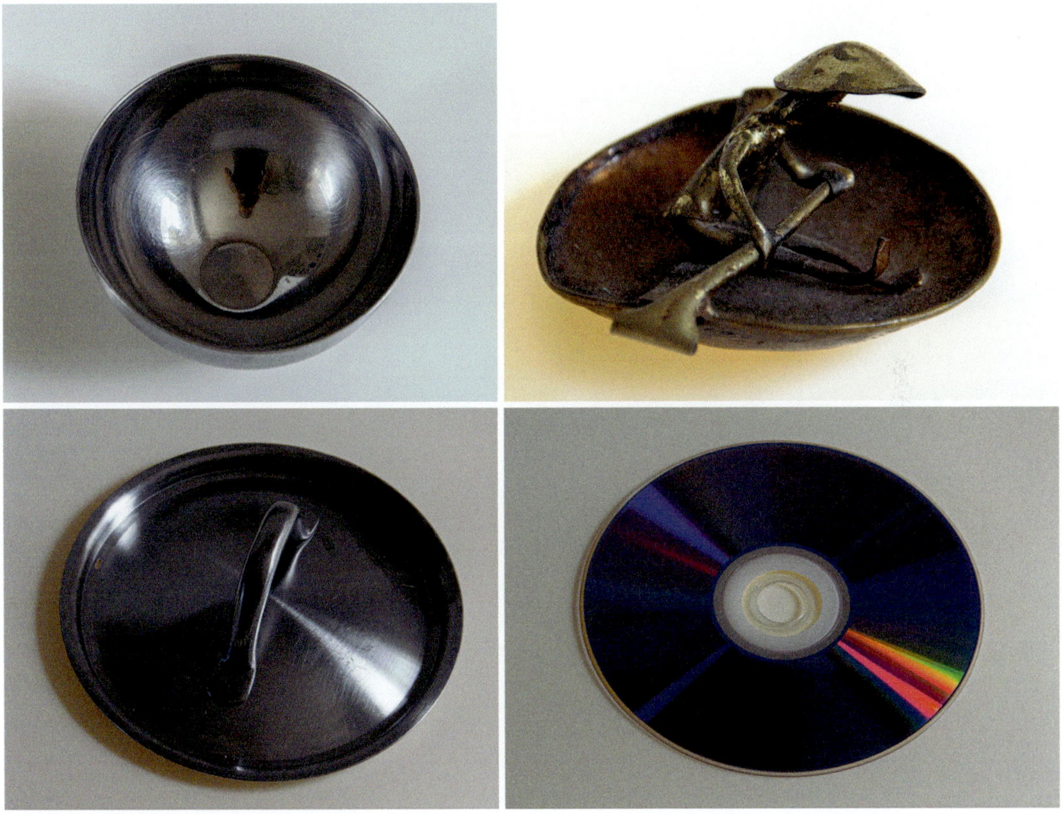

The mirror-like reflection on the stainless steel bowl (top left) is punctuated only be the scratches (notice how they are more visible around the reflection of the bright windows). Decay, wear and dirt occlude much of the original bronze specularity (top right). The circular grooves on a pot lid (bottom left) generate distinct anisotropic reflections. The metallic reflective layer on a CD (bottom right) also shows anisotropic specularity, but the rainbow-like effect (thin-film interference, see Chapter 12) is caused by the transparent plastic coating.

reduction of specularity at straight viewing angles, and reflections are much more consistent across the surface. Mirrors, for example, reflect equally well from any viewing angle thanks to their metallic back coating. Also, because specularity in metals is not mixed with diffuse, the reflected image is much less attenuated (as it usually is in dielectrics), and thus shows much more distinct detail. **Anisotropy**, or the pronounced stretching of specular reflections, is usually the result of elongated or circular grooves in many types of processed metallic surfaces, such as brushed aluminum.

Chapter 10

Lens and Camera Characteristics

In the book's first chapter I wrote about the importance of making digitally created imagery look "photographic" to accomplish photorealism. In this chapter we will examine the most typical lens and camera characteristics and their effect on photographed reality. A good understanding of this subject is vital not only for VFX and visualizations – game engines now support various lens effects, which are used to instill a more cinematic look and enhance photorealism. The term "lens effects" is somewhat of a misnomer, since most lens and camera idiosyncrasies are the results of defects – lenses are not 100% perfect. Yet it is those defects that impart the kind of natural inconsistency to otherwise sterile digital creations. Visual lens imperfections (generally called **aberrations)**, can be divided into monochromatic aberrations (all wavelengths affected equally) and chromatic aberrations (different wavelengths affected independently). Flares and blooms are the result of light being scattered by the glass elements in the lens, while film grain and digital noise are not related to the lens but to the characteristics of the light-recording medium (film or digital sensor).

Defocus

Of all the lens aberrations, defocus is the most significant, because it is used as a story-telling device as well as a visual and artistic instrument in photography and cinematography. It is also one of the most crucial aspects of photorealism and successful integration of 3D and 2D elements. Like atmospheric depth, defocus provides the viewer with vital spatial clues. Moreover, defocus helps establish the sense of an optical medium between the environment and the viewer, and subsequently imparts a photographic feel to digitally created content. Even a subtle sense of depth of field can elevate a CG render to a higher level of believability, while precise matching of defocus is of course a critical step in compositing.

Depth of Field

For every lens, there is a certain specific distance where a point of light is captured at maximum sharpness. That distance is the lens's **focal plane**. In fixed-focus lenses (like the ones used in webcams and surveillance cameras) this distance cannot be changed. However, almost all lenses used in photography and cinematography have a mechanism for moving the focal plane (**racking focus**) manually or automatically by changing the distance between the lens and the sensor/film.

DOI: 10.4324/9781003590293-13

The **depth of field (DOF)** defines the width of the area on both sides of the focal plane where the image is still sharp. Beyond those boundaries, the image starts falling out of focus. A wide (deep) DOF means that even points that are far from the focal plane are still sharp, while at a shallow DOF, only areas that are near the focal plane are sharp, and the rest of the image quickly goes out of focus. DOF is affected by several variables:

Aperture width: smaller apertures (higher f-stop numbers) generate a deeper DOF. A pinhole camera (which is essentially the equivalent of an extremely small aperture, f/100 or higher) produces an almost infinite DOF, or a uniform sharpness at any distance. Such an extremely small aperture is of course not practical for general photography and cinematography, since it lets so little light in. But even commonly used smaller apertures (f/8–f/11) have a decently wide DOF when focused at infinity. Small apertures are commonly used in landscape photography for maintaining sharpness across a wide depth. On the other hand, wide-aperture lenses (f/5 and downwards) produce an increasingly shallower DOF. Lenses with f/1.8 and f/1.4 apertures are popular for portrait photography and product shots because of the way they throw everything in front and behind the subject out of focus.

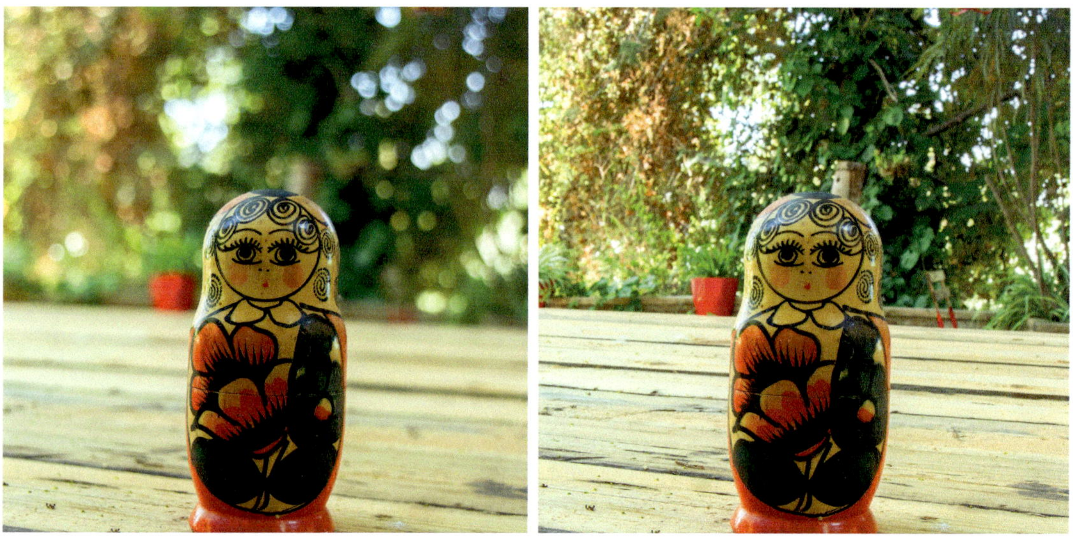

Left: with an open aperture of f/4, the background is thrown out of focus. Right: an aperture to f/22 keeps the background fairly in focus.

Focus distance: the farther the focal plane is from the camera, the wider the DOF. Given the same aperture, focusing on a distant subject results in a wide DOF, while focusing on a subject that is close to the lens produces a noticeably shallower depth of field (a fundamental characteristic of macro photography). Our intuitive perception of scale is tied to DOF exactly because of this principle. We instinctively associate narrow depth of field with small scale. A building feels like a dollhouse and a landscape like a miniature if the DOF is narrow and the focal plane distant. For the same exact reason, miniature models that are meant to appear much larger are filmed with the narrowest aperture possible, to maximize the depth of field (a narrow DOF would instantly give away the scale cheat).

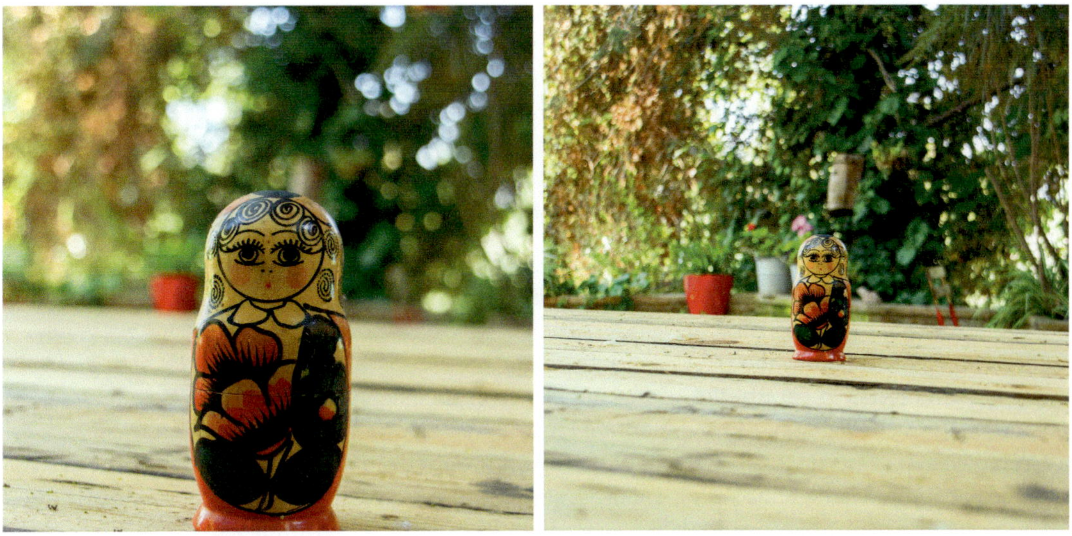

Both images were shot with the same lens and aperture, but the closer focal point (left) generates a narrower depth of field. Moving the focal point a few feet away widens the depth of field and, as a result, the distant background is much sharper.

Now, does **focal length** also affect DOF? In other words, do long lenses have a shallower depth of field than wide lenses? Yes, but to a lesser extent than what it seems at first sight. In fact, much of the apparent difference in depth of field between wide and long lenses lies in the way longer lenses compress perspective and depth. Objects at a specific distance may have a similar amount of defocus when shot with a wide or a long lens. But with the long lens, these objects appear much closer to the focal plane, and their defocused detail much bigger. This creates an illusion of a shallower DOF because it seems like defocus happens over shorter distances. This compression of distance affects not only the appearance of DOF, but also the appearance of atmospheric depth, as objects seem to be much closer than the amount of atmospheric hazing suggests.

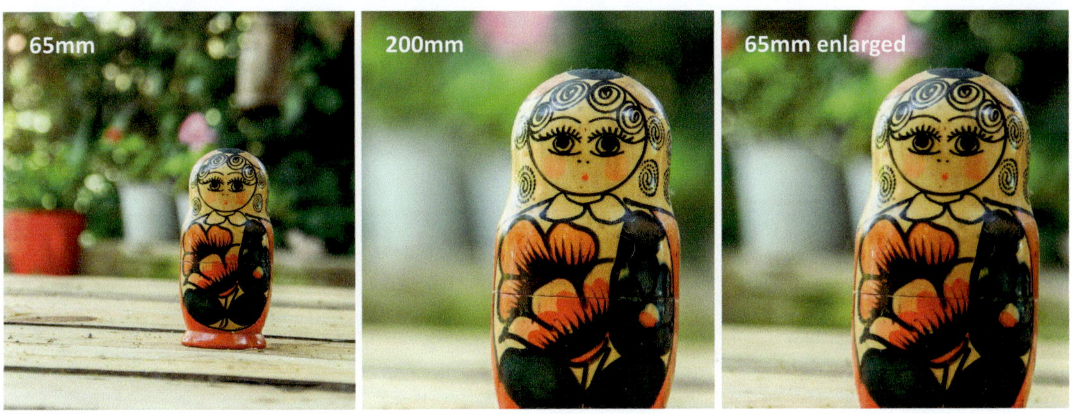

DOF seems to be much narrower in the photo taken with a zoom lens at 200 mm (middle), compared to the 56 mm (left). But enlarging the 65mm shot (right) shows that the difference is actually not that big – the compression of depth contributes much to the illusion.

Bokeh

Defocus blur is characteristically different from other types of blurring. For example, Gaussian blur (the most common blur function) blurs sharp points into soft-edged circles, and averages areas of sharp contrast into a smooth, consistent "mush". On the other hand, lens defocus blurs points into circular shapes with distinctly defined edges. As defocus increase, those circular shapes grow larger, but retain their edge sharpness. In addition, unlike Gaussian blur, areas of contrast are preserved even at high defocus levels.

The term **Bokeh** (from Japanese "boke-aji", or "blur-quality") is used in photography to describe the distinct character and quality of defocus. The shape of a defocused point is not necessarily a perfect circle. It is tied to the physical aperture mechanism of the lens. Simpler lenses use five or six straight blades to control the size of the iris. With such lenses, the resulting Bokeh looks like a pentagon or hexagon. More expensive lenses have nine or more blades, usually curved, to produce a clean, circular Bokeh. The fact that Bokeh is directly related to the shape of the aperture is particularly important in anamorphic lenses, which have an oval, rather than circular, aperture. Anamorphic Bokeh is therefore oval, like a circle squeezed on its horizontal axis. Because of the disproportionate shape of its Bokeh, an anamorphic rack focus has a very noticeable vertical squeeze/stretch effect.

Lens defocus does not grow infinitely with distance. There is a gradual increase in the size of the Bokeh shape from the from the DOG boundary to a point where defocus reaches its maximum level, after which it remains the same. The distance of the gradation and the actual maximum amounts depend on the lens type, aperture width, and focal distance. Another characteristic of lens Bokeh is **highlights bloom**. Unlike Gaussian or quadratic functions, which average colors, the Bokeh of most lenses accentuates highlights by keeping them relatively bright and larger in relation to the rest of the defocused image.

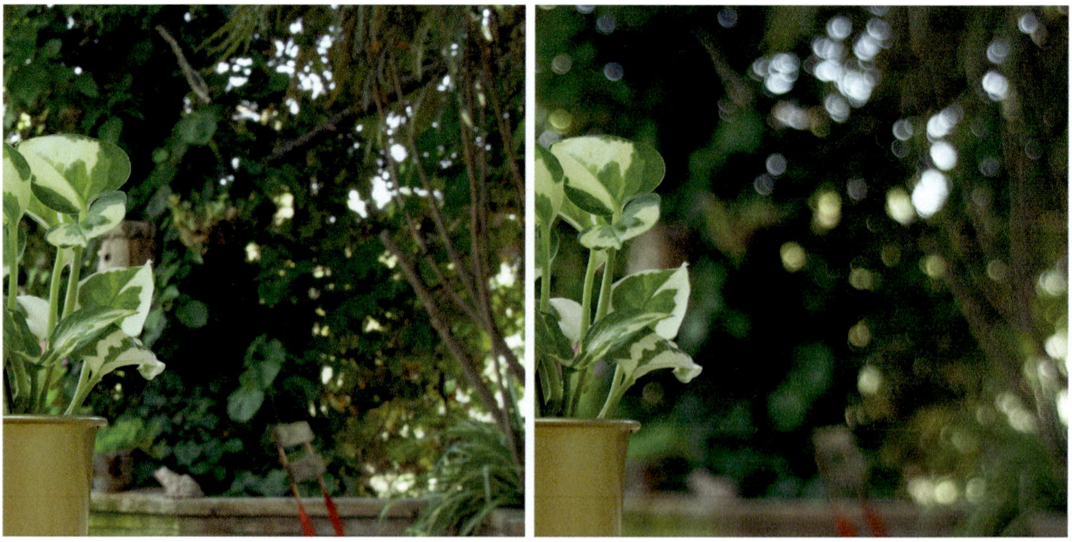

Circular Bokeh, notice how the highlights retain their brightness.

Lens Distortion

The amplified skew, or cornerstone effect on parallel lines is related to the way wide lenses exaggerate perspective, and is not lens distortion. However, when straight lines are bent into curves, this is the result of lens distortion, an aberration that is more typical of wide and zoom

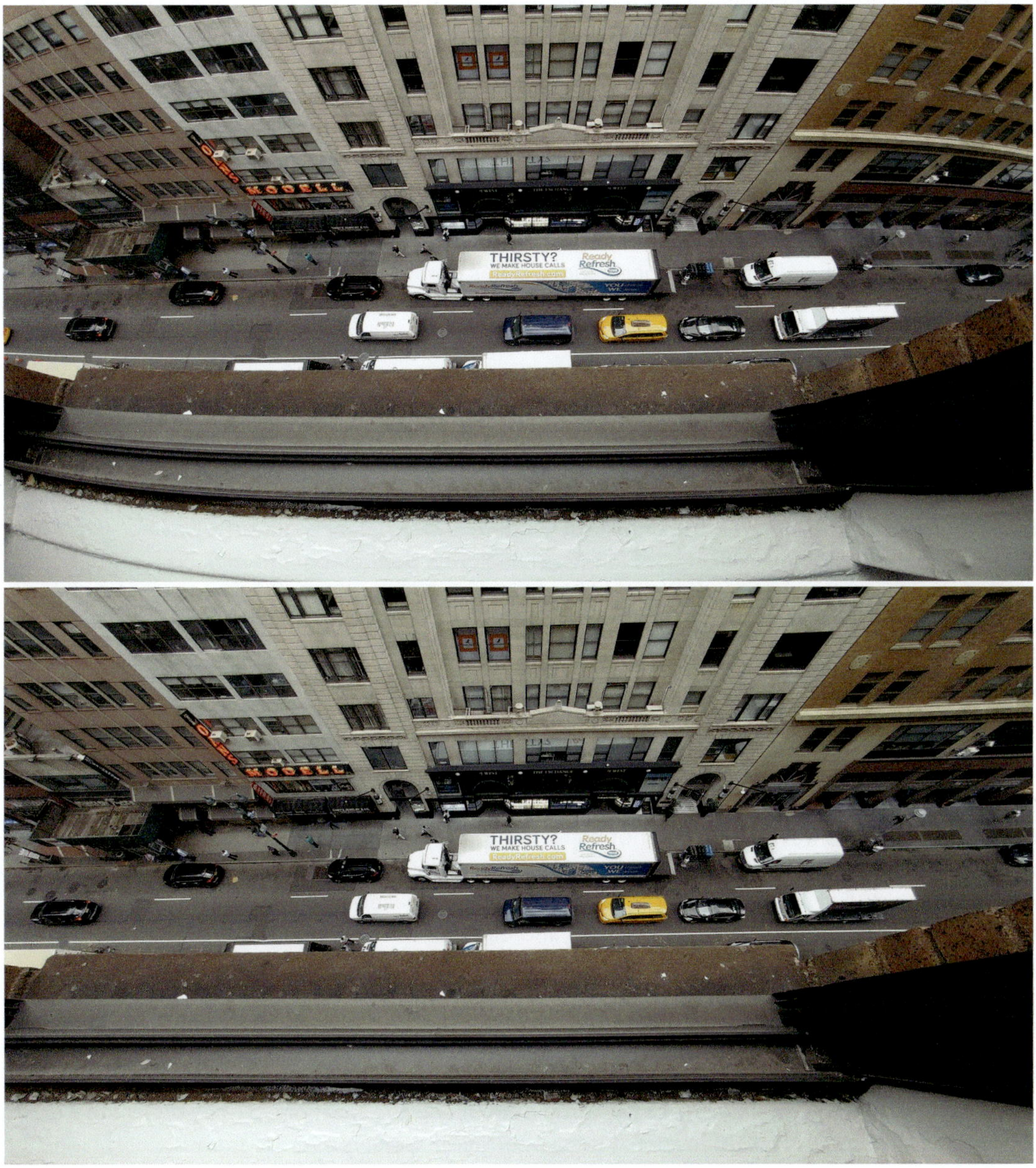

Top: the exaggerated perspective of a tilted-down wide lens generates a strong cornerstone effect of converging vertical lines. But the actual barrel lens distortion causes straight lines to curve, especially toward the edge of the frame. Bottom: when the image is properly undistorted, the curved lines are straightened out but the perspective distortion is kept.

lenses, but varies substantially between lens models. Lens distortion is a monochromatic aberration and affects all wavelengths equally. It is the only aberration that can be relatively easily undone because there is no degradation in the image quality, only displacement of pixels in 2D space. Undistorting footage is indeed a crucial step in VFX, as it allows for accurate camera tracking and integration of CG and 2D elements. However, the original distortion is usually reapplied at the end, and usually there's no attempt to "fix" lens distortion. This confirms the fact that in most cases, photographers and cinematographers treat lens distortion as an integral part of the photographic language, rather than an undesirable defect. Lens distortion is not part of the CG rendering process and needs to be applied as a 2D effect in comp (see Chapter 18). It is therefore not surprising that lens distortion is now offered as a post effect in game engines like Unreal and Unity. While not as dominant an effect as defocus, it is an important aberration that should be applied to final CG renders as part of the photoreal treatment, and must be matched precisely when elements that are composited into live footage for proper integration and tracking.

In a typical lens distortion, lines are either bent outwards (**barrel distortion**) or inwards (**pincushion distortion**), though some lenses generate a mix of both. The exact amount and shape of the curvature vary between lenses and focal ranges but is normally stronger toward the edges of the frame than at the center. The effect of distortion is not depth-dependent, but it is more noticeable on elements that contain straight lines (like buildings).

Chromatic Aberration

Every lens acts as a prism, to a certain extent, shifting and separating different wavelengths. This is of course not a desirable effect, and lens manufacturers do their best to minimize it, but even high-end lenses may exhibit minor chromatic aberration (heavy chromatic aberration is typical of low-quality lenses). All chromatic aberrations stem from the fact that the refractive index of the lens is not equal for all wavelengths. As light passes through the lens, different wavelengths slow down and bend at slightly different rates, and thus hit the sensor at slightly offset positions. This causes some visible separation of color components in the captured image.

There are two main types of chromatic aberrations, depending on whether the wavelengths are separated on the depth (Z) axis or along the focal plane (XY axes). In **longitudinal** (or **axial**) aberration, different wavelengths do not converge at the same distance from the lens, which shifts the focal plane slightly for each wavelength along the Z (depth) axis. In other words, different colors are focused at slightly different distances. Longitudinal aberration appears as purple or green fringing along the edges of objects and high-contrast boundaries, and can happen anywhere in the frame. It is typically more noticeable in areas that are out of focus. In fact, the color of the fringing is different when the area is in front or behind the focal plane (it usually shifts from purple to green).

In **lateral** (or **transverse**) aberration, all wavelengths are focused on the same focal plane and at the same distance, but not exactly at the same point along the focal plane (they are shifted on the X/Y axes). Like Longitudinal aberration, lateral aberration appears as color fringes along high-contrast boundaries. However, lateral aberration never happens at the center of frame, and instead grows stronger toward the frame edges.

While photographers and cinematographers try to minimize chromatic aberration by using high-quality lenses, stopping down, or using post-process treatments, digital artists can (and should) add some artificial chromatic aberrations to the final image. This is usually a subtle effect (unless the goal is to mimic a very bad lens, or to achieve a stylistic effect), but can provide an extra touch of photorealism to CG renders by breaking up their sterile look a bit, and adding some lens "warmth". Obviously, when compositing elements into footage, the artist must match any chromatic aberrations embedded in the footage for seamless integration.

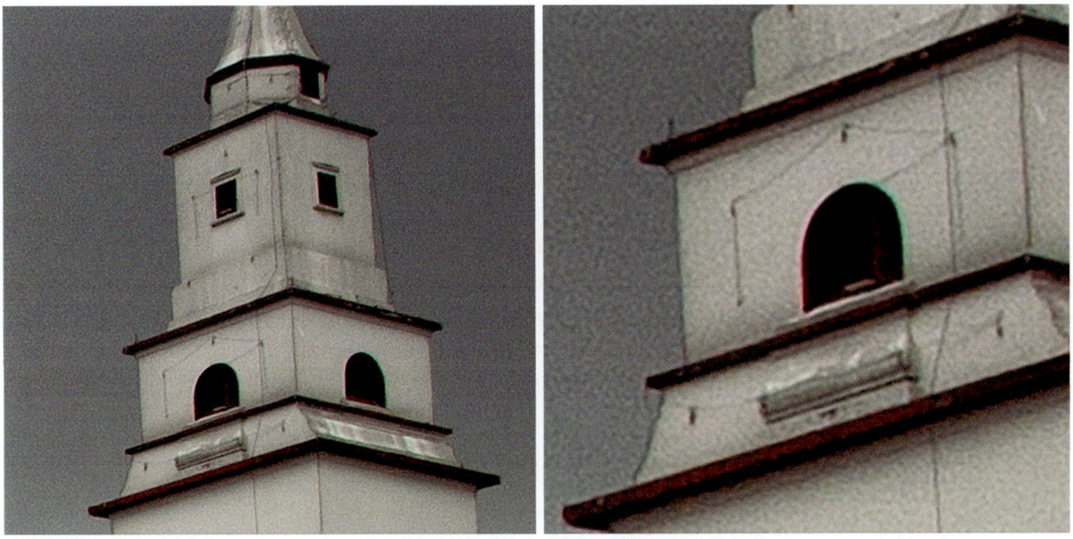

Two extreme zooms on a photo showing chromatic aberration at the edges of the windows, alternating between red-blue on the left and green-blue on the right.

Lens Flares

That tacky wedding-video lens flare effect has thankfully gone out of vogue long ago, and even director JJ Abrahams, the "king of lens flares", admitted to overusing them in movies like and *Super 8* and *Star Trek*.[1] But putting their arguably inflated use aside, lens flares remain an important optical element that, when used judiciously, can add a substantial amount of photorealism to CG renders, VFX composites, and games. Lens flares do not only emulate lens behavior and add warmth and grit, they also help bind together separate elements in the composite into a cohesive environment. The combination of different lighting scenarios and different lenses produces such an immense variety of flares with complex features and behavior, that the subject can easily fill an entire chapter (or an entire book). Within the limited space here, I aim to highlight the main features of lens flares, in hope of providing a base for further exploration.

Flares are not exclusive to glass lenses. They happen in our eyes too. Just bring a small source of light (like your phone's flashlight) into your field of view – you can easily feel the haze-like flaring as the light scatters in your cornea and lens. If (like me) you are also wearing glasses, that hazing is augmented by the scattering in the glass lenses, and additional shapes seem to appear. If you are wearing glasses while driving toward the setting sun, you are experiencing triple flaring, as light is scattered in the windshield, your glasses, and finally your cornea. This is essentially what happens inside a camera lens: at the basic level, light scattering causes a sort of hazy lifting of the blacks. But in most cases, additional separate elements are formed as the light scatters several times through the various glass elements that make up a lens assembly, producing those familiar yet complex chains of stars, circles, polygons, arcs, and various other artifacts. Zoom lenses, which contain more elements than primes, usually produce more complex flares. Here is an overview of the most common flare elements:

Simple Glow

When a camera is pointing toward a source of light, the scattering in the glass generates a glow around the light. When the source is near the center of the frame, the glow forms a circular halo around it. But when the light hits the lens from the side, the steeper incident angle causes a wider, more pronounced scattering that fans out from the light. If the light is relatively strong, it can create a noticeable glow even if the source is out of frame (for example, when the light is just off to the side of the camera). Photographers and cinematographers use lens hoods, flags, or even their hand to block light hitting the lens from the side and reduce flaring. When a flare glow happens in the lens, it lifts the blacks and lower midtones and reduces the contrast, but (like fog) does not brighten the highlights. A subtle flare adds a slight "haze" gradient, a strong flare can completely obliterate the frame.

Diffraction Spikes

Point your phone's flashlight toward your eyes and squint. Notice how, when your eyes are open, the glow is relatively uniform, but as you squint, numerous lines appear, emanating from the light's center – these diffraction spikes are caused by your eyelashes. A similar diffraction may happen when the lens aperture is not fully round. Since the aperture structure is much simpler than a person's eyelashes, the effect is also simpler – usually a distinct star shape, with the number of spikes either matching or doubling the number of aperture blades (for example, 12 spikes caused by a six-blade aperture). The length of the spikes can be uniform, but often, some spikes are longer. This familiar effect, often called **starburst** or **glint,** is generally well-defined when the light faces the lens, but becomes more chaotic and less defined when light hits the lens at an angle.

Simple lens glow (left) and diffraction spikes (right).

Additional Flare Elements

Camera lens assemblies (zoom lenses in particular) comprise of multiple glass elements with varied optical characteristics. When light enters the lens, it is not only scattered by the outer glass, but also by each of the internal glass pieces. This can generate an entire sequence of scattering effects, from glows and diffraction spikes to various shapes like circles, arcs, lines, and polygons (the latter usually matching the aperture shape and number of blades). When the light hits the lens assembly at a relative straight angle, the additional elements overlap in a way that makes them hard to discern. But when light hits the lens from the side, the various flare components spread along the direction of the light, starting from the origin position and ending at the opposite side of the frame, resulting in the classic elongated chain of flare components. The specific types of components and their order in the chain depends on multiple factors: the particular lens model, the angle of the light hitting it, the type of light source (focused or diffused), the amount of dirt on the lens, and the level of wear of the optical and mechanical components – potentially an infinite number of variations. However, the movement of all lens flare chains usually conforms to a simple principle: the elements are aligned along a straight line that starts at the light source and pivots at the center of the frame. For example, if the light moves from the top right corner to the bottom left corner, the entire flare chain rotates, and the end moves from bottom left to top right. Additionally, the brightness and visibility of the flare components vary according to the intensity of the light and its position in relation to the lens.

Dirt and Imperfections

Even the tiniest speck of dust or dirt on the lens becomes substantially more noticeable when light hits it from the side. This factor is often overlooked by artists as they strive to simulate realistic lens flares. No lens is absolutely clean or perfectly smooth. These inherent imperfections not only affect the look of the flare and its different components, they also contribute their own additional components – small circles or streaks that add to the complexity and irregularity of lens flares. Flare components are rarely consistent and change shape and appearance with the smallest movement, angle variation, or light fluctuation.

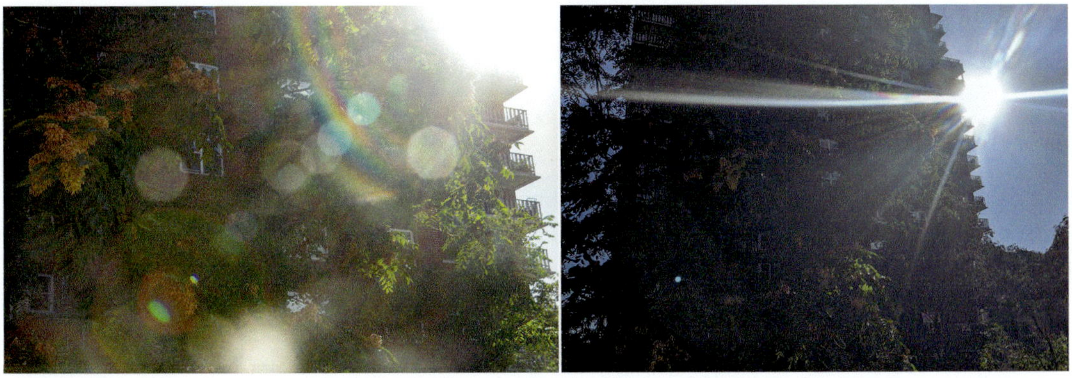

Different lenses produce completely different flares. The photo on the left was taken with a Sigma zoom lens on a DSLR. Notice the complex combination of different elements and the distinct heptagonal shapes generated by the seven-blade aperture. The photo on the right was taken at the same location with a smartphone, and shows quasi-anamorphic diffraction spikes and very few additional elements.

Anamorphic Lens Flares

The oval aperture of anamorphic lenses and the fact that the footage is stretched horizontally makes anamorphic flares distinctly different and immediately recognizable. They are substantially wider on the horizontal axis, typically with one or two prominent, blue diffraction spikes stretching across the frame. This adds a uniquely dramatic visual component to footage and has been used effectively in movies (or, as some may point, has become a bit of a cliché). Anamorphic flares should certainly be used when elements are composited into anamorphic footage but are better avoided with spherical footage or when the context of a CG scene does not call for an anamorphic emulation. For example, adding anamorphic flares to an architectural render may not be the best choice, as it is unlikely that such a render would be captured in real life with an anamorphic lens. This distinct and recognizable effect is best used only where it truly belongs.

Close-up on the elongated diffraction spike of a typical anamorphic lens flare.

Lens Bloom

The definition of bloom is quite fluid, as some use the term interchangeably with lens flare. Most commonly, the term is used to describe a lens glow that happens around bright highlights rather than actual light sources. Since highly specular surfaces reflect the light sources in the environment at relatively high intensity, such areas can scatter in the lens and produce a noticeable glow, or haze. Since lens manufacturers strive to minimize blooming, natural bloom is usually quite subtle and uniform, and rarely displays the complex flare elements associated with strong direct light sources. That said, photographers and cinematographers often exaggerate highlight blooms to achieve a soft, dreamy look, by using lens filters like Glimmerglass, or as a post effect.

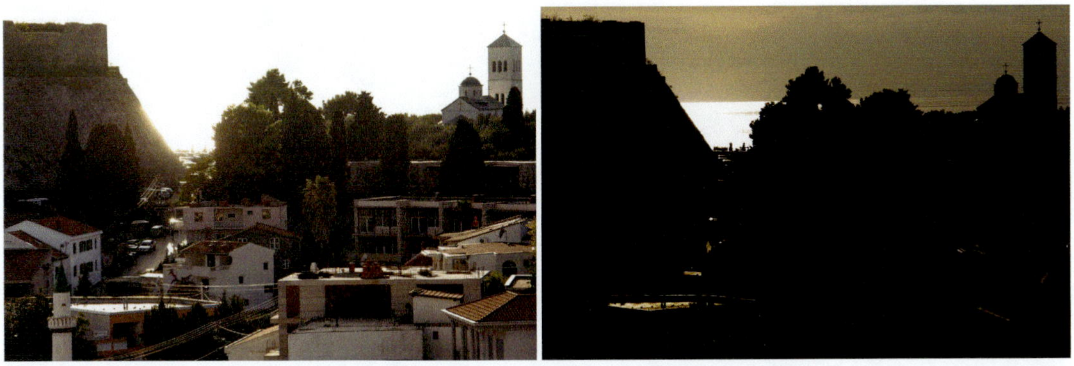

This extreme example of a bloom effects looks almost like a flare (as if the sun was visible in frame). Lowering the exposure (right) shows why: the sun is completely out of frame, but its specular reflection on the water is substantially brighter than anything else in the photo.

Motion Blur

Motion blur is a vital component of photorealism. A fast-moving object without motion blur does not only strobe, it looks extremely unnatural, even in a still image. Admittedly, motion blur does not play the same important role in visualizations and games as it does in VFX. Architectural and product renders rarely involve fast camera motion or fast-moving objects, and although motion blur is available in some games, so far it has not been universally accepted, partly because of technical limitations and partly because gamers tend to prefer visual clarity over optical realism. But in VFX, motion blur is crucial for integration with live action and cinematic believability

Motion blur is not exclusive to cameras – it is also part of our vision. The amount of blur we see depends on the speed of the moving object. However, unlike our vision, camera motion blur is strongly affected by two additional parameters: shutter speed and frame rate. Assuming an object moves at the exact same speed, a longer shutter speed will produce longer motion blur. Sports photographers use wider aperture lenses and fast shutter speeds to minimize motion blur, for example. Movies are shot at 24 frames per second. But higher frame rates can reduce or eliminate motion blur. Interestingly, attempts to shoot entire films at high FPS have so far received mixed responses. The critical reaction to Ang Lee's 2016 film *Billy Lynn's Long Halftime Walk*, which was shot at 120 FPS, is a good example. While some critics praised its unprecedented clarity, others lambasted the effect as un-cinematic. As Vox magazine's Emily VanDerWerff put it: "At 120 fps, things look a bit like higher definition video – or, as many people put it, like a soap opera".[2] Over a hundred years of movies have left a mark on our cinematic perception, and the reduced motion blur of high frame rates can feel quite unnatural.

The word "blur" in motion blur is somewhat deceiving because the effect is markedly different from standard Gaussian blur or defocus. It should better be thought of as a trail that connects pixels between two adjacent frames, and unlike regular blur, it does not fade off softly but often ends abruptly at a defined line. While motion blur certainly smears details in the image, the smearing is always directional, and contrasted features in the image can be clearly seen as streaks. However, a complex fast motion like that of a shaky hand-held camera can produce an equally complex motion blur pattern that is not easily emulated with 3D or 2D software.

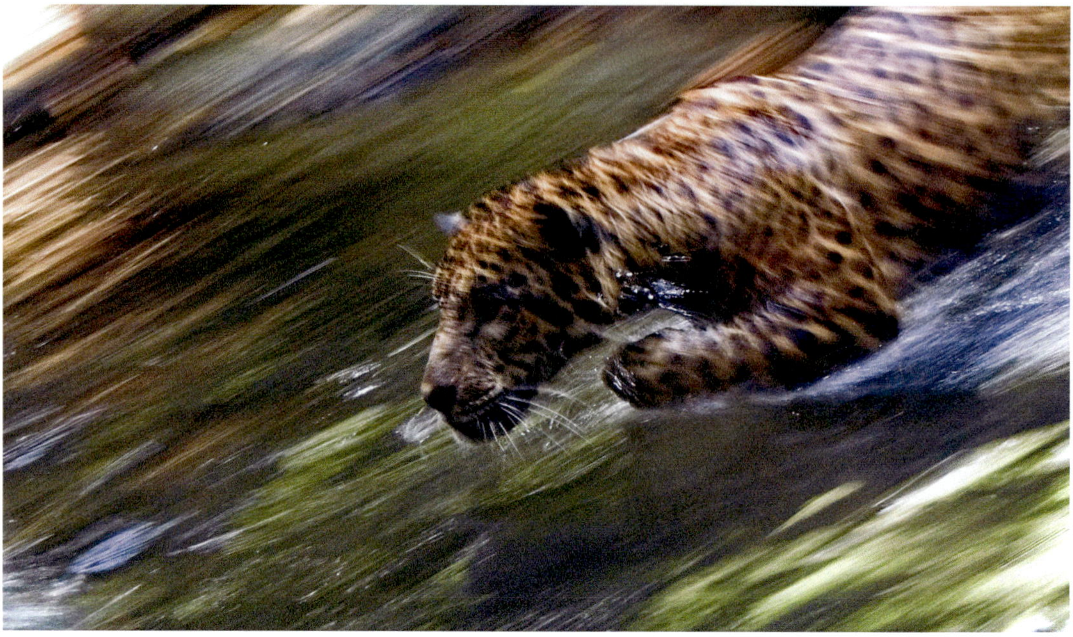

The camera's motion is clearly felt in the pronounced streaking, while the leopard's own motion generates a different streak direction on its back. Notice the general sharpness of the streaks – motion blur is more stretching than blurring.

Grain

Film grain is the result of the small particle of silver halide reacting to light photons. In digital cameras, noise is usually the result of the sensor's reaction to low light and high ISO settings, and is generally less prominent than film grain. Despite their differences, film grain and digital noise (which we will also call "grain" here for the sake of simplicity) are the very final layer of photorealism in digital imagery. Because grain happens uniformly over the entire image, it has an important role in compositing as a glue that binds together various elements. Grain adds a photographic feel to CG renders and is crucial for integrating CG elements with footage. Missing or mismatched grain, even if hard to detect without close inspection, can nonetheless cause composites to fall apart, especially when they are color-corrected.

The exact appearance, scale, intensity, color variation, and randomness of grain and noise vary greatly between different film stocks and digital camera models. But there are some basic characteristics that apply to all types:

- Grain covers the entire image uniformly, but its visibility depends on the underlying colors. Generally, grain is more pronounced in the darker areas, and less visible in the brightest.
- Grain (both film and digital) is never equal in each of the RGB channels. Digital sensors are more sensitive to green and red light (same with the green and red layers in film), so the blue channel is usually considerably noisier.
- Grain is random, and changes on every single frame. It is never static.

Matching grain is an important (and often neglected) part of compositing and will be discussed in Chapter 18.

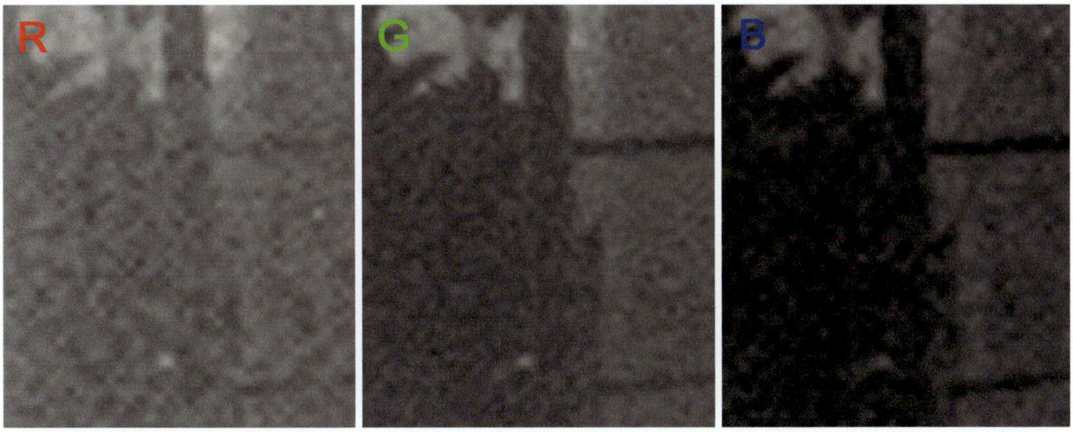

An extreme close-up on footage captured by an Alexa digital video camera, showing different grain patter, size, and intensity on each RGB channel.

Notes

1 https://www.theverge.com/2013/9/30/4788758/j-j-abrams-apologizes-for-his-overusing-lens-flares.

2 *Billy Lynn's Long Halftime Walk* is all technology, no movie, Emily VanDerWerff, *Vox Magazine*, 2016.

PART 3

THE CG WORLD

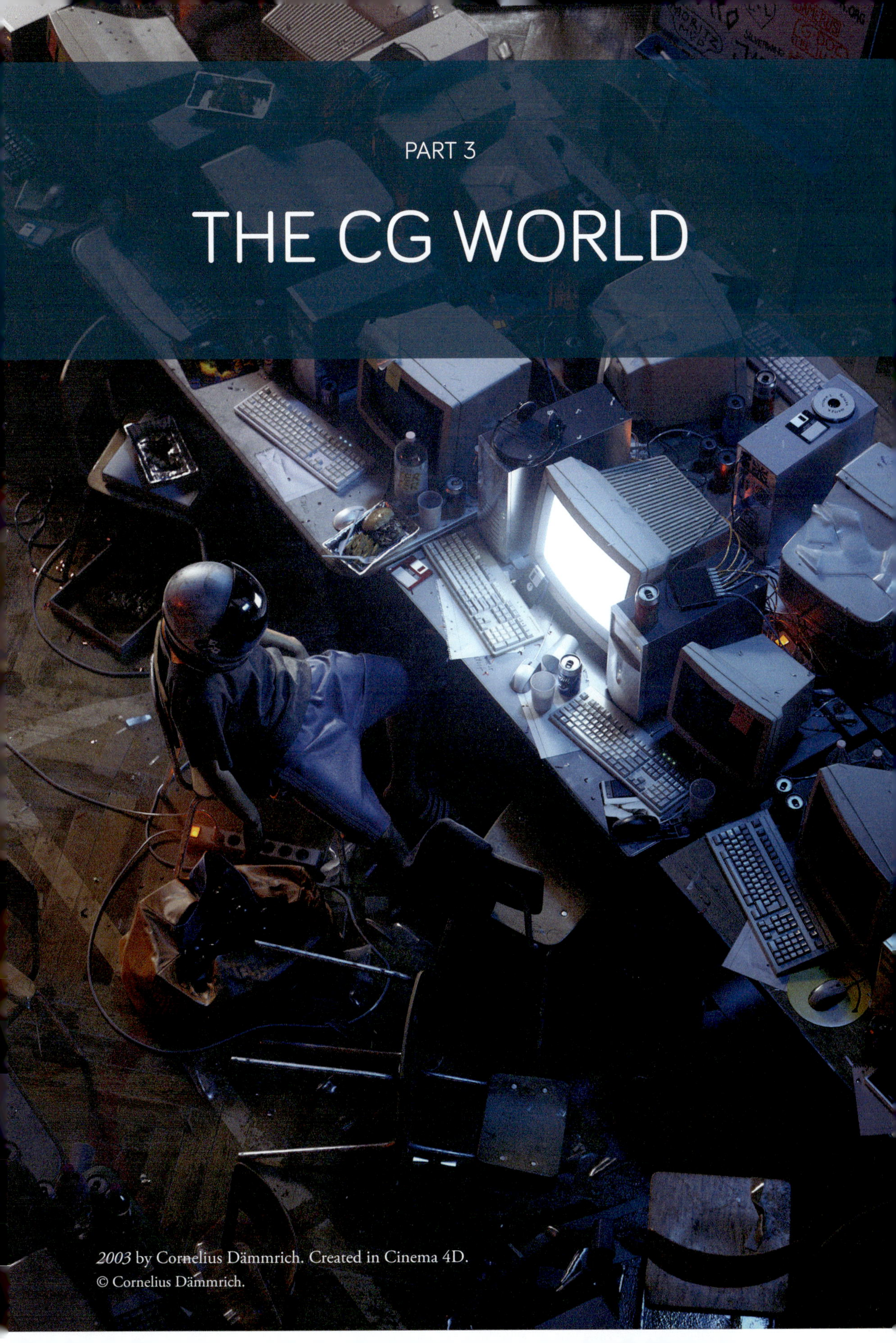

2003 by Cornelius Dämmrich. Created in Cinema 4D.

Here is the good news: it is easier today, more than ever before, to render convincingly photoreal CG imagery. Physically-based rendering, BRDF shaders, IES lights, an extensive selection of smart lighting, shading, and texturing tools, the widespread availability of scanned 3D models and textures, ready-to-use HDRIs, and advanced real-time technologies – all these factors contribute to the relative ease of generating believable CG content. What once required equal amounts of trickery and ingenuity to work around the many limitations of simulated reality is nowadays a much simpler affair: follow real-world paradigms in texturing, shading, and lighting – and you should end up with photoreal results.

And now the not-so-great news: the world has gotten used to physically plausible renders. Whether you create VFX for films and TV, architectural visualizations or product renders, your clients and your audience are expecting a level of photorealism that would have been considered very challenging, if not unattainable, just ten years ago. And talk about games . . . yes, real-time rendering does not yet allow quite the same photoreal results as offline rendering. But the emphasis here is on the word "yet". Technology is consistently marching forward, and the gap is narrowing. With every new game that pushes the envelope of realism a little more, gamers are expecting the next one to go even further.

These lofty expectations certainly put the onus on CG artists, who must keep up with a higher than ever threshold of photorealism. The fact that contemporary shading and lighting works so much like the real world requires not only a deep familiarity with photography and physical light interaction, but also a mastery of contemporary CG tools. Lighting and shading are of course crucial for photoreal rendering, but so are textures, and to some extent modeling as well. To better emphasize how each of these tasks contributes to the final quality of photorealism, I chose to organize this part of the book in the reverse order of the CG process – starting with rendering and lighting, then shading, texturing, and ending with modeling.

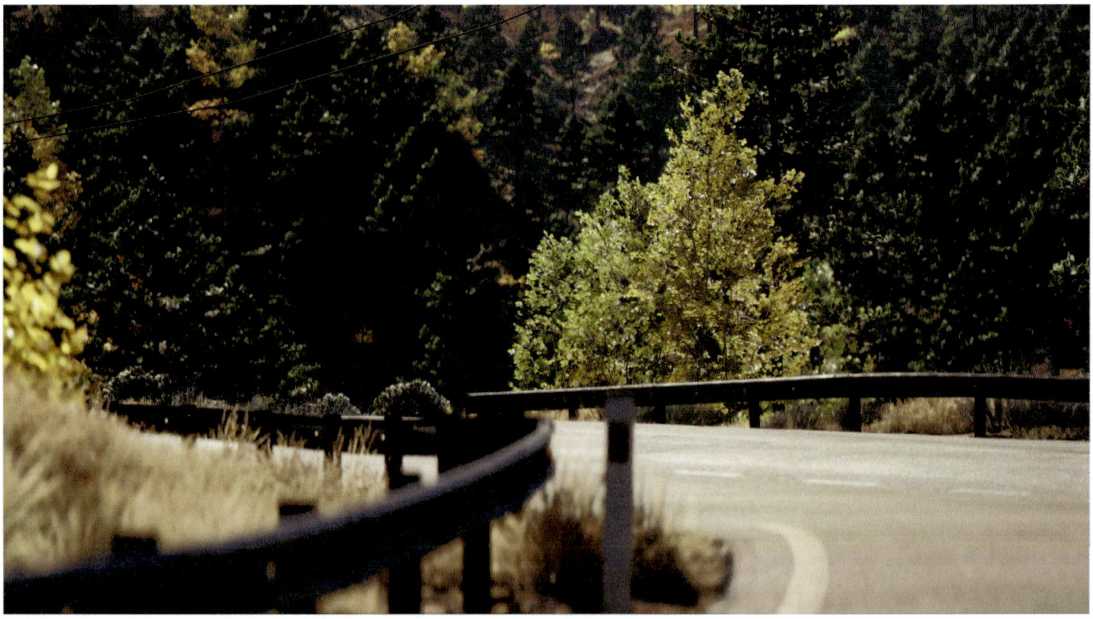

PWG Ecosystem by Gökhan Karadayi. Unreal Engine, Procedural World Generator.

Chapter 11

Rendering and Lighting

When 3D computer graphics started evolving in the 70s, the theory of light was already well established. The knowledge was there, the laws and paradigms were there, the formulas where there, the works of Newton, Huygens, Fresnel, Planck, Einstein, et al., were there. What wasn't there? Computers that could handle the immense computational challenge of simulating physical light behavior and surface/light interactions. It is easy to forget just how much computing power has increased since the early days of CG rendering. Here's an example: the CDC 7600, one of the most powerful super-computers of the 1970s, a behemoth that occupied an entire room, had a computing power of 36 million megaflops. By comparison, the Samsung Galaxy S6 phone had a computing power of 36 BILLION megaflops. A 2015 consumer phone was still a thousand times (!) more powerful than the 70s strongest super-computer.

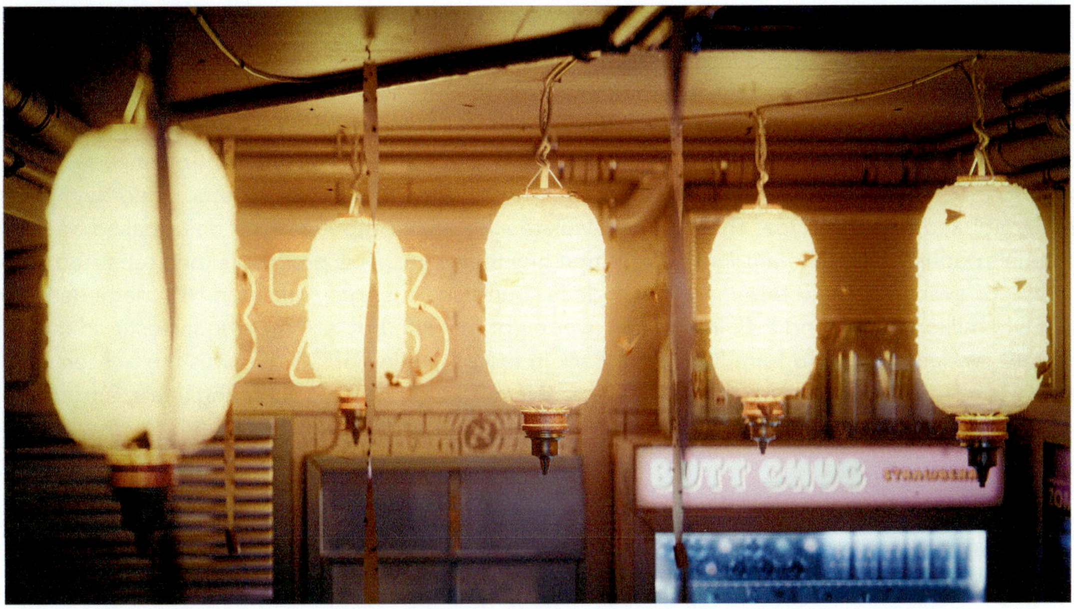

Detail from *You Touch, You Buy*, by Cornelius Dämmrich. Cinema 4D, RizomUV, Octane Rrender.

DOI: 10.4324/9781003590293-15

The pioneers of CG rendering in the 70s and 80s had a mere fraction of the computing power we have today. With such extremely limited means, a true physical simulation of light was not even an option. The first CG light emitters, shading models, and rendering algorithms were never intended to faithfully emulate real-world lighting. Instead, they mimicked light behavior by way of crude approximations, using smart compromises that allowed CG artists to produce some surprisingly realistic imagery despite the hardware limitations of the time. Today's CPUs and GPUs are fast enough to support the type of physically-based rendering that was a distant dream 30 or 20 years ago. But to better understand the contemporary tools we have at our disposal, it is important to see them in a wider evolutionary context, especially since many tools that have been with us for decades are still being used today.

From Scanline to Path Tracing

A render engine has one, and only one task: to determine what color a pixel is. At the end, after the scene is modeled, textured, shaded, and lit, it all comes down to this one question: what color is this pixel? And what color is the pixel next to it? And the one next to it? Even though the answer is just three numbers (RGB) per pixel, the road to it is a long and winding one, because those three numbers depend on a multitude of factors. To obtain an absolutely precise answer, the ideal render engine would need to examine everything in the CG scene that may, in one way or another, affect the RGB value of each specific pixel: all the light sources, every interaction of photons with any possible surface in the scene, the various material properties of every surface, and so on. Such an ideal renderer (let us call it **fully unbiased**) would be the closest thing to a true light interaction simulator. And while we are closer to this ideal renderer today, such a prospect was not a viable option in the past, and so rendering methods had to follow a much-simplified process.

Scanline/Rasterized Rendering

A scanline renderer gathers all the information about the geometry in the scene, then transforms that 3D information into a 2D projection based on the camera's viewing plane, a process called **rasterization**. Depth is sorted via a **Z buffer**, to ensure that only the front-most surface is rendered and keep the ones behind it hidden. The color of every pixel is determined by the shader applied to the surface. The shader calculates the color output at each point based on the material properties and the amount of light that hits that point. This process is then repeated until a color for every pixel is determined, which does sound like a lengthy affair (an HD image has over two million pixels, and a 4k image has over eight million), but is in fact the fastest and most efficient rendering method for a 3D scene. Back in the beginning of CG, this was the only way to render in a feasible amount of time. Nowadays, this rendering process is blazingly fast, and forms the foundation of contemporary game engines, enabling smooth refresh rates at high resolutions.

Raytracing

Scanline/rasterization is fast, but very limited in its ability to generate photoreal renders. It cannot provide a truly accurate answer to the "what color this pixel is?" question because it ignores several key factors: much of the light-surface interaction depends on the surface's surrounding environment

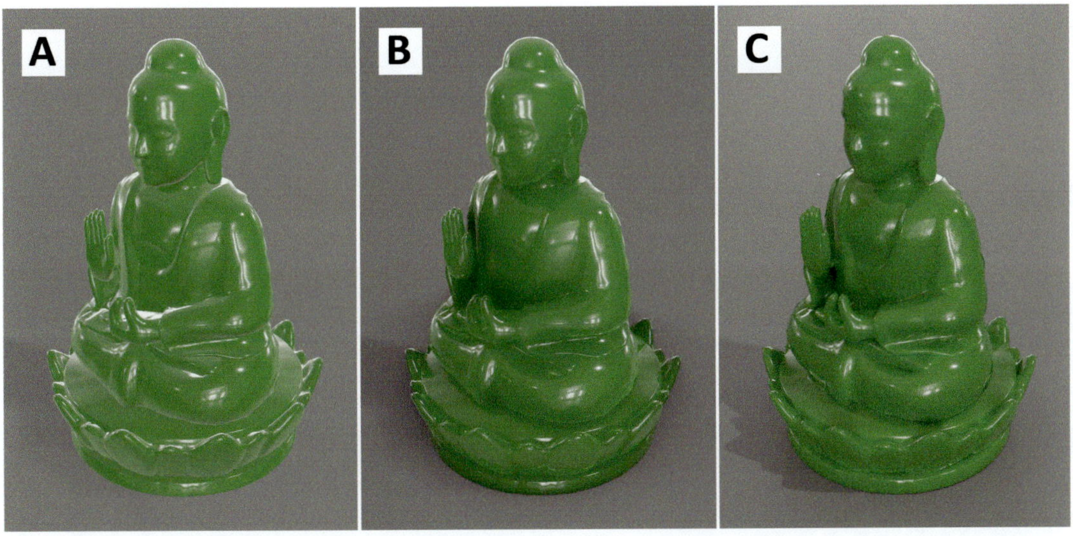

These examples compare Blender's real time *Eevee* renderer, and its offline full-on raytracing engine, *Cycles*. A: Eevee simple real-time render, B: Eevee real-time render with ambient occlusion, screen-space reflections and subsurface scattering approximation (more accurate), C: Cycles render with path-traced reflections, true subsurface scattering and GI (most accurate). These lighting and shading aspects will be discussed in this chapter and the following one.

(as discussed throughout Part 2). The exact color of every point in the scene depends on specular reflections (what exactly is being reflected at this point?), transmission and refraction (what do we see through the surface at this point?), and, most importantly, bounce light (how is the environment contributing to the lighting at this point?). If we really want to find out how light behavior is affecting the color of every point in the scene, we need to start following photons around: shoot multiple rays from each source of light, and follow their path as they bounce around the scene and interact with surfaces. This, however, would be an extremely inefficient undertaking. As I mentioned in Chapter 1, the camera only sees photons that directly hit the sensor. If we started tracing light from the source, we would be spending an eternity following photons that will never even be seen by the rendering camera, and therefore will have little or no effect on the image.

Raytracing follows the path of light backward, starting from the camera and eventually ending at the light source/s. This ensures that only rays relevant to the rendered image are traced, substantially speeding up the process. Yet despite this smart approach, ray tracing is still painfully slow in comparison to scanline rendering. Imagine a single ray being traced from the camera to the first point of surface interaction. This single point may be affected by multiple lights, and reflect other surfaces, which in turn may also be affected by multiple lights and other surfaces, and so forth. One single ray needs to be split into multiple rays at the first point of interaction, and each of these rays may be further split into even more rays at subsequent interaction points, multiplying the ray number exponentially. In raytracing, rendering a single pixel may require millions of calculations (as opposed to just a few in scanline rendering), yet the added realism is well worth the longer render times. Indeed, ray tracing became a mainstay of offline rendering and elevated CG to a new level of photorealism by generating true specular reflections, realistic refractions, and accurate shadows. But raytracing remained too prohibitive for real-time rendering until recently: in 2018, Nvidia unveiled their RTX technology, and since then ray tracing is making a steady foray into games, VR applications, and real-time visualizations.

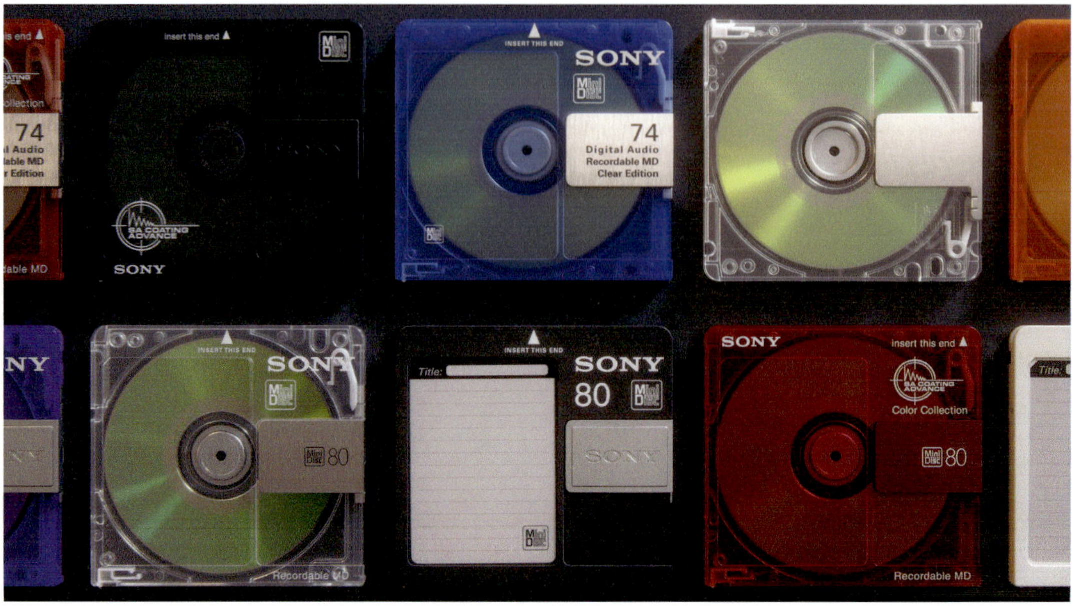

Minidisc by Raphael Rau. Cinema 4D, Octane Render.

The Challenge of Global Illumination

Accurate reflections, refractions, and shadows are important for realistic renders, but the contribution of bounced light, or **global illumination (GI)** is the most crucial ingredient in the photorealism recipe. Only global illumination can faithfully reproduce all the intricacies of light interaction, and its effect, though often quite subtle, makes a tremendous difference. Global illumination is substantially more complex to calculate than other components and was not initially incorporated in raytracing renderers. In the history of CG, many brilliant animations, visual effects, and visualizations were created without any global illumination. Instead, they were masterfully lit by artists who used simple point and spotlights to create the illusion of reflected light and create a convincing ambience. Several methods were consequently developed to cope with the challenge of global illumination, either by "faking it" (ambient light emitter, ambient occlusion) or by using more efficient calculation methods (radiosity, photon mapping).

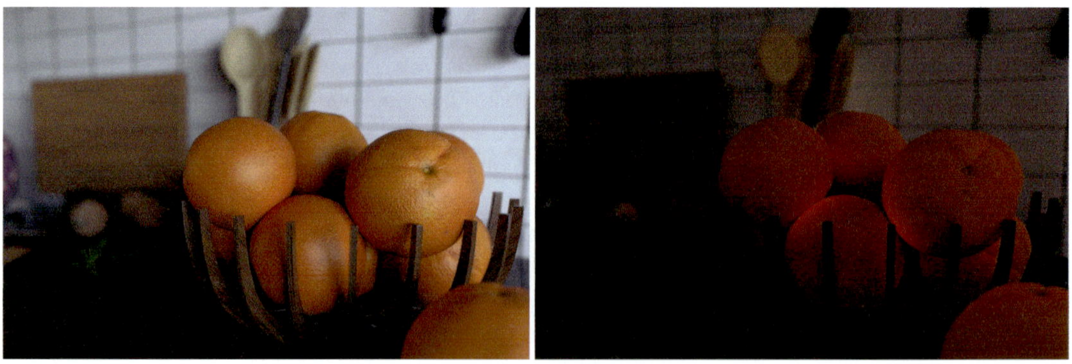

The image on the right shows only the indirect, or reflected light component of the path-traced render on the left. Notice the strong inter-reflectivity between the oranges and between the cooking utensils and the wall.

AMBIENT LIGHT EMITTER

The ambient light emitter was used from the early days of CG as a quick and cheap way to crudely simulate the effect of reflected light or diffused sky radiation. Though in some applications (Maya, for example), the ambient light emitter has an optional basic directionality, it is not really a light at all, but rather a multiplier that uniformly raises the overall color gain of the scene. The ambient light was originally used to "fill in" for missing reflected or scattered light in both exterior and interior scenes. However, it produces an utterly flat and unrealistic "light" with no shading or occlusion, and although it can still be found in some rendering software, there is little reason to use it nowadays for offline rendering (and it is quickly becoming obsolete in real-time rendering as well).

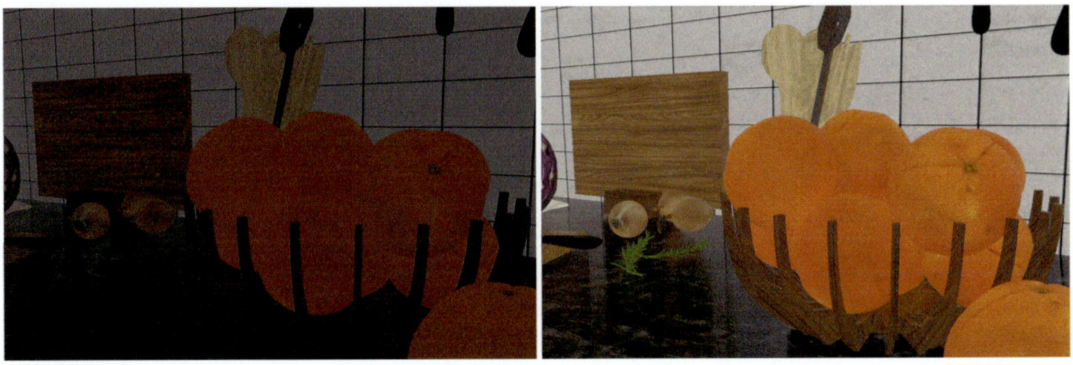

The old ambient light was a rather poor substitution for indirect illumination, as it only acted as a simple brightness multiplier.

AMBIENT OCCLUSION

Ambient occlusion (**AO**) is not a global illumination solution – it does not calculate reflected light at all. But it is a much-improved take on the crude ambient light emitter, and can be used as a very efficient "cheat" for GI. Ambient occlusion assumes multi-directional ("ambient") light emanating from a consistently white sphere around the scene, and then checks to see how much of that light reaches every point in the scene. Naturally, areas that are more exposed to the theoretical light dome are more illuminated than areas that are partially occluded, resulting in the soft shadowing characteristic of diffused sky radiation on a fully overcast day. This process is significantly faster

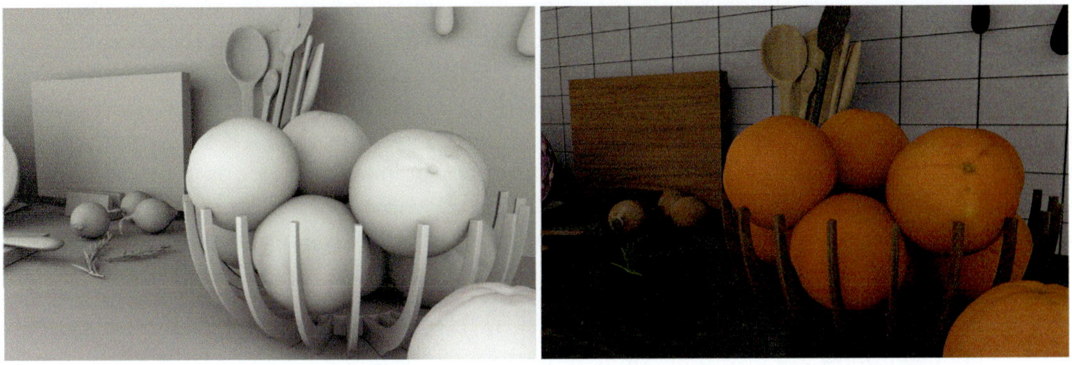

An ambient occlusion pass (left) can be combined with the flat ambient light shown in the previous figure to generate a "poor man's global illumination". This approach is used especially in games and other real-time applications.

than calculating a multitude of bouncing light paths and can add a substantial level of depth and photorealism by replacing flat ambient light with a shaded one. It is thus commonly used in game engines and real-time display of CG applications, and is an important component of texturing for enhancing bump/normal maps (see Chapter 13). In addition, an ambient occlusion render pass is used for adding contact and proximity shadows in the compositing stage (see Chapter 16).

RADIOSITY AND PHOTON MAPPING

As the first effective solution for GI, radiosity was an exciting development in CG rendering that was quickly adopted by ArchViz artists. Radiosity is view-independent and calculates reflected light energy over the entire scene. It divides the surfaces into small patches, and then checks each pair of patches for their **view factor** – essentially how much they "see" each other. Patches that are in close proximity and/or facing each other bounce more light between them, as opposed to patches that are distant or at oblique angles. Although radioisty was, for a while, the most efficient way of calculating the contribution of reflected light to a scene, it was limited by the fact that it could only calculate diffuse light (ignoring the contribution of specular reflections), and was detached from the main raytracing process. Also, its reliance on dividing geometry into patches made it prone to errors and artifacts along hard edges and sharp angles.

Like radiosity, photon mapping works as a separate, complementary process to standard raytracing, however photon mapping is view-dependent. In this 2-stage solution, photons are first shot from a light source, and their interaction with surfaces is traced and recorded in a photon map. In the second stage, the renderer uses the projected photon maps to determine the radiance value of the surface at each point. This value is then combined with the direct illumination values obtained by raytracing. While photon mapping requires less computing power than path tracing, faster hardware made path tracing the preferred choice for fully integrated ray-traced global illumination.

Path Tracing

Almost every contemporary **physically-based renderer** (**PBR**) uses a method called **path tracing**, which represents, to date, the most holistic (and physically accurate) approach to raytracing. A path tracing renderer treats all light equally regardless of whether it is direct or reflected, incoming or outgoing. In conjunction with **BRDF shaders** (which will be discussed in Chapter 12), path tracers integrate global illumination seamlessly with diffuse and specular reflection and refraction. Although this requires heavier calculations then older GI approaches, reasonable rendering times are still possible thanks to the **Monte Carlo** approach. Monte Carlo is a wide term that applies to many fields, not just CG rendering, and refers to algorithms that speed up computationally intensive processes through random sampling. The concept is similar to statistical surveys: to find out which ice cream flavor is the most popular in the USA, for example, the surveyor would never interview the entire population (which will give the most accurate result but will take forever), but a much smaller selected group instead (which will give a reasonably accurate result in a manageable time).

A Monte Carlo path tracer uses a similar principle by tracing a limited number of random rays, in effect **sampling** the environment at various locations and for various light interactions, instead of methodically tracing every single possible ray. The quality and accuracy of the image depends on the number of samples. A low **sampling rate** results in noisy renders,

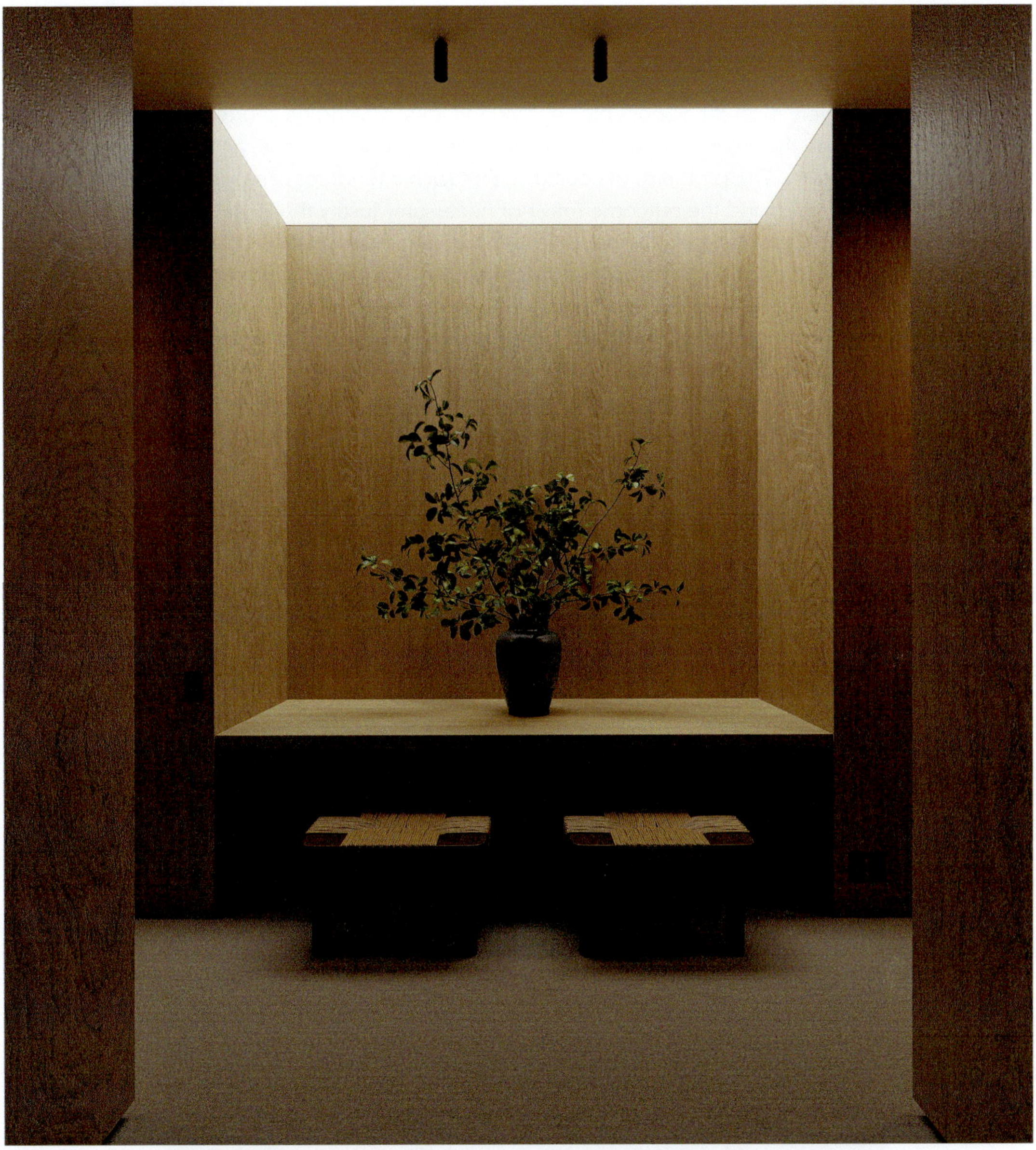

Simple Interior by Javier Wainstein. 3D Studio Max, Corona Renderer.

© Javier Wainstein.

because not enough information is gathered. In contrast, overly high sampling rates are counterproductive, as rendering times are increased for no substantial visual benefit. Each different CG scene therefore has its own sampling "sweet spot" of optimal quality/time setting. Reaching it is made easier by assigning separate sampling rates to different lighting components (diffuse, specular, subsurface scattering, etc.). Usually, the noisiest components are the indirect ones (like indirect diffuse or indirect specular), and higher sampling rates can be set specifically for these components while keeping the direct illumination sampling values lower.

Unbiased Versus Biased Rendering

The "ideal render engine" I mentioned earlier is considered fully unbiased, because it diligently and meticulously checks every pixel for all possible light interactions, making no shortcuts on the way. This is bound to provide the most accurate "render solution" to a given scene, and, supposedly, the most photoreal results. The problem, of course, is that such a renderer can take a painfully long time to produce the image, and the outcome may not necessarily justify the long wait. By contrast, a biased renderer uses various shortcuts to speed up the process, making biased choices as to what is more important and what can be ignored. Does that mean that the result of biased renderers is less photoreal? Put in the correct perspective, the debate between "biased" and "unbiased" is rather pointless. The aforementioned "ideal renderer" does not exist, and even renderers that are considered unbiased still use an approximation of physical light behavior. Path tracing relies on sampled (not absolute) results, and often uses **importance sampling** that focuses the samples in areas that are likely to contribute more to the final output. This in itself is a biased process (physical light does not make such choices). In other words, all renderers are estimators of reality, not reality itself. Take, for example, two of the most powerful (and widely used) render engines: Arnold and V-ray. The first is considered unbiased, and the other biased. Both produce consistently excellent results and have a proven track record as production workhorses. Is one better than the other? Redshift, another popular renderer, is known for its blazing speed, which it owes to an array of smart shortcuts and optimizations. The fact that it is a clearly biased renderer does not necessarily make it inferior. Most contemporary PBRs are capable of similarly realistic results, whether they are considered biased or unbiased.

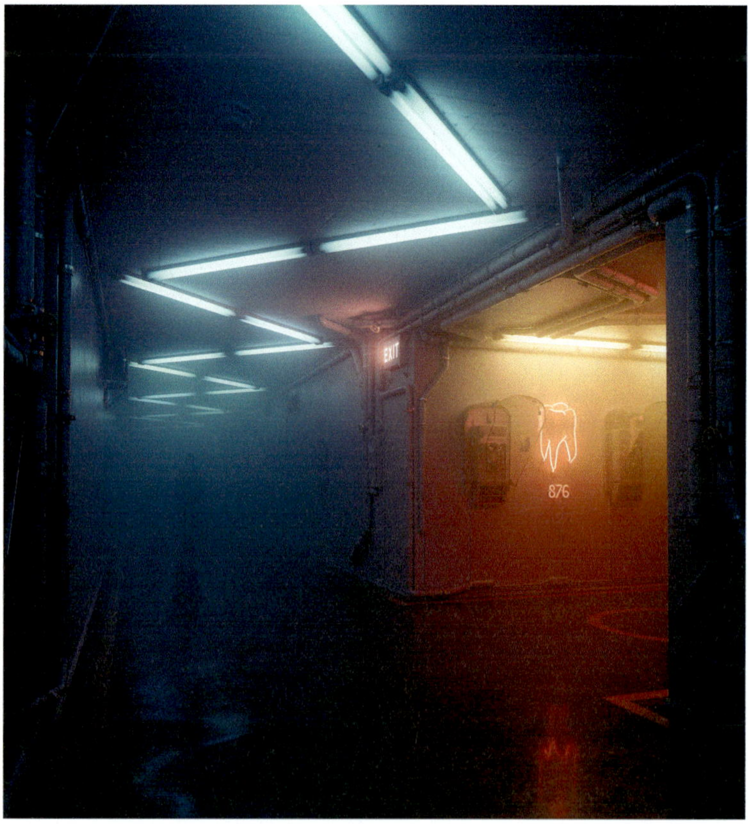

Delpher 876 by Cornelius Dämmrich.

Traditional Light Emitters

Point/Spot Lights

As its name suggests, the point light generates light from an infinitely small point, and a CG spotlight is essentially a point light with a restricted, adjustable cone of emission. The fact that the light emanates from a single point makes these light emitters considerably faster to calculate then area lights, and virtually immune to noise issues. But point/spot lights are physically incorrect because real-world light always originates from a physical, measurable surface. Point and spot lights therefore have several key limitations:

- Being infinitely small, point/spot lights cannot produce true, accurate soft shadows with an umbra and penumbra that relate to the shape, surface scale, distance and angle of the light (soft shadows can only be "faked" through a global softness parameter).
- Since point/spot lights have no scale, they do not exist as visible entities in the CG environment, and therefore cannot appear in specular reflections. This goes against the very nature of specularity, where (as discussed in Chapter 4) light sources play a crucial role. The traditional representation of point/spot lights in specular shaders like Phong and Blinn is merely a "cheat", not a true reflection of a light-emitting entity.
- Point lights can only emit light in a spherical fashion. Elongated fluorescent tubes, rectangular fixtures, neon signs, and all other non-spherical light fixtures are hard (if not impossible) to emulate.

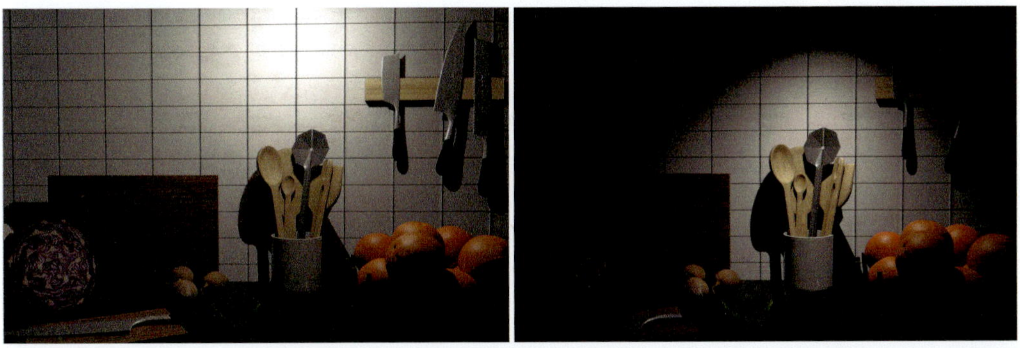

Point light (left) and spot light (right). Notice the perfect, harsh shadows and strong contrast.

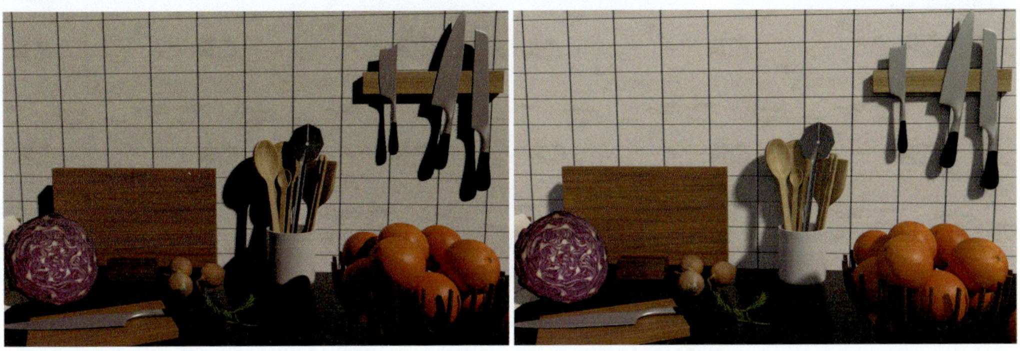

Directional light (left). On the right, with a bit of shadow softness and additional sky dome for diffuse sky radiation (see the following).

Directional Light

Like point and spot lights, the directional light has no surface or size, and light emanates from an infinitely small source. But unlike point/spot lights, the position of a directional light has no effect (only its direction), and it casts parallel rays and shadows, which makes it a basic but nonetheless effective simulator for sunlight. As described in Chapter 6, the sun is so distant, that its light on earth (for all practical purposes) is parallel. The directional light is therefore still widely used in both real-time and offline rendering. Most render engines offer a shadow softness parameter (sometimes called angle). This is necessary to simulate the effect of the atmosphere (and time of day) on the softness of sun shadows.

Contemporary Light Emitters

Area and Mesh Lights

As computing power increased, **area lights** became the choice emitters for most offline rendering purposes. Their advantage over point lights is clear: area lights generate light from a surface that has a physical shape and scale, and thus solve the three crucial limitations of point lights: accurate soft shadows, true visibility in specular reflections, and non-spherical illumination. The visual characteristics of light fixtures are often derived not from the bulb itself but from the wider surface of scattering mediums like lampshades, reflectors and diffusers, and area lights are well-suited to mimic such characteristics. Their disadvantage is that numerous rays must be shot from the surface toward every angle, as opposed to a single ray in point lights. This can introduce noise issues, especially in the shadows, which requires higher sampling rates and increases the render times.

Area lights usually come as a rectangle, disk, or cylinder – essentially the most common shapes of lighting fixtures. However, since PBRs use a holistic approach to lighting, any object with an emissive shader effectively becomes a light emitter. As a more efficient alternative to emissive shaders, most PBRs let you convert any mesh into a **mesh light**. This retains the shape of the geometry but adds the functionality and additional controls of area lights. Mesh lights are particularly effective for creating neon signs and other irregularly shaped light fixtures.

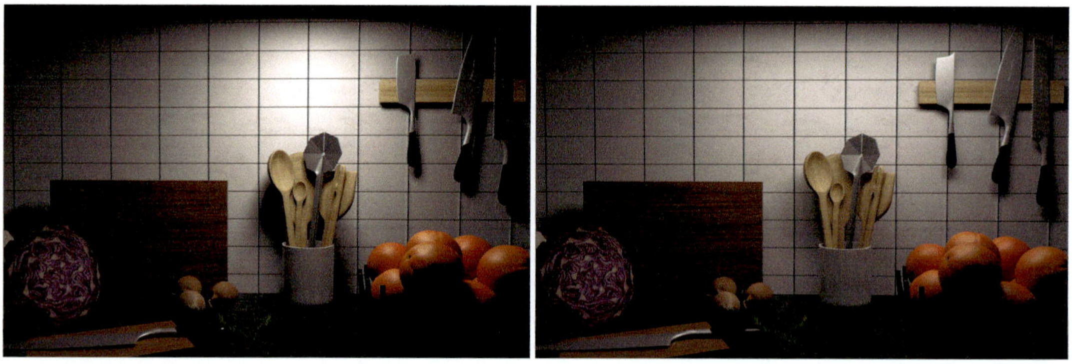

A small circular area light (left) produces a result that is similar to point light, but with much softer shadows and a more natural light spread. With an elongated rectangular area light (right) shadows are much softer and the image is more pleasant to the eye.

Image-Based Lighting

As mentioned in Chapter 3, 16-bit HDR images can store a very wide range of luminance values, and thus capture real-world lighting at its fullest. **Image-based lighting (IBL)** constitutes an integral approach to CG lighting, not only because it uses real-world lighting snapshots instead of artificial emitters, but also because it creates a full environment around the CG scene, which can be used for specular reflections, as well as transmission and refraction. IBL is of course extremely useful in VFX, because it can capture the lighting on the set, which can then be applied to CG elements for optimal integration with the footage. It is also great for replicating real-world studio lighting for product visualizations, and of course to generate believable outdoor lighting (diffuse sky radiation in particular). It is important to remember that despite its effectiveness, standard IBL is not a three-dimensional environment – it is merely an "egg shell" that surrounds the CG scene. Therefore, it is only good for distant or background lights that do not interact with elements in the scene or the camera (for example, it cannot replace street lights in a driving car scene, interior lights in an architectural walkthrough, or any light in a playable area in a game).

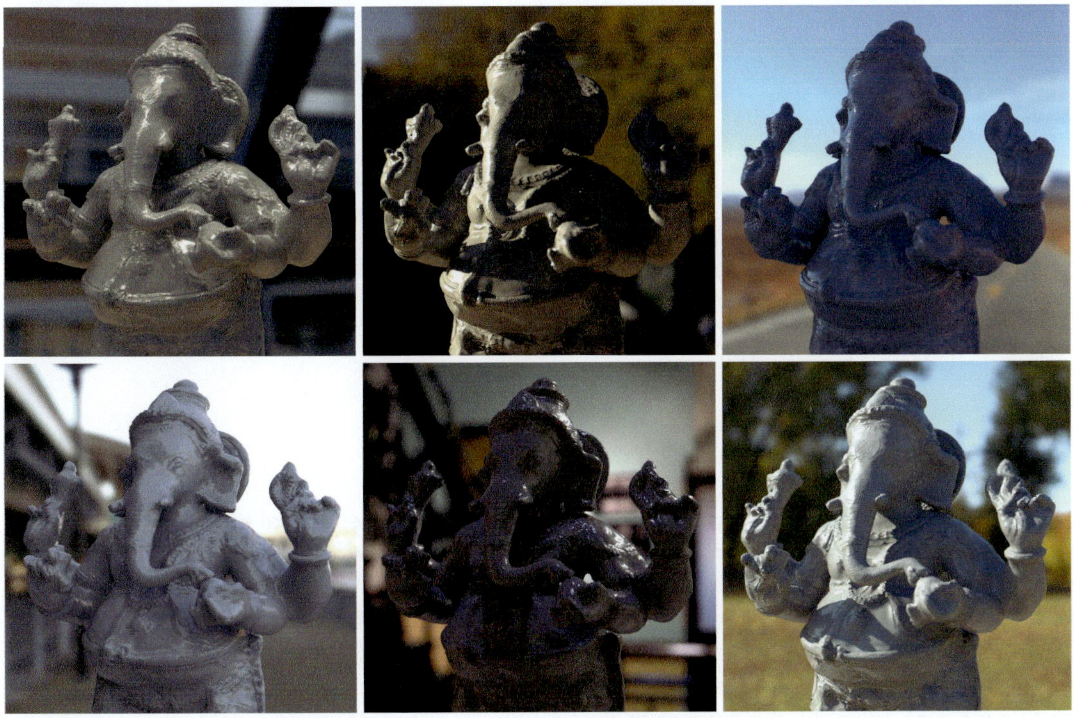

The same gray-shaded object, illuminated by six different spherical HDR images.

To successfully serve as an IBL source, an image needs to fulfill some key requirements:

- It needs to be a consistent, uninterrupted 360 panorama. This is usually done by shooting the environment with a very wide (or fish-eye) lens and rotating the camera three or four times (depending on the lens) on a nodal head to complete a full circle. The images are then stitched together into one spherical panorama that can be applied to a spherical **environment light** (also called **sky light**, **dome light,** or **sky dome light**).

- Even the best cameras cannot capture a full dynamic range in one shot. Therefore, several images must be taken for each angle, each at a different exposure. Usually, five or seven exposures are taken, both below and above the optimal exposure, and these are merged using dedicated software to produce a 16- or 32-bit image with an extremely wide dynamic range.
- If used for lighting only, HDR panoramas should be slightly or moderately blurred, to avoid any sharp peaks or artifacts that can create render issues. However, a sharp version should be used for specular reflection, and care must be taken to fix broken seams and clean up unwanted elements such as passing people. In most scenarios, only the upper hemisphere is used, as there usually is some kind of ground geometry in the scene that blocks the lower half of the HDR panorama.

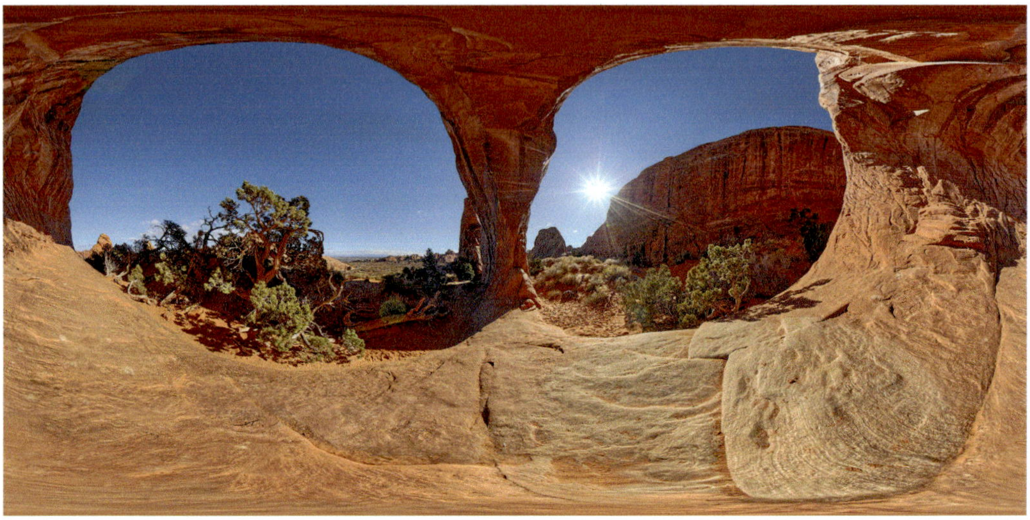

A spherical HDR image from the sIBL archive.

Procedural HDR Environments

Instead of using photographic captures of real-world panoramas, spherical environments can also be created from scratch. This is an intriguing subset of IBL that covers a few different approaches: some programs (such as HDR Light Studio) let you design a virtual studio lighting environment by dropping 2D light elements that emulate lighting fixtures into a 2D spherical environment map. These are especially useful for product visualization. For outdoor lighting, many render engines offer a procedural sky map. These physically-based sky simulations let you adjust the sun's position (or pitch/azimuth), the amount of sky turbidity (higher amounts simulate an overcast or hazy sky with less direct sunlight), sky color, and the sun's intensity and scale (for sun appearance in specular reflections). The color and gradation of the sky automatically shifts according to the position of the sun, simulating the effects of Rayleigh and Mie scattering. Finally, dedicated landscape-creation applications like Terragen and Vue have built-in atmosphere simulators that, in addition to providing procedural sun and sky interaction and volumetric clouds, apply a three-dimensional atmospheric volume to the entire scene that interacts with the objects and generates depth-based haze and fog (even crepuscular rays). Procedural IBL environments thus offer control and flexibility that is not possible with HDR images, and they do not require shooting, editing, and stitching photographs. But they often lack the realistic detail and irregularity of real-world environments, and, of course, cannot be used to match the set lighting in VFX).

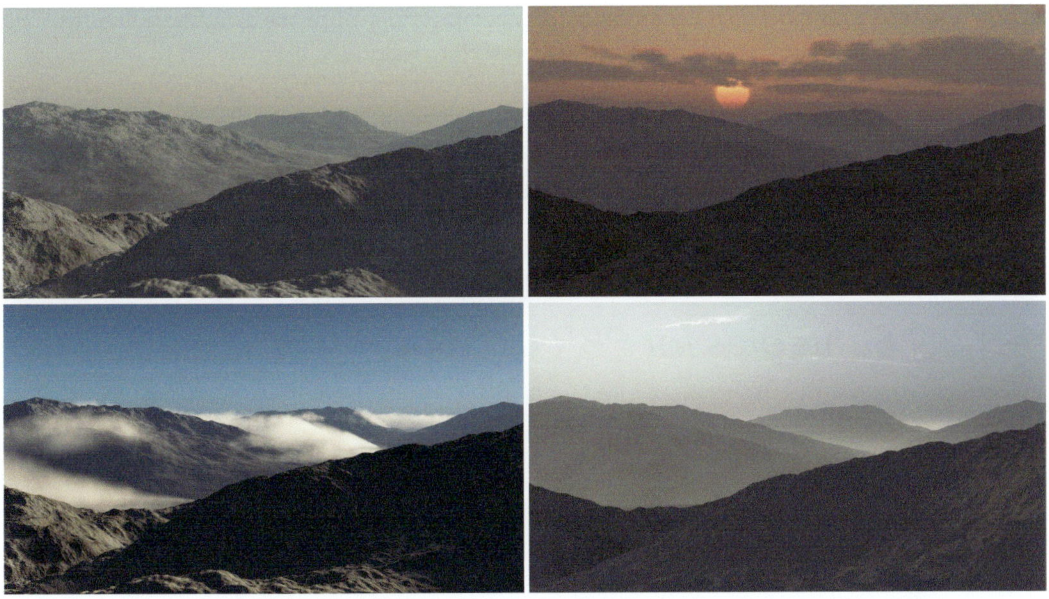

Unlike standard IBL, Vue's atmosphere model is three-dimensional and truly volumetric. Tweaks to sun position, fog, haze, clouds, and other parameters can drastically change the look of a scene.

Photometric Lights

The Illuminating Engineering Society (**IES**) instituted a digital format for storing the characteristics of specific real-world lights. CG photometric lights can read IES profiles and

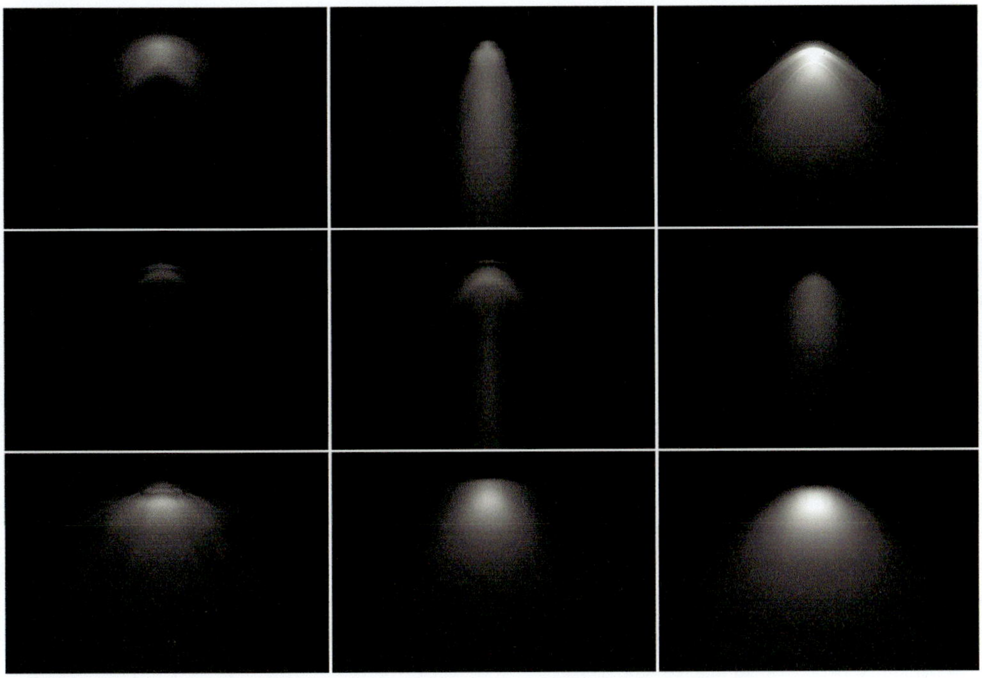

Nine different IES profiles, showing various light emission and spread characteristics.

closely simulate the lighting distribution and color temperature of physical lighting fixtures. This is of course extremely valuable for architects and interior designers: not only do photometric lights accurately emulate the proposed lighting at the visualization stage, but they allow experimentation with different lighting in a virtual environment, and pre-visualizing the results as part of the lighting design process. IES profiles can be directly obtained from different lighting manufacturers, and using a photometric light is as easy as dropping it in the scene and loading the IES profile.

Essential Strategies for PBR Lighting

Scale Matters

Scale is arbitrary in the virtual world of CG, where one unit can be an inch or a mile, a millimeter or a kilometer. In the past, when light emitters were limited to point, spot, and directional, and when global illumination was not an integral part of the raytracing process, there was little importance to the scale of surfaces or the light sources themselves. But with physically-based renderers, scale matters: lights have a physical size and accurate quadratic falloff, and all surfaces and textures contribute to the global illumination. A light source with a one square inch surface emits considerably less energy than a ten square feet light, and a one-inch white cube contributes much less to the global illumination than a ten feet cube. PBRs follow real-world paradigms and work best when real-world scale is preserved in the scene. While the intensity of lights can be increased or decreased to compensate for wrong scale, doing so tips off the physical balance of scene. Moreover, the luminance of texture maps cannot be tweaked in the same way as CG lights, which can further push the lighting away from a photoreal equilibrium. It also goes without saying that photometric lights lose their accuracy when scale is not coherent. So, whichever base unit is set in the renderer's global parameters, it is important to ensure the elements in the scene (including area lights) match the scale of their real-world counterparts.

Natural Daytime Lighting

In the past, achieving convincingly realistic daytime light was a challenge, as it required careful placement of fill lights to simulate diffused sky radiation and occlusion shading. With contemporary PBRs, it is often as simple as dropping a spherical HDR image into the scene and hitting the render button. The combination of Monte Carlo-based fully integrated global illumination and BRDF shaders ensures that light energy emanating from the sky dome is distributed and propagated accurately within the scene. However, sometimes there is a need to further control the balance between direct sunlight and sky radiation. Thus, it is a common practice to erase the sun from the spherical HDR image with a clone-paint tool, and use a directional light instead. While the light's angle, intensity, and hue should roughly match the captured sunlight in the image, there's still some leeway that allows more precise control over intensity and even exact angle. As mentioned previously, most PRBs offer some form of built-in physical sky simulation which can be used to quickly generate accurate sky radiation with varied hued and intensity based on the position of the sun. But such simulations usually lack the real-world detail and variance that is captured in an HDRI, so are better used as look development tools for experimenting with different lighting conditions. There are numerous online libraries of high-quality sky, environment, and interior HDRIs (many for free), so finding a suitable one for a desired lighting is easier today than ever before.

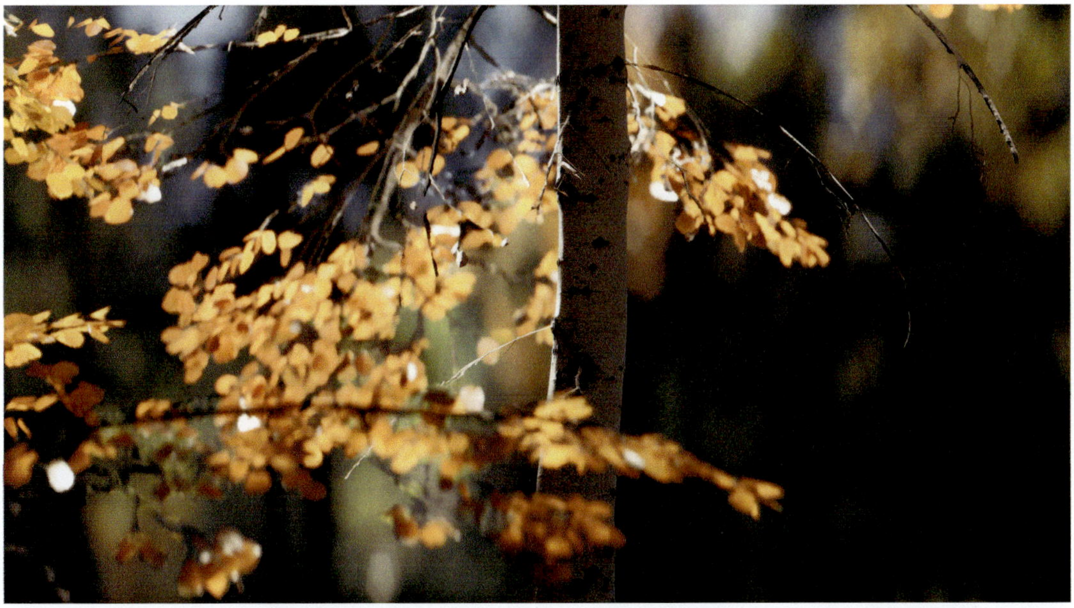

PWG Ecosystem by Gökhan Karadayi. Unreal Engine, Procedural World Generator.

Skylight Color

The color of diffuse sky radiation is not as blue as the sky itself, but HDR images of clear skies tend to cast an overly saturated blue light, because the light color is derived directly from the deep blue pixels in the image. Some desaturation of the HDR image is useful to achieve a more balanced and realistic result. Also, even though HDRIs have a wide dynamic range, their brightness often needs to be boosted because the merged exposures do not represent the true luminosity of the captured environment. Finally, the color (and especially the white balance) of HDR images is subjective and depends on the camera type and camera settings. When shooting HDR photos, it is very helpful to include a color card (ideally at the bottom of the frame), so that color balance can corrected and optimized for lighting purposes.

Ambient Occlusion for Outdoor Scenes

As mentioned earlier, ambient occlusion can be used as a cheap and fast alternative to global illumination, especially for real-time rendering. This "cheat" is particularly effective in exterior scenes because the role of bounced light is much less critical outdoors than it is indoors. In an enclosed space, the contribution of bounced light is crucial for the render result (without GI, a room lit by a single window would be much too dark). In exterior scenes however, a large portion of the incident light is reflected back up toward the sky and contributes little to the illumination of the environment. The ambient occlusion alternative works especially well for natural scenes with a lot of small detail and variations (trees, plants, etc.), but is less effective for large, uniform surfaces (such as building exteriors) where the lack of true light reflection and color bleeding is more apparent.

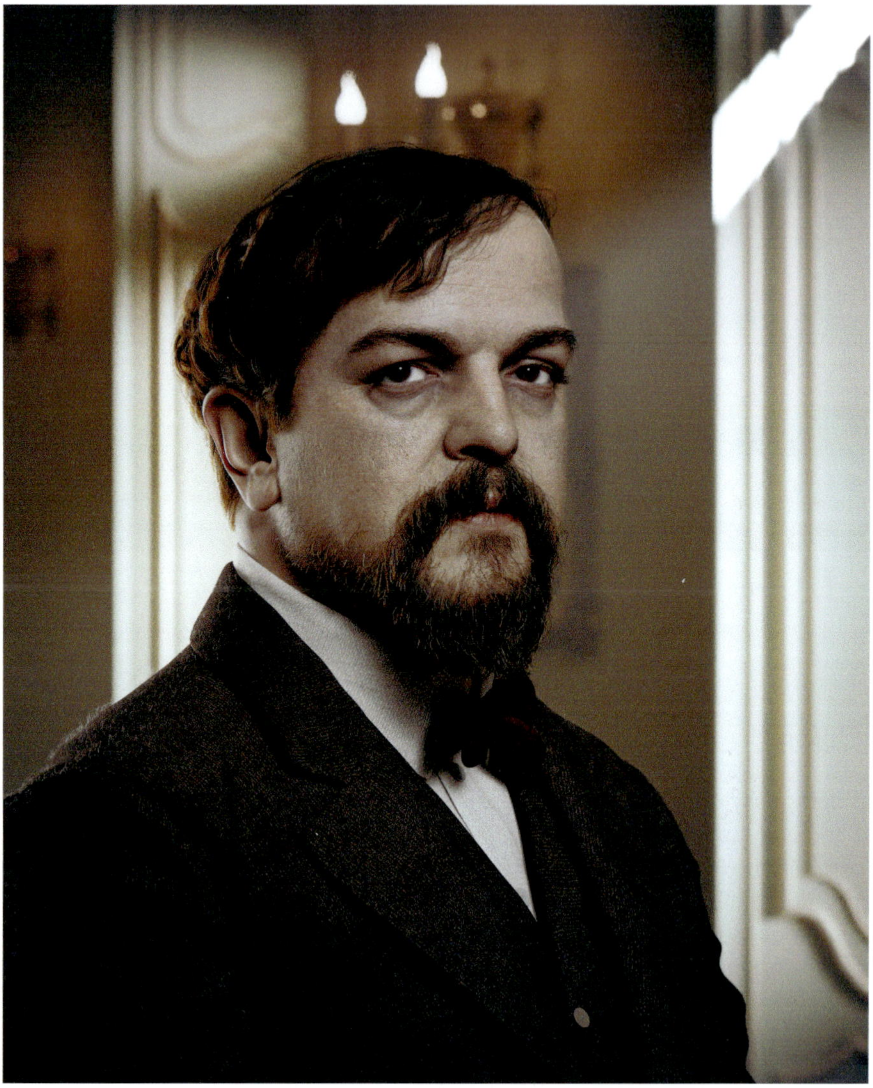

Claude Debussy by Hadi Karimi. Maya, Substance Painter, Zbrush, Xgen, Arnold.

Daytime Interior Scenes

Indoor settings that are illuminated by daylight can be problematic: while the exterior environment is a full 360 degrees hemisphere, the light that reaches the interior space through the window(s) is restricted to small areas. Most of the sampling is thus wasted on parts of the environment that do not contribute to the lighting, while the lack of samples in the critical regions causes noisiness. Some rendering applications offer a **light portal** tool, basically a plane that is placed in front of the windows and focuses all the exterior light sampling toward that area. Another often-used option is to ditch exterior IBL altogether, and instead place large, bright area lights in front of the windows. This is how interior sets are lit in real life, and it works particularly well when the view out the windows is a simple 2D background image rather than a 3D environment (as it often the case in architectural interior renders).

Man-made Lighting

Chapter 7 provided an overview of the principles of light intensity (Lumen/Lux), and Kelvin color temperature, as well as a rundown of the most common emitters. As mentioned, the type of light modifier (lamp shade, diffuser, reflector, filter, lens, barn doors) often affects the characteristics of the light more than the lamp itself. Rectangle, disk, and cylinder area lights work for most light fixtures, with mesh lights supporting emitters of more complex shape. Some render engines provide an option to set light temperature in Kelvin units, and photometric lights offer an even closer approximation to real-world fixtures. However, the believability of the scene (especially interior renders) depends not only on the right shape, intensity and color of the light, but also on its interaction with the different elements in the light fixture, so even when using photometric lights, it is often necessary to model and shade fixture elements like casings, lamp shades, and diffusers separately. While this may slow down the rendering, the added level of realism achieved through physical fixture elements is worth the price, especially in interior architecture and design.

Case Study: Product Lighting

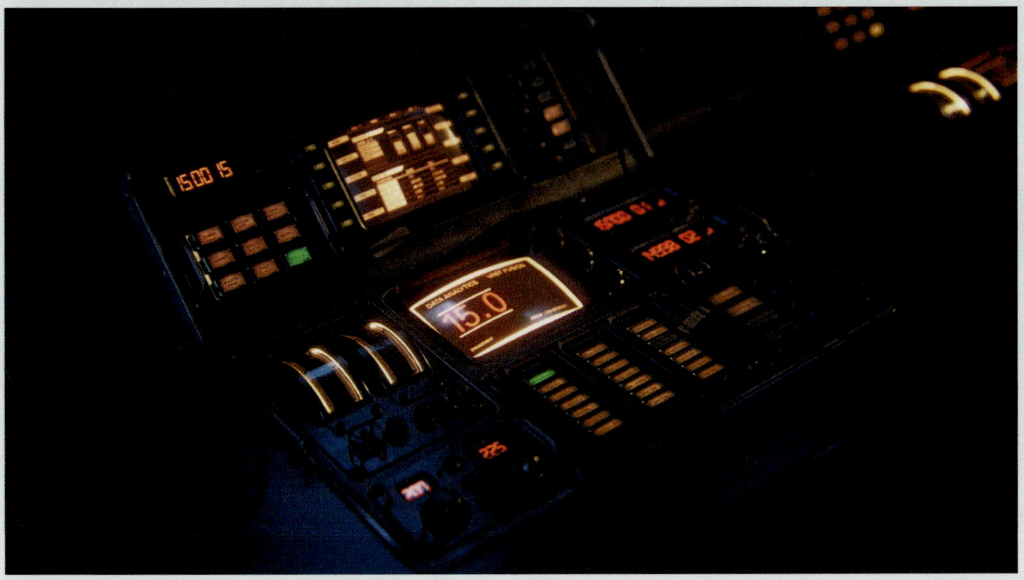

Console by Raphael Rau. Cinema 4D, Octane Render, Photoshop.
© Raphael Rau.

Raphael Rau's render of an audio mixing console is a masterful example of effective CG lighting. First, he blends the ambient light with a warm center spotlight and then adds a cooler sideways window light. Three additional reflector lights add dimension and definition. Finally, the various panel lights are added, each with its own specific tint and intensity. The carefully balanced result is a complex and highly realistic representation of a dark control room space. and subtly emphasizes the console's features and contours without ever going over the top or breaking the photoreal quality.

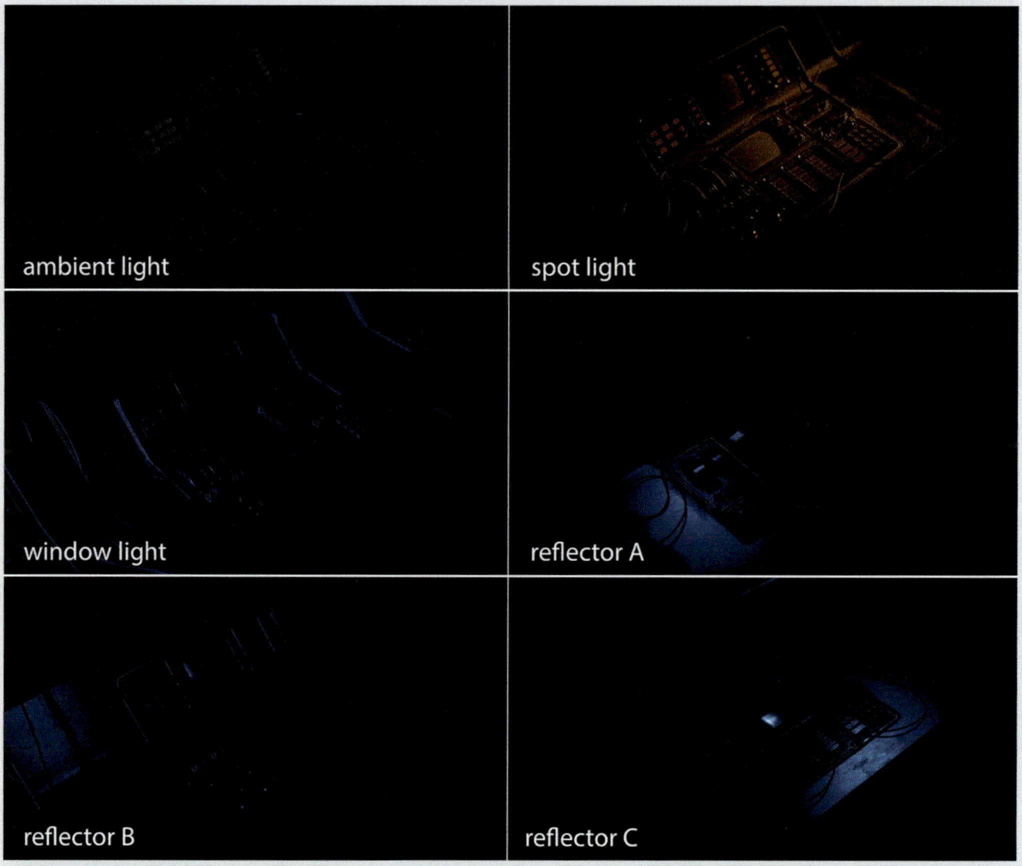

Console lights breakdown.

© Raphael Rau.

Chapter 12

Shading

A Brief Overview of Shader Evolution

The most fundamental function of a shader is to calculate the amount of illumination at every point on the surface. The term "shader" itself comes from the traditional art technique of shading areas of partial light occlusion to simulate a three-dimensional look in a 2D painting. Like all other aspects of CG rendering, the development of shaders has been tightly linked to the growth in computing power. A **Lambertian reflectivity model** calculates only diffuse reflections, a much simpler process than calculating specular reflections, because there's no need to reference the environment. Earlier shading models combined a basic Lambertian diffuse component with a simplified specular component.

The most primitive shading model in the evolution tree was the **flat shader.** It checked the angle of each face on a CG model relative to the light source, to determine how much it receives. Since each face was shaded separately, rendered surfaces had a distinctly faceted look. While this could work for a cube, for example, it was unsuitable for any curved, smooth surface. The next evolutionary step, the **Gouraud model** (first published in 1971), calculated the color value at each point of a triangle face, based on the normal vector of that point, and then interpolated those values linearly between vertices. This produced a smooth shading across the surface. While the Gouraud model was good enough for Lambertian shading (and faster to render than subsequent models because it only calculated the color on vertex points), it did not work well for specularity – any highlight that happened to fall at the center of a polygon was simply not be "seen", since only vertices were calculated.

Moving on – Bui Tuong Phong published a paper about his light reflection model in 1975. The **Phong** model provided a better alternative to Gouraud shading but required heavier calculations. Unlike Gouraud, which interpolated the color value between vertices, the Phong shader interpolated the vertex normal vectors first, and used this averaged vector value to assign a color for each pixel. This ensured a per pixel accurate shading, and solved the specular limitation of the Gouraud model. The Phong model was thus more suitable for shiny surfaces. A few years later, Jim **Blinn** developed a modification to the Phong model that reduced rendering time by calculating half-way normal vectors. The Phong/Blinn model became the pillar of CG shading for many years and was used in numerous animation films, VFX, visualizations, and games. However, it is now made obsolete by the more capable **BRDF** model.

DOI: 10.4324/9781003590293-16

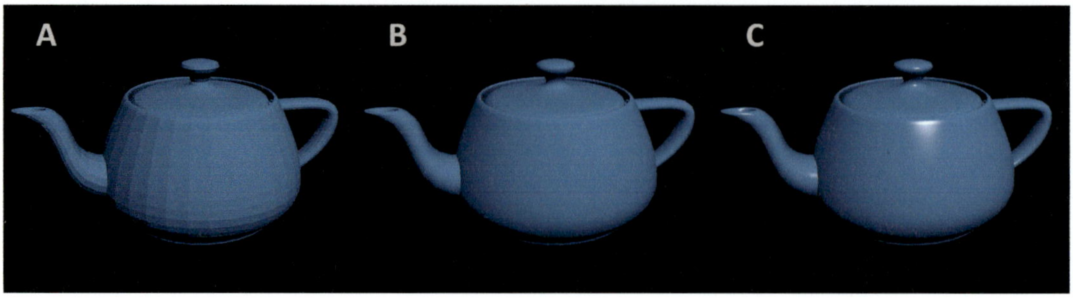

A: faceted lambert shader, B: smooth lambert shader, C: Blinn shader.

The BRDF Shading Model

Most physically-based renderers use the **Bidirectional Reflectance Distribution Function** model as the basis for their surface shaders.[1] This latest step in the evolution of CG shaders provides the most effective simulation of realistic surface-light interaction, and represents a number of important advantages:

- BRDF shaders use microfacet models like GGX or Cook-Torrance to accurately emulate surface roughness not only for diffuse reflections, but also for specular reflection and transmission.
- BRDF shaders calculate true isotropic and anisotropic specular reflection for dielectric as well as metallic materials.
- Fresnel effect is always factored in reflection/transmission, and is based on either simple refraction indices (for dielectrics) or complex refraction indices (for metals).
- BRDF shaders behave according to the law of conservation of energy. The specular, diffuse and transmission components are always balanced to prevent any situation where the amount of light leaving the surface exceeds the amount of incident light. For example, specular intensity is automatically reduced with higher roughness values so that the same amount of reflected energy is maintained over the wider specular spread.
- Multiple specular layers are possible within a single material (for example, many BRDF shaders have a second specular layer for clear coat and glossy paint effects).
- True depth-based transmission color is possible (color gets more pronounced as light travels deeper through the surface).
- Complex effects like transmission scattering, dispersion, and subsurface scattering are possible.

This impressive list shows why a BRDF model acts as a sort of "super-shader" – it is capable of realistically simulating almost any type of solid and liquid surface and provides a comprehensive and flexible foundation for physically-based rendering. While BRDF shaders in various rendering engines differ in some respects, they largely share a similar feature set, which makes it much easier for artists to transition from one application to another, or to create cross-platform materials and texture sets. The following is an overview of the most common BRDF features and how to approach them from a photoreal perspective:

Diffuse

Most BRDF shaders include (in addition to the basic diffuse color and diffuse weight parameters) a **diffuse roughness** setting. This uses a microfacet distribution function to simulate the variance in diffuse reflectivity and light falloff on rougher surfaces. Small roughness values are suitable for materials such as coated plastic, water, and glass, while higher values work well for fibrous or grainy materials like concrete, plaster, sand, or paper.

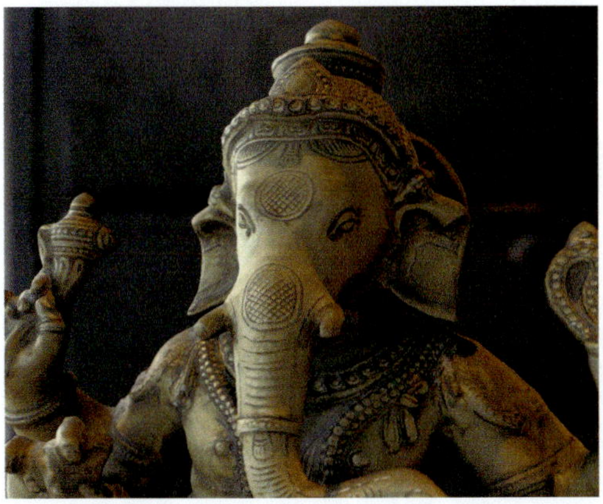

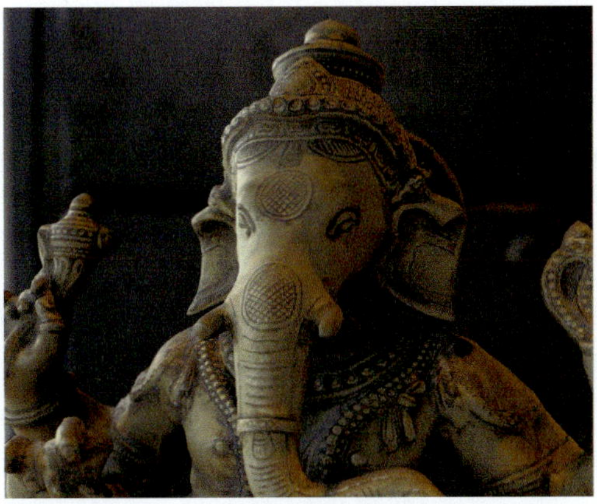

Left: no diffuse roughness, right: maximum diffuse roughness. The difference is subtle, but when diffuse roughness is high, the surface reflects less in the illuminated areas and the diffuse reflection is more evenly spread out.

Dielectric/Metallic Toggle

This parameter (also called "**metallic**" or "**metalness**") is found in all BRDF shaders and has a vital role in the photoreal simulation of materials. While real-world materials are either metallic or dielectric (there is no in between), in most applications this toggle is a range slider, to accommodate texture maps that drive this parameter and may have intermediate values. As noted in Chapter 9, there are some fundamental differences between dielectric materials and metals, so toggling from dielectric to metallic changes some key aspects of the shader:

- The model for specular reflection is changed to properly represent the characteristics of metallic specular reflection, and simple index of refraction is switched to complex index of refraction for correct metallic Fresnel effect.
- Since metals do not reflect diffusely, the diffuse color parameter effectively controls the specular tint. Setting the diffuse color to yellow, for example, causes all specular reflections to be tinted yellow (as in gold). This is consistent with the principle that the perceived color of a metal is derived from its specular tinting.
- In some applications, certain parameters that are no longer relevant are either grayed-out or ignored. These may include: diffuse roughness, specular intensity/weight (the metallic model assumes 100% specularity for metals), index of refraction, and transmission (again, as the assumption is that metals are never transparent).

Specular

Since BRDF specular reflection maintains energy conservation, the rougher the specularity, the lower its intensity. So, by simply varying the **specular roughness**, one can effectively cover the full range, from highly reflective materials to the dullest surfaces. The **specular intensity** (or specular weight) control is therefore redundant and should normally be kept fixed at 100%.

The **index of refraction** in BRDF shaders does not only affect refraction. It is also linked to the Fresnel effect of specular reflections. As discussed in Chapter 9, a higher IOR increases reflectivity at facing angles. The IOR of most liquid and solid dielectric materials lies between 1.3 and 2.4, so the difference in the Fresnel effect is rather subtle. While it is possible to achieve a crude pseudo-metallic look by raising the IOR much higher, it is completely unnecessary – switching the BRDF to the metallic option is a much better way for simulating metals, for all the reasons previously described.

As in older shader types, BRDF shaders include controls for **anisotropy** amount and direction. But unlike Phong or Blinn shaders, energy conservation in the BRDF model means that the amount

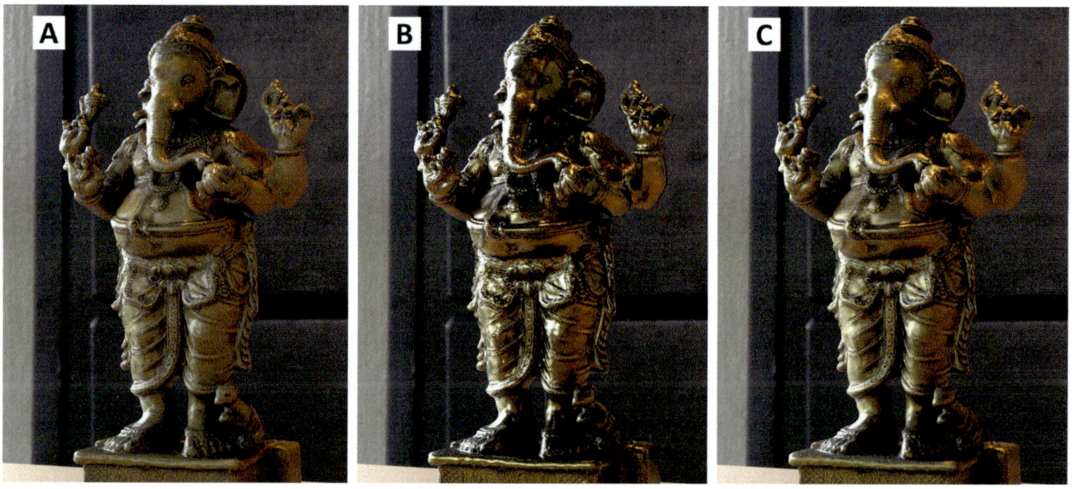

A: shader set to non-metallic (dielectric), B: shader switched to metallic (notice the increased reflectivity and luster), C: roughness is increased, but the metallic feel is maintained.

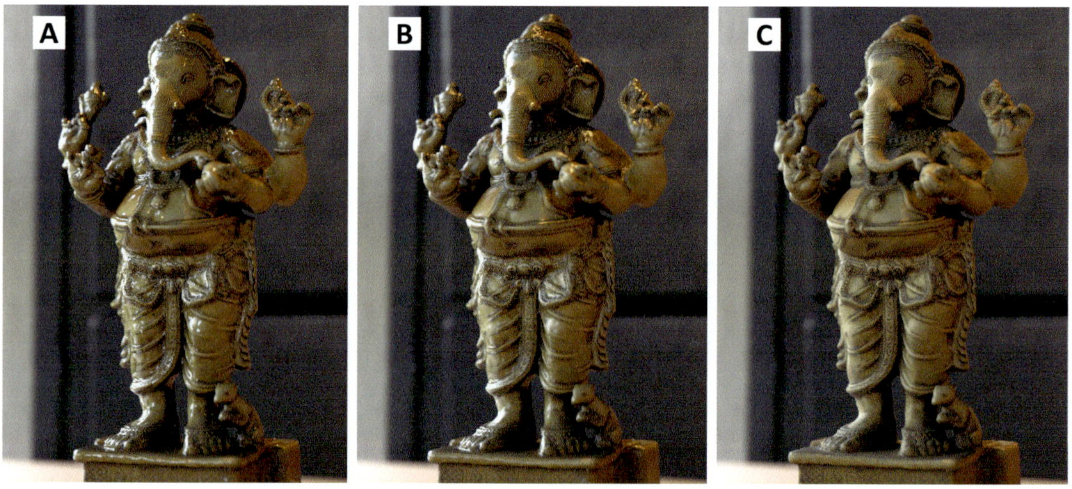

A: a dielectric shader with specular roughness at 0.05 looks smooth and shiny, B: with roughness increased to 0.3, specular reflection is still visible, but the surface feels duller, C: with roughness at 0.7, the surface looks completely matte. Note that specular intensity has remained at 100%.

of stretching is also affected by the specular roughness parameter (the higher the roughness, the farther anisotropy stretches the highlights). For example, a roughness setting of 0.05 would yield minimal stretching, even if anisotropy amount is set to 100%. This prevents a situation where extremely bright highlights are stretched over a wide area, resulting in an albedo value that exceeds one. This is also a more photoreal approach: real world anisotropic highlights are almost always the result of tiny elongated grooves or bumps on the surface (for example, brushed metal), and thus are rougher by default.

Coat

Most BRDF shaders come with an optional second specular layer called **coat** (or **clear coat**). This additional layer helps simulate a very wide range of materials that have a thin, clear coating, lacquer, or resin finish, like many types of plastic, lacquered wood, and coated metals. Coat is also effective in simulating wet surfaces, especially when modulated by a texture. Usually, the main material has a relatively higher specular roughness, while the coat layer provides a shinier, sharper specularity. The combination of a duller base surface and a glossy outer layer creates a complex look that is simply not achievable with a single specular setting.

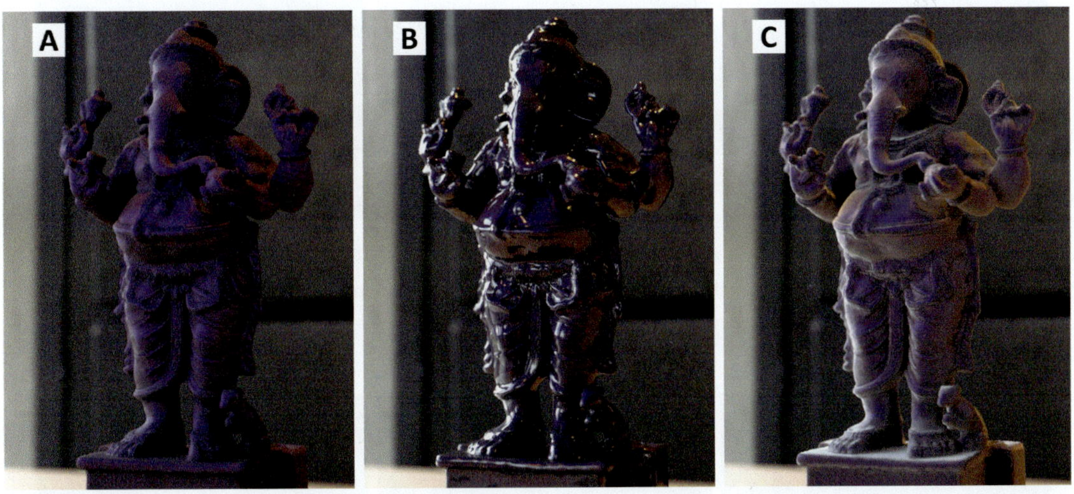

A: rough dielectric base layer, B: clear coat with a low roughness value is added, C: sheen effect (see in the next section).

Transmission

Glass, liquids, gemstones, and other transparent (and semi-transparent) materials were never easy to emulate with Phong and Blinn shaders. BRDF shaders, with their adherence to energy conservation, accurate Fresnel effect, and microfacet transmission distribution, can generate extremely photoreal transmission, refraction, and scattering. Exact transmission parameters differ between applications, but the following description should cover the most common ones.

Most shaders provide control for **transmission weight** (or **transmission amount**), but it is generally better to keep it at 100%, and avoid using in-between values. This ensures that diffuse color has no effect on the appearance of the material (some applications automatically disable diffuse color when transmission amount is set to 100%, others use the diffuse color as the transmission color). Just like with specular reflections, the **transmission roughness** should be used instead to control the amount of transparency, or how clear or "cloudy" the material is. In most BRDF shaders, the transmission roughness is tied to the specular roughness, so increasing the roughness respectively reduces the clarity of the material's transparency. This simulates real-world material properties, where clear transparent materials are also quite shiny, while half-transparent, cloudy materials are duller. However, some BRDF shaders also offer a separate transmission roughness parameter. This is useful when a material needs to be only semi-transparent but still maintain sharp highlights (for example, murky water).

In transparent and semi-transparent materials, **transmission color** (**Fog color** in some render engines) replaces diffuse color to define the base hue of the material. There is a significant

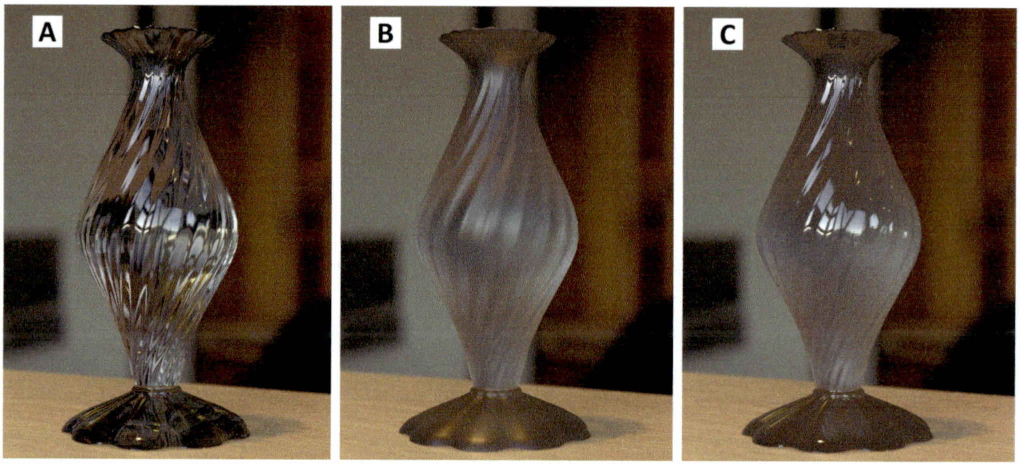

A: 100% transmission with low specular roughness (clear transparency) B: 100% transmission with high specular roughness (cloudy transparency), C: 100% transmission with low specular roughness but high additional transmission roughness (cloudy transparency with sharp highlights).

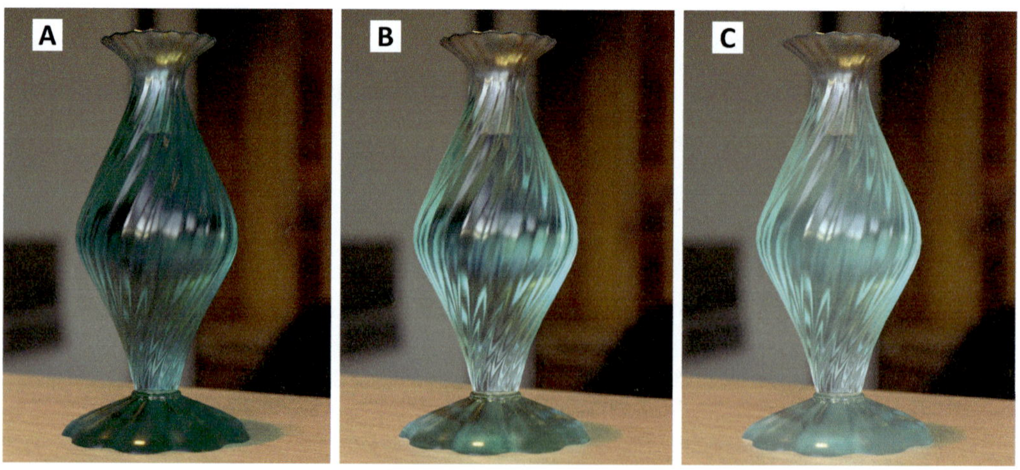

A: high transmission depth result in deeper coloring in thicker areas, B: low transmission depth, C: the same depth amount with increased scattering, causing the transmission color to spread more evenly.

difference here: diffuse reflection happens at (or just below) the surface, while transmission color happens as light passes through the object. Transmission color varies with depth – brighter and less saturated in thin areas, darker and more saturated in thick areas. BRDF shaders faithfully recreate this variance of color, which greatly increases the photorealism of transparent and semi-transparent materials. Some shaders offer a separate control for **transmission depth**, which can be used to increase or decrease the apparent thickness of the transmitting material regardless of the real-world scale of the object.

Transmission scatter, another parameter found in some BRDF shaders, is especially handy for emulating thick, murky liquids like honey or certain gemstones. Increasing scattering causes light to spread more evenly through the volume, which works particularly well when specular/transmission roughness is fairly high (clear liquids have minimal scattering). A **scatter anisotropy** parameter allows for a more precise control of the direction of light scattering.

Subsurface Scattering

Most diffuse reflections involve at least some minimal light scattering just below the surface. For practical purposes, many dielectric materials can be emulated in BRDF shaders without using the more computationally intensive subsurface scattering. However, even the smallest hint of scattering can substantially enhance the look and feel of a wide range of non-metals: many types of plastic, foam, stones, rocks, clay, plants, thick liquids, and more. The enhanced photorealism is often worth the extra rendering times, especially for close-up renders where even a subtle effect can make a big difference. Of course, subsurface scattering is a must for materials that are characterized by a deeper penetration of light under the surface: wax, skin, marble, thick leaves and stems, certain types of plastic, gemstones, and minerals, etc.

When the subsurface scattering amount is set to 100%, the diffuse color loses its role, and instead, the **subsurface scattering color** becomes the material's base color. The most important parameter for controlling subsurface scattering is **radius,** which defines how deep the scattering happens

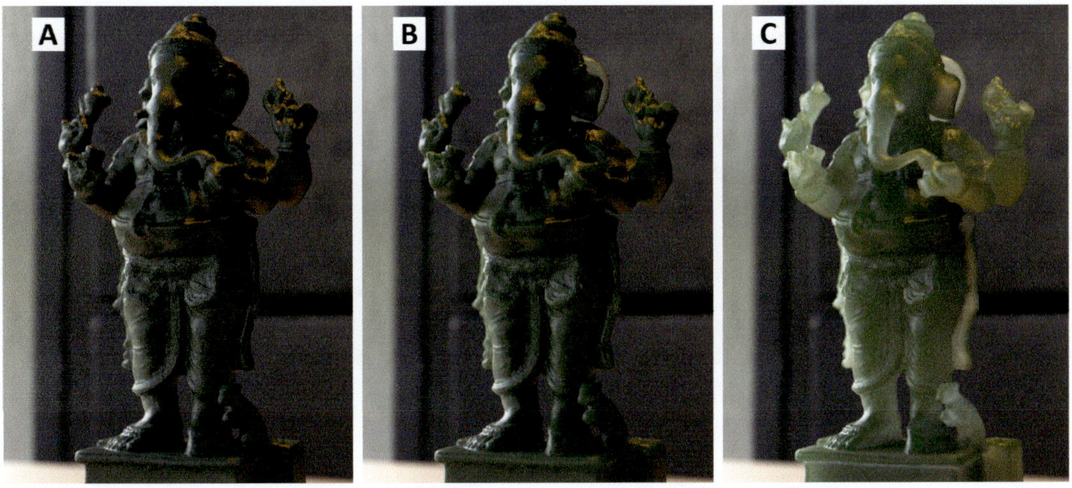

A: no subsurface scattering, B: subsurface scattering with a small radius (notice the "waxy" feel), C: subsurface scattering with a large radius, resulting in a quasi-translucent look.

under the surface. A small radius generates a subtle, almost imperceptible effect that nonetheless greatly affects the "feel" of the material, giving it a softer, milky look. This is particularly suitable for marble and some types of plastic and stones. A wider radius creates a noticeable sense of translucency, important for materials like wax. As the radius is increased even further, the material may start feeling almost semi-transparent. However, transmission and subsurface scattering are fundamentally different: no matter how large the subsurface scattering radius is, the material always remains opaque (you cannot see through it). Subsurface scattering and transmission, therefore, serve different purposes, and are normally mutually exclusive. That said, unusual and interesting shading effects can be achieved by lowering the subsurface amount below 100%, and creating a mix of diffuse, transmission, and subsurface scattering, with each component assigned its own color map.

Emission

When emission is above zero, the surface acts as a light source. Render engines offer better light emitters (like area and mesh lights), which are optimized for the task and have dedicated controls. But emission is useful for self-illuminated surfaces that contribute little light to the scene (like distant light sources, stars, or neon signs) and effects like phosphorescence and bioluminescence. It is not advisable to use emission as a way of making a surface brighter or make it stand out in the render. Doing so breaks the law of conservation of energy and damages the photoreal coherence of the render. It can also slow down the render because the surface is now treated as an additional light source in the scene. Emission should only be used for self-illuminating surfaces.

Additional BRDF Features

Beyond the basic features describedpreviously, different BRDF surface shaders may have one or more of these extra features:

SHEEN

Although this word is often used to describe glossiness or shininess, in CG terminology it relates specifically to the characteristic velvety effect of microfiber in textiles (and other fuzzy surfaces). Sheen is not specular reflection but rather a soft luster that is related to the facing direction of the microfibers. It is therefore a distinct effect that is particularly useful for emulating garments, upholstery, rugs, carpets, bed linens, and even skin.

THIN-FILM INTERFERENCE

Thin-film simulates the diffraction that happens when light hits a very thin, clear layer like the wall of a soap bubble, an oil slick, or some types of glass coating. This wave interference creates a rainbow-like appearance that shifts pattern and color depending on the angle of the surface and the light.

DISPERSION

The separation of frequencies as light is refracted in transparent materials is the cause of chromatic aberrations in lenses. The amount of dispersion in different types of glass is indicated by their **Abbe number** (also called the **V-number**). The higher the Abbe value, the lower the dispersion. Some render engines provide a dispersion feature that is especially useful for accurate visualizations of glass and jewelry. The Abbe number of most glass types is somewhere between 100 and 20, so the effect of dispersion is normally very subtle and hardly perceptible. It is possible to achieve prism-like results by pushing the Abbe number below 20, though the results become more stylistic than photoreal.

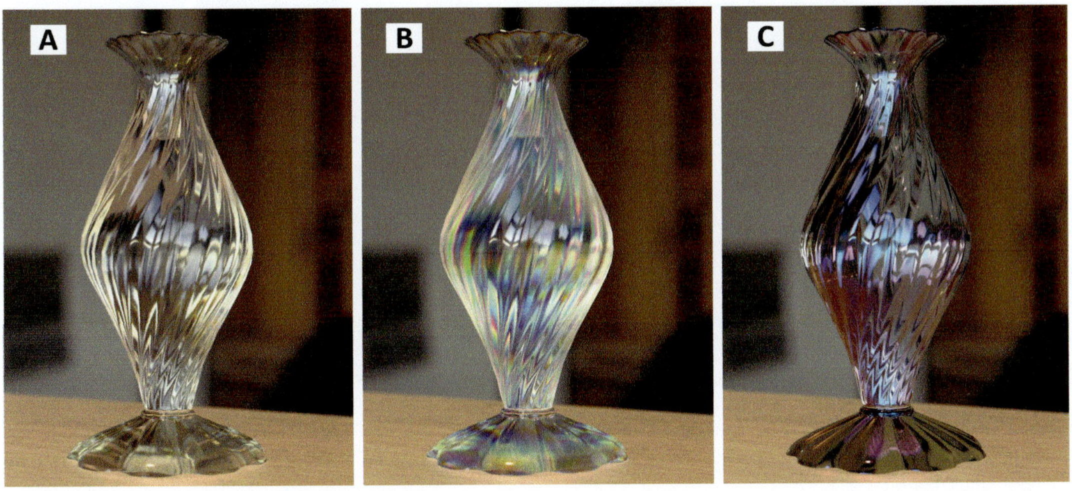

A: an Abbe value of 30 produces a subtle, realistic dispersion that makes the glass feel slightly yellowish, B: an Abbe value of one generates and exaggerated prism-like effect, C: thin-film interference is a completely different effect, suitable for soap bubbles and oil slicks.

Other Common Shaders

Since a shader is merely a description of how light interacts with a surface, shader-writers can create an almost endless variety that ranges from specific material types, to shaders that simulate complex systems like grass or ocean waves. While it is impossible to cover every type of shader here, the following is a brief overview of some of the most common shader types available in rendering packages:

Car Paint Shaders

Cars are a ubiquitous subject in many types of renders, from VFX and commercials to product visualizations, so it is not surprising that most rendering packages offer a dedicated car paint shader. Generally, such shaders are an elaboration on the clear coat principle, and offer extended control over the color and specularity of the base and coat layers, as well as an emulation of the tiny flakes that are a characteristic feature of metallic car paints. These additional features help create a rich, multi-layered look that is useful not just for cars, but for many other types of painted surfaces.

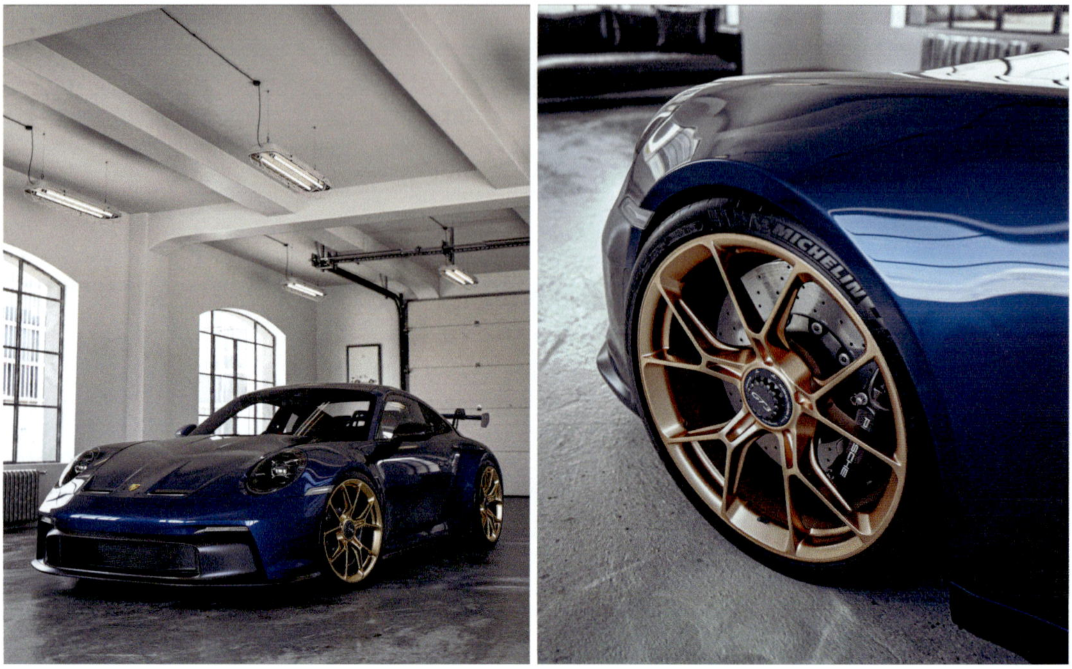

Porsche 911 GT3 by David Baylis.

3D Studio Max, Unreal Engine © David Baylis.

Volumetric Shaders

Voxel-based fluid simulations are used to simulate volumes of small particles like fog, smoke, dust, and fire. To successfully emulate their unique scattering and absorption characteristics, the shader needs to have specific material attributes that are quite different from solid or liquid materials. Most physically-based renderers offer volumetric shaders with dedicated controls for features like density (or thickness), scattering, opacity, and even emission and temperature (for light-emitting volumes like fire).

Hair/Fur Shaders

Because hair is a collection of numerous thin strands, it reacts to light in a markedly different way than solid or liquid surfaces as well as volumetric entities. Hair and fur display a distinct combination of reflection, transmission, scattering, and sheen that is difficult to simulate using standard surface or volume shaders. While in the past, some VFX companies developed their own proprietary hair and fur shaders, nowadays most renderers offer a hair/fur shader out of the box. Unlike standard BRDF shaders, which share similar parameters across different applications, features, and terminology in different hair shaders is quite varied. However, most hair shaders provide detailed control over melanin (natural hair color) or dye (artificial color), as well as glossiness, backlighting and sheen (or glint) amounts. Randomizing color and other parameters is important for achieving a complex, photoreal hair/fur look, and most hair shaders offer some control over randomness and subtle variations.

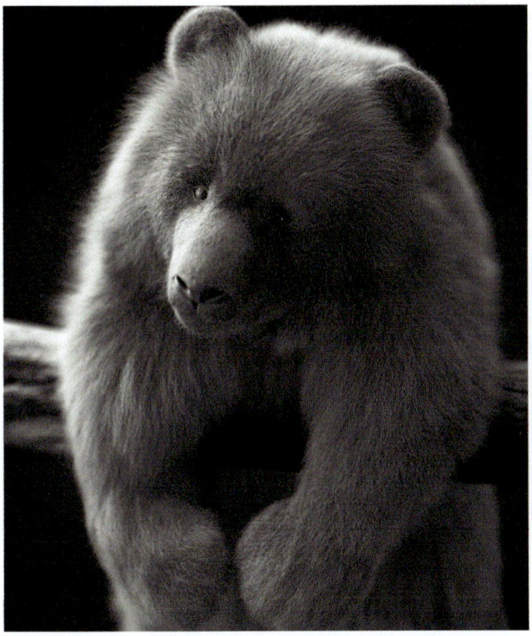

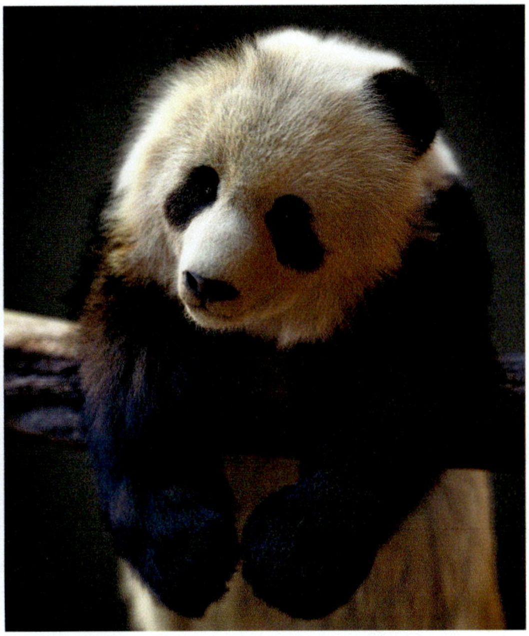

Hanging Panda by Massimo Righi. Maya, XGen, Arnold.

Case Study: Emulating Plastic Materials

When we say that "the CGI looks like plastic" we usually mean that it lacks detail and feels artificial. But plastic materials abound in man-made objects, and it is, in fact, a true challenge to emulate them realistically – exactly because plastic is usually smooth and consistent. Electronic devices and appliances use a lot of plastic parts, which often appear dull and unrealistic in CG product visualizations.

In Raphael Rau's renders, plastic always feels tangible and realistic. I asked Raphael how he approaches plastic surfaces. "If the plastic is opaque or dark, or rather far away from the camera, I just use the typical mix of a diffuse Lambertian base with a reflection mixed by a Fresnel term" says Raphael. "If the plastic has translucent or transparent characteristic, the trick is to use a refractive material model, like you would for glass, then add the right amount of subsurface scattering".

Using Cinema 3D as his main application, with Octane, Redshift, and Cycles render engines and Fusion for compositing, Raphael applies what he calls a "half procedural shading approach" to surfacing. He explains:

> While I use texture maps, those are usually tileable 2k maps that I create myself from photographs. I often use triplanar projections as well as some distortion techniques or a mix with 3D noises to hide the tiling. Then I also introduce what I call "curvature-dependent shading". I create vertex, curvature, or ambient occlusion maps to apply

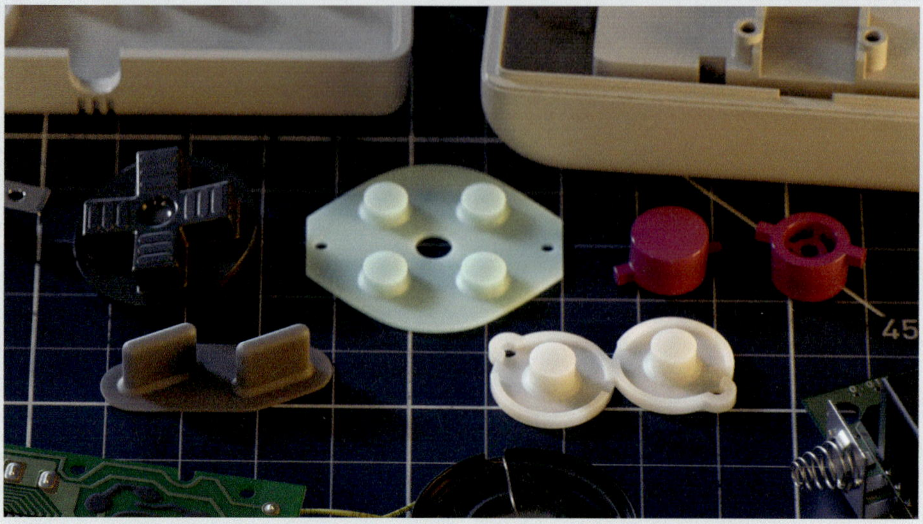

Gameboy by Raphael Rau. Cinema 4D, Octane Render.

usage characteristics to the shader (for example, add glossiness to exposed edges with some bump and scratches, or rougher attributes to dusty cavities and concave areas.

"I use dirt maps or other surface variation to create shader irregularity that can be applied to different channels of the shader to break up the inherently clinical consistent CG surface look" continues Raphael. "In product viz, this is usually kept to minimal levels, but in my still life renders, I am free to set shader irregularity as strong as I would like it to be".

Note

1 Some 3D applications use the term BSDF (Bidirectional Scattering Distribution Function). In general, the principles of BRDF discussed here apply to BSDF as well.

Chapter 13

Texturing

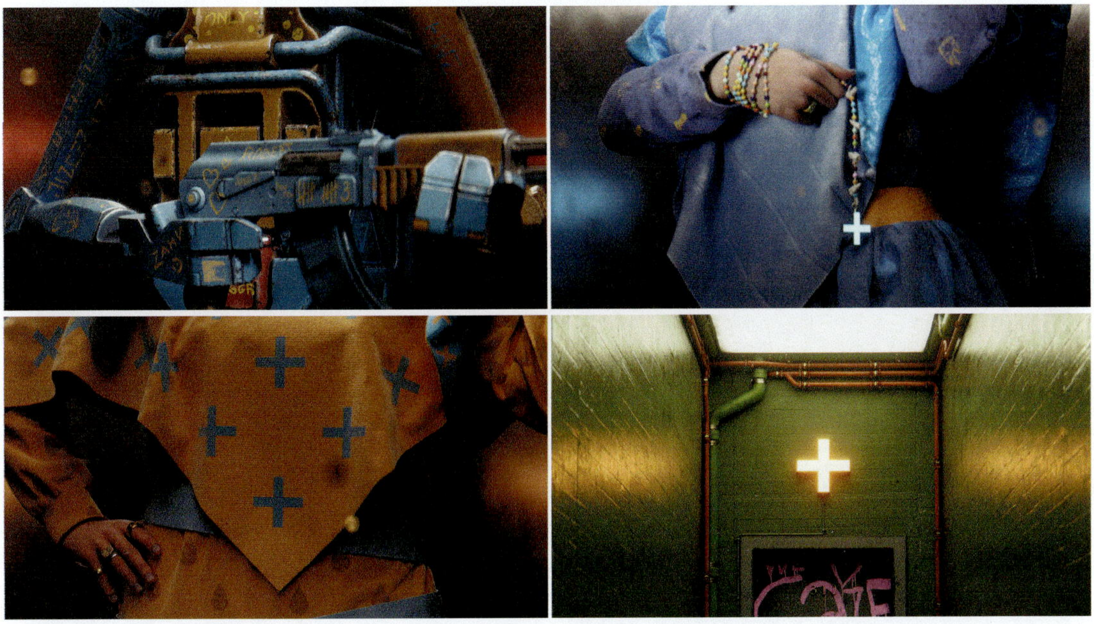

Plus by Cornelius Dämmrich. Cinema 4D, DAZ Studio, Substance 3D Painter, Octane Render.

Textures add surface detail that is crucial for believable imagery. One can say that texture mapping has greater impact on photorealism than modeling, shading, or lighting, because our eyes are instinctively drawn to color variations and patterns that define surfaces. Simply switching the color texture on a cube can instantly transform it from a stone wall to a concrete slab, a cereal box, a metal container, or a book. But texture mapping is of course much more than just colors and patterns. By definition, a texture is something that you can feel, not just see. A wooden table feels different than a towel, and the smooth surface of a plastic box feels different than the surface of a stone. So, while we cannot physically touch surfaces in a CG render, it is the role of texture maps to visually recreate that "feel" by modulating shader properties like bumps, displacement, ambient occlusion, roughness, metalness, or transmission. Texturing can substantially raise the level of believability but can also do the extract opposite: no matter how good the models,

DOI: 10.4324/9781003590293-17

lighting, and shaders, common texturing mistakes like noticeable repetition, implausible grime maps, or over-the-top bumps can easily ruin a render. This chapter focuses on the contemporary PBR texturing workflow, and aspects of texturing that are most pertinent to photorealism,

PBR Texturing

Physically-based rendering is a wide term and does not necessarily indicate a specific category or method of rendering. But it is often used in CG talk to describe a texturing workflow that is based on a set of maps that drive the most common shading parameters of BRDF shaders (diffuse color, specular roughness, dielectric-metallic definition, bumps, displacement, and ambient occlusion). Depending on the material, this base set may be augmented with additional maps (like transmission, opacity, and subsurface scattering). PBR texturing now works with the vast majority of renderers, including most game engines, and this makes it easier to create materials that can be used in different applications. In this section we will look at the role of each texture type within a PBR map set, and discuss ways to maximize the texture's contribution to the photoreal appearance of the material.

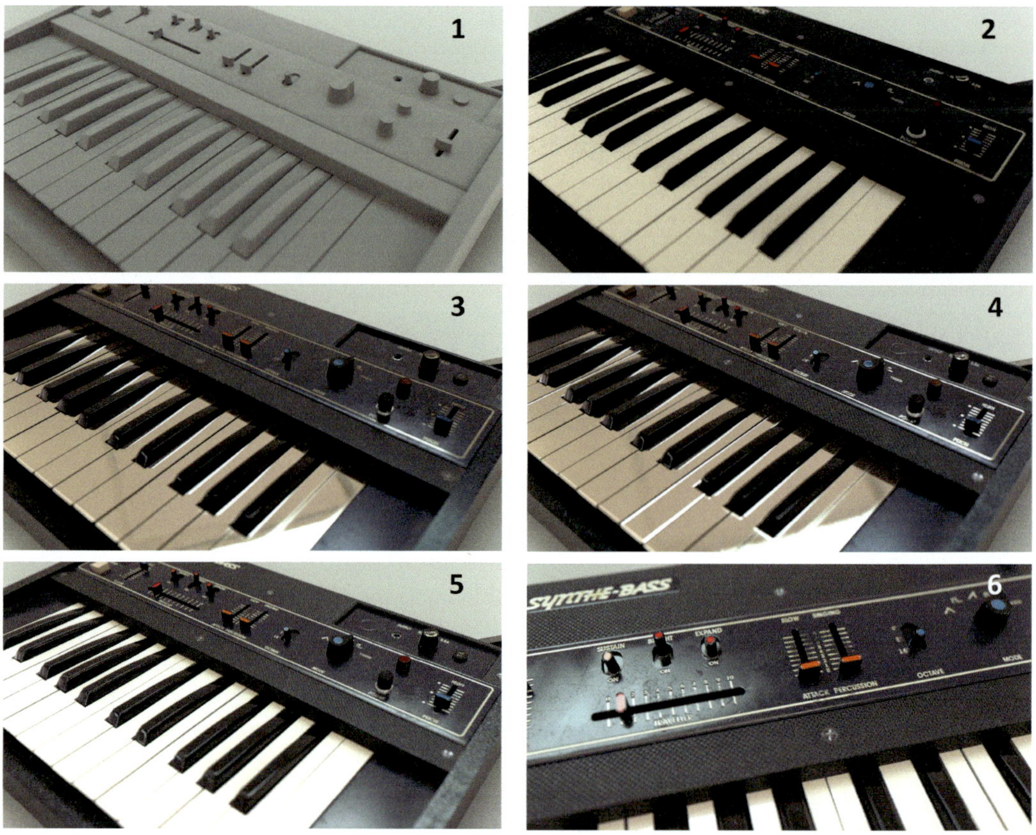

Stages of PBR texturing. 1: untextured model, 2: base color texture, 3: specular roughness map, 4: bumps/normal map, 5: metallic/dielectric map, 6: closer angle, showing how the roughness and bump maps are used to add subtle imperfections.

The Linear Workflow

A few words about the linear workflow: there are no color spaces in nature – the interaction of light and surfaces is universally linear. Physically-based renderers must operate in linear space for light interaction to be calculated properly. Unfortunately, different display technologies make it necessary to add color curves to digital images. By default, computer images are saved in sRGB color space (they have a sRGB gamma curve baked in) so that they display correctly on computer monitors. But when sRGB images are used in linear color space, they are much too dark. Global illumination in PBRs depends on the color values of textures in the scene, so overly dark textures completely skew the lighting solution, forcing the artist to boost the light energy of scene emitters, which throws the render off the photoreal path. Initially, texture artists had to manually convert images to linear color space, but nowadays most 3D and rendering applications provide a color management system that automatically linearizes images (reverses their color space curve), based on the specified source color space and format. While most color maps are based on sRGB images, many of the secondary maps described in this section are not feeding color data into the shader. For example, the color values of a normal map define XYZ vectors, and a roughness or metallic texture is a zero-to-one grayscale map. To work properly, non-color maps should always be set to linear (raw) color space at input to maintain their values.

Base Color Map

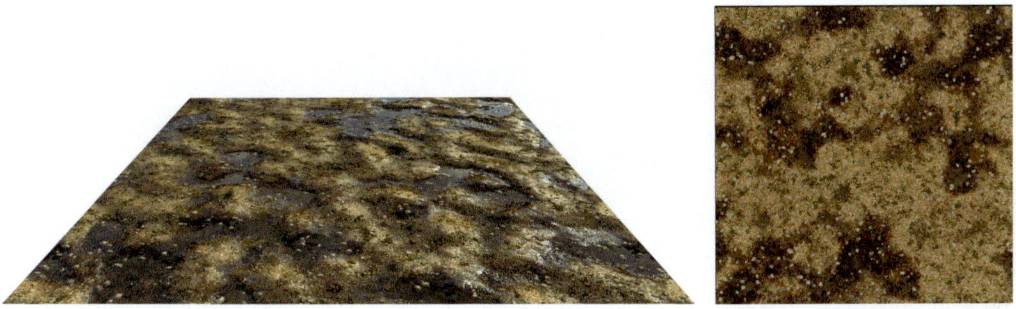

The color map for a wet ground material, showing the different sand, mud, water, moss, and pebbles colors.

Also called **diffuse color** or **albedo**, the base color texture is the most important in the PBR map set, for two main reasons: first, our eyes instinctively focus on color patterns to make out what we see. Therefore, color textures are the front-line "delineators" of the material and are scrutinized more than, say, bump textures, because we are so sensitive to implausibility in color. Second, color maps in a PBR system do not only represent the material's diffuse colors – they define the way a surface reflects light and how it affects the environment. Whether you are using photographs, procedural functions, texture painting, or some combination thereof, the base color map has a crucial role in the final render result. Consequently, there are some key issues to be aware of:

DIELECTRIC/METALLIC DIFFERENCES

The base color map has a different role in metallic materials. In dielectrics, the base color texture represents the color of the diffuse reflection color (the diffuse albedo), and defines which wavelengths are absorbed and which are diffusely reflected. In metallic materials, the base color map affects the specular reflection color (metals derive their colors from tinting the specular reflection). This is important to understand in the context of PBR safe colors described next.

PBR SAFE RANGE

Physically-based rendering abides by the law of conservation of energy. Since the base color texture defines the albedo values of the surface, extreme bright or dark values may cause the surface to react unnaturally to incoming light by reflecting or absorbing too much of it, in essence breaking the rule of conservation of energy. While there is no precise definition for a "good" color range, the common safe range for diffuse color maps lies around 30–240 (using a standard 0–255 sRGB scale), while the range for metallic color is around 180–255. Most texture authoring applications provide tools for checking the brightness of color textures. Ensuring that values fall within the PBR safe range can eliminate lighting and rendering issues down the line.

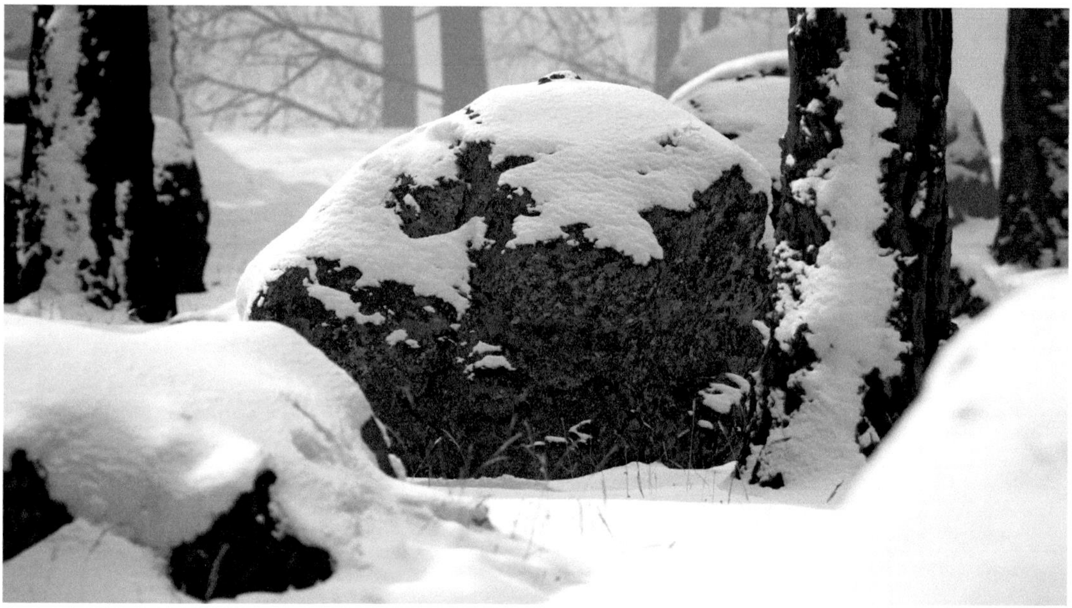

Snowy Forest by Gökhan Karadayi. Unreal Engine.

CONTRAST AND SATURATION

Regardless of PBR safe ranges, extreme contrast and saturation in base color textures should also be avoided. Color textures need to react well to any CG lighting setup in the scene, and sharp differences in luminance values, as well as pitch-black areas or overblown highlights, may cause parts of the object to pop out unnaturally in the render. Extreme saturation is also problematic, as it makes it hard to balance the lighting in the scene (later in this chapter we will discuss ways to avoid these issues when shooting and sourcing material for textures).

AVOIDING OVERLY BUSY COLOR

There is a tendency to cram all the texture information into the color map. While sometimes this is necessary, it often leads to an overflow of color information on the surface, and a noisy, unrealistic feel.

Take a look around you – many surfaces have a rather flat color, and some of the variations we detect are due to roughness or reflectivity, and not necessarily the base color. It is therefore good practice to delegate some of the detail to secondary textures like specular roughness, bumps, or even displacement.

Roughness Map

In a PBR/BRDF setup, it is assumed that the specular amount is constant, and only the specular roughness varies between shiny and dull materials. A roughness map is necessary for complex, layered materials. For example, on a partially-painted wooden beam, the roughness value should be higher where the underlying wood is revealed, and lower in the painted areas. In a wet pavement material, roughness is higher in the dry parts. But simple single-layered materials can also benefit from differences in specular roughness, which can hint at areas of smudges or dust. A roughness map is a great way to add subtle imperfections to smooth, clean surfaces like glass windows or polished metal.

Extracting a roughness map from photographs can be tricky since the color values in a photo do not necessarily represent specular roughness. Texture authoring applications offer some smart functions to simplify the process, but sometimes it is better to paint the roughness values manually when using a photo as a base texture.

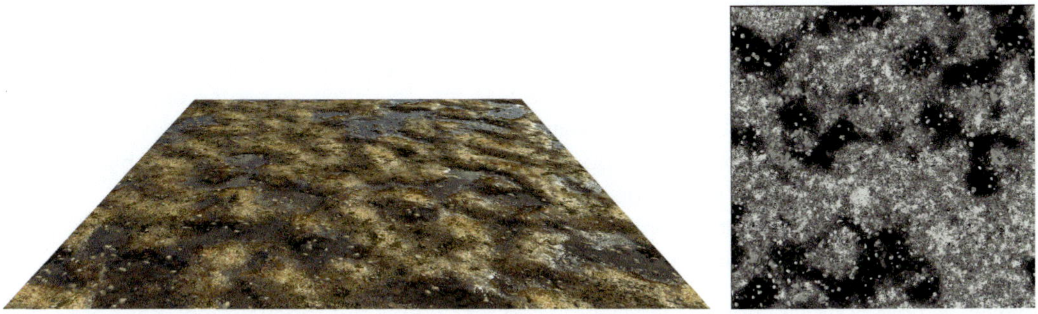

The roughness map is brighter in the sand and pebbles (rougher surfaces), and much darker in the water puddles (reflective areas).

Metallic Map

The metallic (or metalness) map is a grayscale texture that defines which areas in the material have metallic properties. This is an essential map for any layered or mixed surface that combines

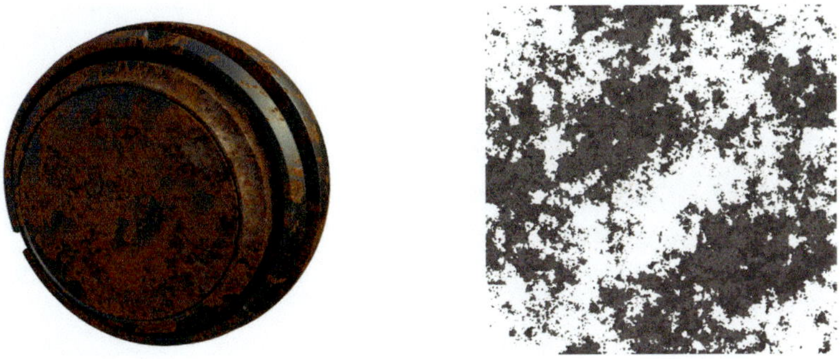

Rust areas are darker (dielectric) in the metallic map of this simple rusty metal material.

metallic and dielectric elements. Rusted metal is a classic example – while the clean non-oxidized areas are metallic, the rust itself is not. Without a metallic map, the rusty parts would look strangely reflective and metallic, and the material's layered feel will be lost.

With procedural or layered textures, it is usually easy to derive the metallic map's values from the relevant layer/s. When processing a single photo, it is of course harder. Often, the metallic map can be derived from differences in hues. For example, in a photo of a rusty surface, the rust may have a more saturated warm tone then the underlying metal, and a similar difference in hue is likely to happen with painted metal. Regardless of the method used to create the metallic map, it's important to crank the contrast and values of the map so that they are close to either zero or one, and avoid too much gradient gray shades: areas should be either metallic or dielectric, and the transition should be fairly sharp.

Bump (Normal) Map

From the very beginning of CG, bump maps were used to create the illusion of surface roughness and small-scale detail. The emphasis here is on "illusion" and "small scale", because unlike displacement, bumps do not affect the actual geometry, and are nothing more than a shading trick of the eye. Bumps are therefore somewhat of a double-edged sword: when used correctly they can add a remarkable amount of subtle detail to any surface by breaking up specular reflections and adding dimension and depth. There is, however, a certain tendency to push bump mapping too far in an attempt to increase realism by emphasizing surface features. This often causes the opposite result, because real-world surface roughness is generally only noticeable up close, and quickly loses its definition with distance. When bumps are too strong, their prominence from a distance harms the realism of the image. And when viewed up close, exaggerated bumps reveal their "fakery", especially along the outline of the object. It is therefore a good practice to combine bumps and displacement, relying on displacement for the larger features and using bumps only for smaller detail.

Originally, simple grayscale height maps were used for bumps, but today normal maps are preferred, because they use RGB to represent a three-dimensional relief rather than just an up-down value in a grayscale map. Traditional game modeling relies heavily on normal maps, which are first extracted from a detailed, high-poly model and then reapplied as bumps/displacement to a low-poly version of the same model. This technique also works well in VFX and visualization, especially when dealing with scanned or sculpted models that contain a massive amount of surface detail. That said, bump maps are a crucial component in practically every material, as very

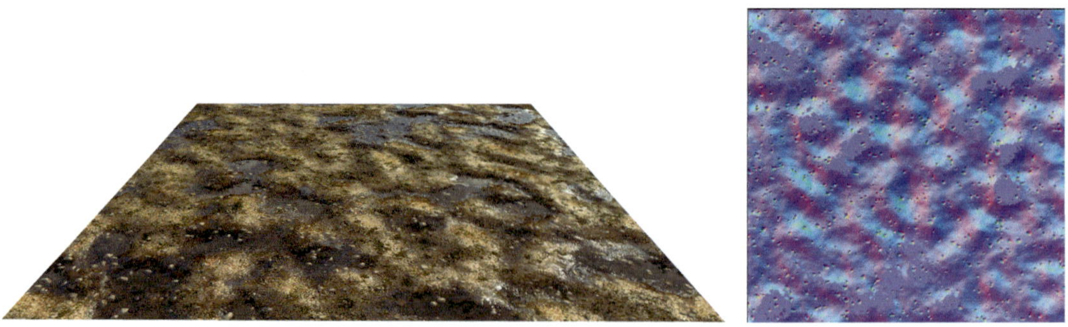

The normal map uses RGB to represent XYZ relief values.

few are perfectly smooth. The relationship between bumps and specular reflections is particularly important. A subtle bump texture that is hardly felt as surface roughness can nonetheless break specularity in ways that hint to fine irregularity. For example, even a clean glass window can feel more realistic with a few scratches or some unevenness of the glass. When used well, bump mapping can add a level of complexity to evenly-colored surfaces, and thus contribute to a richer and more photoreal look.

Generating a normal map from a photograph can be quite challenging, because dark and bright values do not necessarily represent height. For example, the roughness pattern of a rug may be completely different from its color pattern. The process of generating bump maps requires image manipulation skills in order to isolate the roughness-relevant color differences from the rest (material-authoring applications like Substance Alchemist have some useful tools to help with the process). Sometimes, when there's not enough information in the color image to produce meaningful height detail, it makes more sense to create the bumps through texture painting or procedural generation, separate from the color image.

A fully procedural workflow greatly simplifies the creation of normal maps because height detail is separate from color detail. But even with procedural textures, it is important to choose the right balance between larger and smaller features. When the bump values of a tiny dent and a deep fissure are similar, the material looks overly noisy and the amount of detail feels unnatural. Large-scale and small-scale bumps maps need to be sensibly mixed with this relative balance in mind.

Ambient Occlusion (AO) Map

In physically-based rendering, ambient occlusion is calculated as an integral part of the global illumination process. However, the renderer can only detect light occlusion based on geometry – it cannot calculate occlusion for bumps. The AO texture is thus a useful companion to the bump (normal) map. It reinforces the illusion of nooks and crannies on the surface by shading cavities and partially occluded areas in the material itself, adding depth to an otherwise flat texture. The AO texture is usually applied as a multiplier to the base color texture, so darker (more occluded) values in the AO map darken the colors while white areas have no effect.

In games (which rely heavily on normal maps for surface detail), AO maps are a necessity, but they are useful in any type of render situation where surface detail is conveyed through bump maps. In a procedural workflow, the AO texture is directly derived from the height information. With

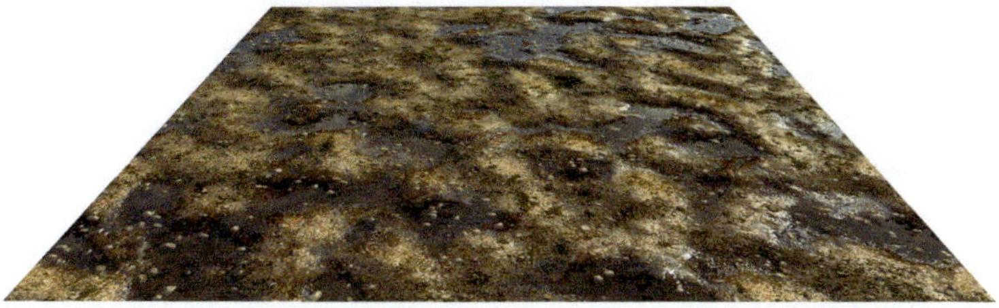

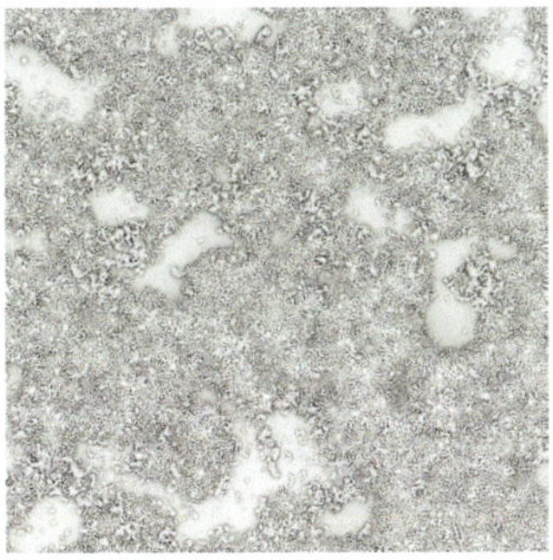

Areas that receive the least amount of light are darkest in the AO map (under pebbles, for example).

photos, the process is tied to generating a grayscale bump map: the same values that are used to define lower (recessed) parts in the bumps are used to define a stronger occlusion. As with bump maps, AO textures should be kept subtle: over-cranked occlusion can tip the lighting balance in a scene by shifting too much emphasis toward small-scale shading detail. In the final render, the darkest values in the AO texture should not be darker than true geometry-based occlusion in the scene.

Displacement (Height) Map

Unlike bumps, displacement effectively modifies the underlying geometry. This makes the displacement texture somewhat unique among PBR maps because it is, in fact, a modeling tool (rather than a shading tool). For displacement to work properly, the affected geometry must have enough subdivision at render time. The finer the displacement detail, the denser the subdivision must be. For this reason, displacement maps are used when height detail is prominent enough in the context of the scene. For example, it is quite an overkill to use displacement for a carpet texture (bump will suffice), unless the carpet is viewed at extreme close-up. On the other hand, displacement works best when height differences are not too extreme, and the geometry is not overly stressed. Displacement and bumps work well together, especially when displacement is used for the largest features in the material while bumps fill in the smaller detail. Therefore, when generating displacement maps from images, it is advised to mask out (or blur) small features in the image and keep only the larger detail.

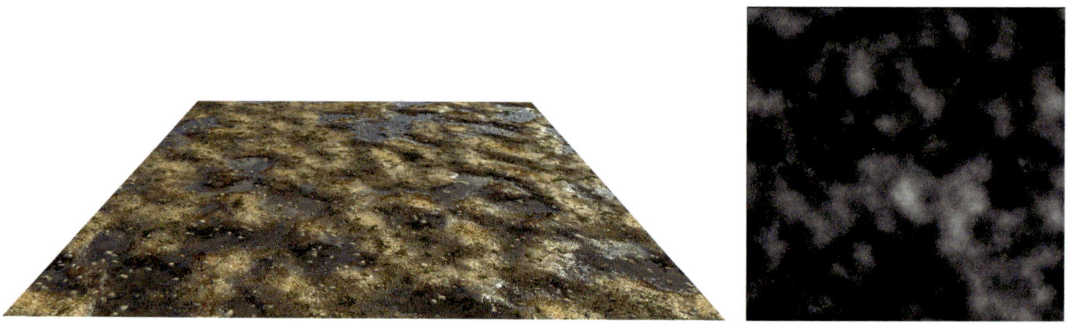

The displacement map is a simple altitude map, usually representing larger surface features.

Transparency Map Versus Opacity Map

A transparency map controls the transmission amount in the material – it is useful for varying the transparency of a surface, or to differentiate between opaque and transparent areas in a layered material. A transparency map does not affect reflection and bumps, which may still be visible in fully transparent areas (for example, on a clear window glass). However, sometimes the goal is to simply make parts of a surface invisible. In such a case, an opacity map should be used. The opacity map does not control any physical transmission properties of the material, it simply causes the surface to completely disappear wherever the opacity value is zero. Opacity maps are mostly used as "cookie-cutters" for thin elements like leaves or twigs, for creating objects like fences or nets without actually modeling the fine detail, or as a mask in layered materials.

Texture Generation Workflows

Essentially, there are three basic workflows for creating a texture map:

- Using one or more photographs (captured, scanned, or sourced).
- Generating the map procedurally, using a dedicated procedural texture authoring application.
- Hand painting the map, either in a 2D application (usually with the unwrapped UVs as a guide), or directly on the model, using a 3D paint application.

There is no superior methodology, since the choice of workflow depends on the specific material, model, and context. Each of these texturing approaches has its merits and limitations, and often the most practical and efficient solution is a combination of two or three approaches. In the following sections, we will examine these texturing workflows and look at ways to maximize their advantages and improve photorealism.

Image Textures

The main advantage of photo-based textures is of course the fact that they are inherently photoreal. Image textures retain all the complexity and subtlety of the real world and can add a remarkable amount of realism to even the simplest CG scene. Photographing or sourcing photos for textures is a more straight-forward process then procedural generation, and image textures carry the kind of intricate, natural detail that is hard (and sometimes impossible) to emulate artificially. The increased popularity and availability of quality 3D scans has redefined image

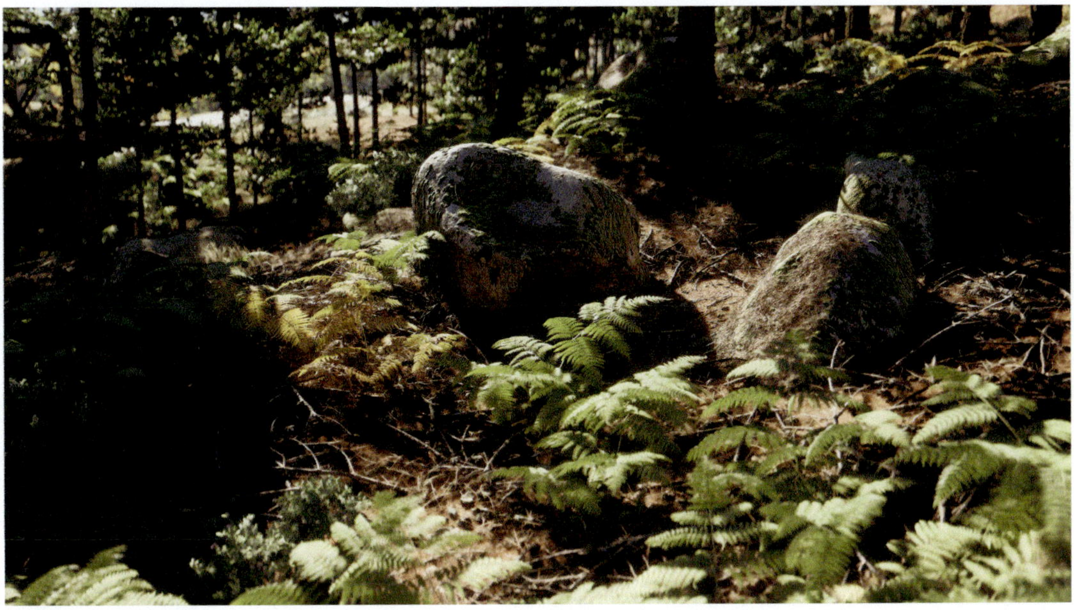

PWG Ecosystem by Gökhan Karadayi. Unreal Engine, Procedural World Generator.

© Gökhan Karadayi and ScansLibrary.

textures as an inseparable and organic part of the model. This further strengthens the role of image textures as the most effective means of connecting CG to the real world.

That said, image textures have some significant disadvantages:

- Photographs are two dimensional, so image-based textures need to be wrapped around objects. This requires meticulous UV mapping and applying multiple textures to different sections of the mesh to avoid stretching or tearing.
- Image textures often need to be tiled to cover large surfaces. While there are many tools that facilitate tiling, photographs may need substantial retouching work to achieve seamless tiling (and some simply don't tile well).
- Lighting is baked into every photograph, and photos with strong directional lighting can be extremely problematic for texturing. This substantially limits the choice for sourced images and forces some strict limitations on shooting photographs for texturing.
- Generating secondary maps (like specular roughness, bumps or ambient occlusion) from a photograph can be very challenging, because the color information does not necessarily represent height, occlusion, or specularity.

Shooting and Prepping Photos for Texturing

CG artists have access to an enormous selection of free and commercial generic photographic texture maps, and the higher quality ones are often preprocessed for optimal lighting and seamless tiling. But many situations call for specific textures that can only be obtained by photographing or scanning surfaces. This section offers some advice for minimizing the issues listed previously, while shooting and prepping materials for texturing.

Resolution

Texture size is important – a resolution that is too low may result in degraded, soft, pixelated, or aliased look, which can hurt the believability of the material. 2K texture maps are the norm for generic materials, but not every texture needs to be 2K. For example, a texture map that is intended to cover a small area or to be tiled multiple times (like a single bolt or a single bathroom tile) can, and should be, smaller. On the other hand, textures that are meant to cover a large surface with no tiling (say, an entire brick wall) may need to be 4K or even higher, depending on the final render resolution. When shooting photos for textures, it is best to use the highest available resolution, to allow cropping and further editing down the line.

Baked-in Lighting

The rule is simple: avoid direct light. For example, direct sunlight produces harsh shadows, bright specular reflections, and extreme differences in luminosity between lit and shadowed arears. This does not only limit the texture's usability in different CG lighting scenarios, it makes it extremely hard to generate subsidiary maps like specular roughness, normal, or displacement. Outdoor photos and 3D scans should always be captured under overcast sky or in full shadow. Indoor photos should be lit as flatly and diffusely as possible, avoiding any noticeable effect of direct light sources, or strong cast shadows. Even a gradual color change due to soft shadows can make it hard to tile the texture. Inconsistencies in lighting and luminosity must be corrected and "flattened" in an image-editing software or a dedicated texture authoring application.

Angle, Perspective, and Focus

Ideally, texture photos should be taken at a straight angle. For the same reason, it is also better to use medium or long lenses to minimize the distortion of wider lenses. When shooting large or tall objects, it is not always possible to find a parallel vantage point (for example, shooting the trunk texture of a large tree from the ground). Perspective and distortion can be corrected in an image-editing application. Sharpness, on the other hand, cannot be restored in post, so photographs should always be captured as sharp as possible, by avoiding a shallow depth of field (using a smaller aperture and keeping the focal point at a reasonable distance).

Exposure

Brightness and contrast can be easily changed as the image is prepared for the texture. But muddy blacks or clamped highlights cannot be salvaged. It is important to ensure that enough detail and color information is captured at both extremities of the color range. Photos should always be saved as raw in the camera, never JPG, to preserve the full dynamic range and allow for extensive image manipulation in the texture prep stage. If possible, three or five different exposures per angle can really expand dynamic range and help produce a texture with consistent lighting and clear detail.

Optimizing Photographs for Tiling

As mentioned, avoiding directional or uneven lighting goes a long way toward smoother tiling. Color and luminosity across the texture should be equalized, to avoid a noticeable repeated gradient. Another area that needs careful attention is repetition: singular or unusual detail (like a knot in a tree bark or a lone bolt in a concrete wall) can add a great deal of realism if the texture is used as is, but will look conspicuously repetitious once the texture is tiled. Such detail can be easily spotted when tiling of the image and should be removed (usually with a clone-paint brush).

Procedural Textures

Procedural texture maps are generated from scratch using a combination of different mathematical functions like noise, fractal algorithms, and pattern generators. 3D applications have always included a basic selection of procedural functions (for example, Maya's Brownian, cloud, or marble functions). But procedural material creation received a serious boost in recent years with the introduction of dedicated material-authoring software like Substance Designer, which provides powerful procedural tools coupled with a streamlined system for exporting materials as texture map sets for contemporary PBRs.

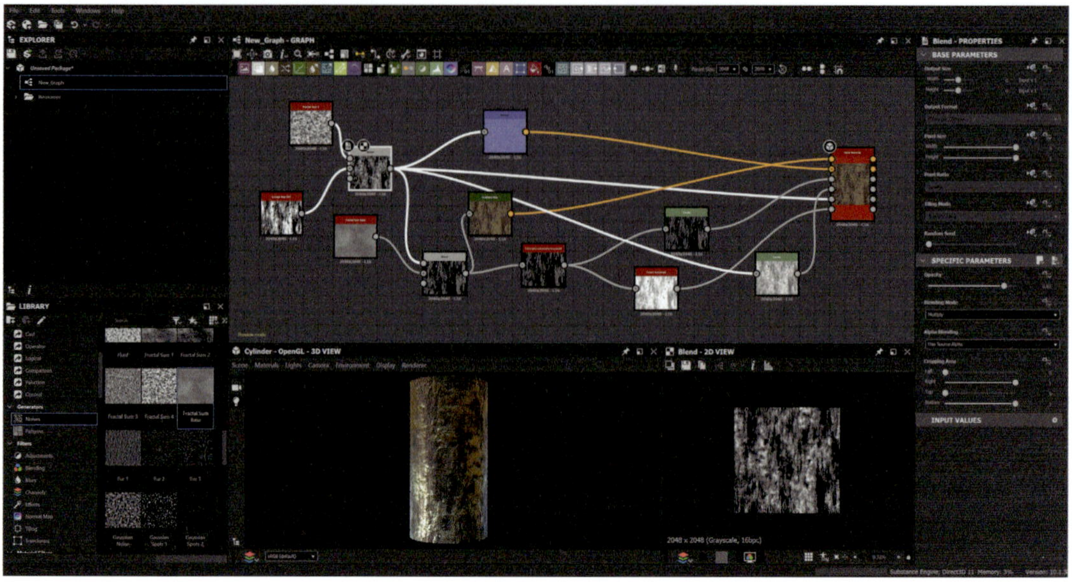

A simple procedural material node graph in Substance Designer. Complex materials can have much more elaborate graphs.

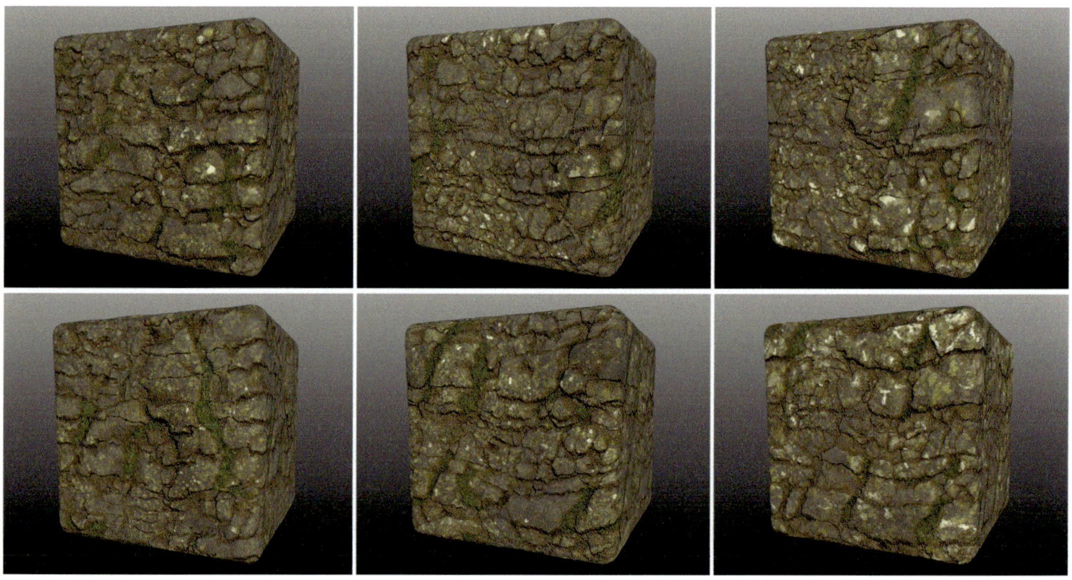

Six variations of a procedural stone wall material, all similar yet each uniquely different, generated with a single mouse click to change the random base seed.

Procedural textures have some significant advantages over image-based textures:

- They are fully editable and controllable, with full flexibility at every level and over every aspect.
- A single "master" texture can be easily and quickly randomized or tweaked to create a range of variations on the same look.
- Procedurally-generated maps are not hampered by baked-in lighting and work with any lighting setup.

- A procedural workflow makes it easy to generate secondary maps for bump, occlusion, roughness, etc.
- Tiling is easy and seamless.

Being a mathematical product, procedural textures lack the inherent real-world realism of photographic textures and tend to feel somewhat mechanical and generic. It is of course possible to create intricately rich textures with a purely procedural workflow, but successfully implementing the typical real-world imperfections and detail requires meticulous work and expertise. However, this downside must always be weighed against the power of procedural textures. Take, for example, a leaf texture: while it is possible (but not easy) to emulate such a texture procedurally, it makes a lot more sense to just scan or photograph the leaf (or find a photo of it) to get the exact, real look with very little effort. But what if this texture needs to be applied to thousands of leaves on a tree? This is where a procedural workflow shines, because it can easily generate multiple variations of the same texture, avoiding the cloned look of a repeated image textures.

Combining Workflows

Texturing workflows have changed in the past few years. Look at games for example: the introduction of new texture authoring applications opened up the vast possibilities of procedural

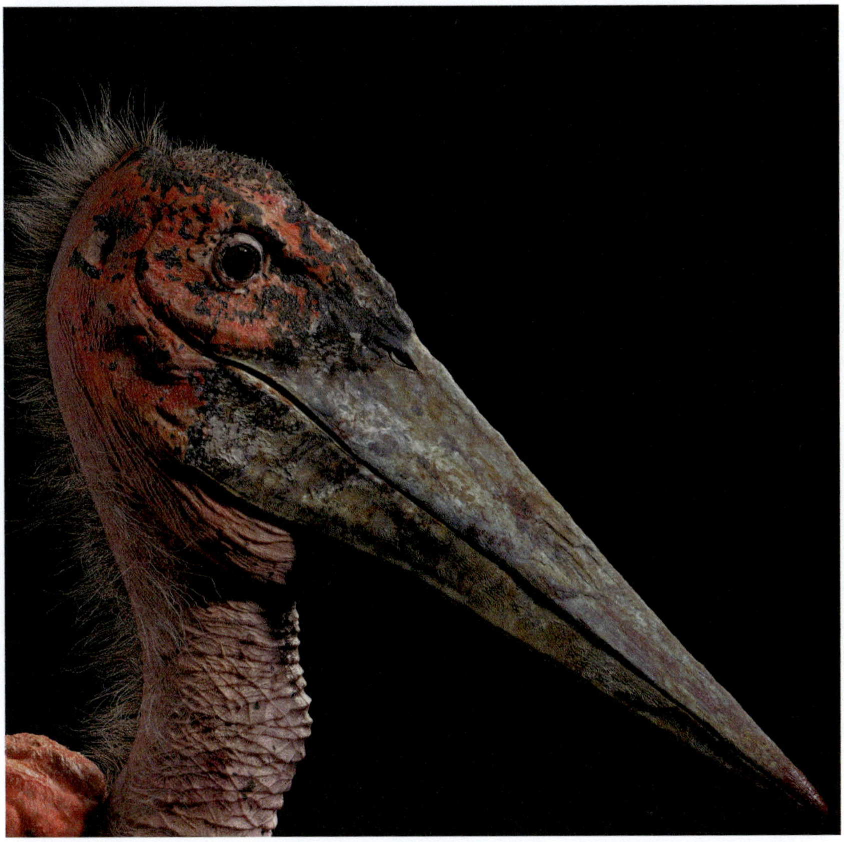

Marabou Stork by Jose Iuit. Zbrush, Maya Xgen, Mari, V-Ray.

texturing. Expansive 3D environments in many games require a huge library of textures and a fast workflow, and the fact that a single procedural texture prototype can be easily changed and repurposed is a boon. Procedural texturing brings the promise of infinite variation, but ironically, the results are often quite the opposite: many games suffer from a sort of generic "procedural look" that makes everything feel nearly identical. It is almost as if you're seeing the same rusted metal/peeling paint material everywhere. Like any other aspect of digital art, texture authoring is a craft that requires creativity, imagination, and attention to detail. The best results can often be achieved by mixing images, procedural functions, and hand painting. Here are some tips for combined workflows:

- Texture maps that needs to cover wide areas (for example, ground textures) often suffer most from repetitious tiling. One way to dodge this issue is to tile the base image texture a few times first, then use the tiled result as a base, to which additional large-scale procedural and/or hand-painted maps are added. This helps breaking (or masking) the repetition while preserving the underlying photographic quality.
- Procedural functions are great for creating patterns and tiles for pavement, walls, structures, etc., and can be combined with a base image map that provides the fine color detail and variation. In general, using images for the color and procedurals for bump and displacement makes it much easier to produce effective secondary maps. For example, a brick wall can be created by first generating the brick pattern procedurally, then adding color detail for the bricks and the mortar from image textures. The procedural pattern is then used for clean and precise bumps, AO, and roughness, while the diffuse map benefits from the realism of photographic detail.
- One way to avoid a "procedural look" is to mix several different noises or grunge maps (using overlay, screen or multiply) to break up the familiar patterns of Perlin noise and other commonly used functions. Varying the scale and intensity of the noise layers can help creating a rich, natural texture. Another option is to break up procedural patterns with image textures. This can add irregularity and randomness to function-based patterns.
- Texture authoring applications offer smart procedural functions for applying dirt, wear, and tear in certain areas, based on the height, ambient occlusion, or curvature values of the underlying layers. This is useful for quick, basic dirt/gunge layers, but specific, topology-based wear and dirt should be hand-painted on the actual object using a texture paint application.
- Hand-painted wear and dirt effects look best when added in physically-correct areas. Finger smudges and spots normally happen in the areas of higher interaction (such as doorknobs, touch buttons, or touch screens, levers or handles); Scratches appear around areas of friction, or areas that are exposed to the elements; Dust usually collects in areas of least interaction, under objects, and in cracks and nooks; Water marks follow paths downward along the curvature of the surface, and so on. Topology-specific texture painting can truly boost the believability of texture maps by tying the features to the model in an organic, interactive fashion, and is a good way to avoid the previously mentioned generic grime look.

Chapter 14

Modeling

It goes without saying that a disproportionate human figure or a grossly inaccurate car model reduces the believability of a render. But assuming 3D elements are modeled correctly and match their real-world counterparts, does modeling really affect photorealism? Much of this book is focused on light and its interaction with surfaces (and, consequently, color). Modeling is thus a unique subject in the context of this book – it has nothing to do with light and color. Or does it? This chapter is not an overview of modeling techniques – there are many books about that subject. Instead, this chapter focuses on two specific modeling aspects that particularly affect photorealism: modeling for lighting, and procedural generation of complex natural detail.

Modeling for Lighting

Modeling is traditionally split into two main types: technical modeling and sculpting. The first include mostly man-made objects, from furniture and cars to spaceships and buildings, while sculpting is most commonly used for modeling creatures of all sorts, from humans and animals to monsters and aliens. Both categories require accuracy and an understanding of the subject, whether it is mechanics, construction, or anatomy, and both benefit from careful studying of reference material, images, or blueprints. The main difference (at least in the context of this chapter) is that humans and animals are sculpted as one continuous surface, and are naturally curved and flowing, while technical models, just like their real-world counterparts, are usually built from multiple separate pieces. The vast majority of those pieces are derived from simple geometric forms: a table, for example, is essentially five boxes (or one box and four cylinders), and a building is one big box made out of multiple smaller boxes (doors, windows, balconies, stairs, railings, etc.).

Herein lies the issue of light interaction, and the need to model for lighting: in the mathematically perfect world of CG, box-like objects have perfectly sharp corners, but in reality, that is hardly ever the case. The edges of the five boxes that make up a table have been cut and sanded, leaving the corners slightly rounded. As discussed in Chapter 5, rounded corners and edges act as "highlight magnets", displaying noticeably stronger specular reflections. Such beveled corners can be found practically everywhere, and even when they are extremely thin, rim highlights still

DOI: 10.4324/9781003590293-18

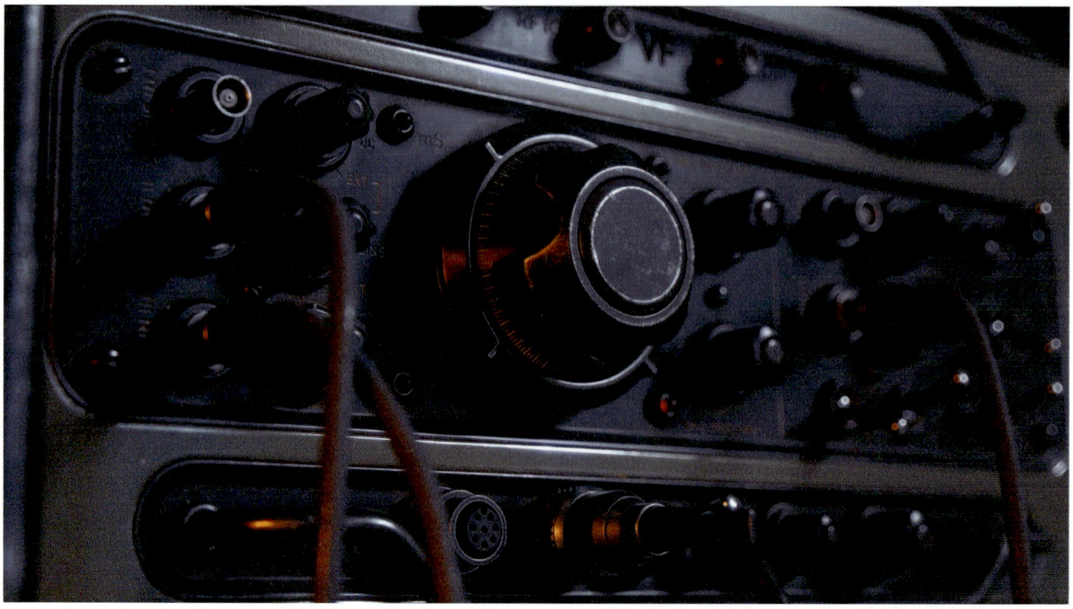

Meters by Raphael Rau. Notice how almost every edge is slightly beveled, catching highlight and giving the render a rich, photoreal feel. Cinema 4D and Octane.

© Raphael Rau.

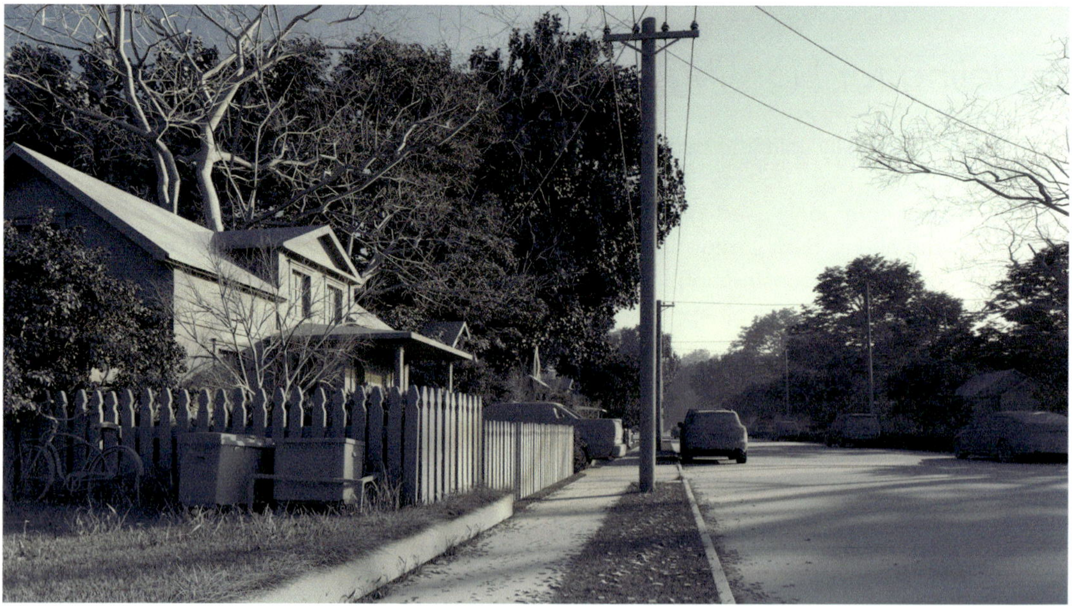

Even with no textures or shaders, Mohamed Raof's street model already feels realistic and believable because of the fine modeling detail. 3D Studio Max.

© Mohamed Raof.

contribute substantially to the way objects catch light. When a CG model has perfectly sharp corners, the effect of rim highlights, which we normally take for granted, is sorely missing – renders feel dull, flat, and "CG". Modeling for lighting means addressing sharp corners and edges by judiciously applying bevels or chamfers. That extra bit of detail is extremely important for photorealism and is usually well worth the additional modeling time (and poly count). That said, the necessity of beveled edges must be weighed for each particular case. Take, for example, two scenes: an architectural render of an office building and a render of an interior apartment. In the first, beveling the corners of the building or various architectural elements would probably be an overkill – the building is too large and too far from the camera for rim highlights to even be noticeable. But the apartment interior could certainly benefit from beveled edges on furniture and even wall corners. Beveling in games and other real-time rendering is often limited by poly count restrictions, but rim highlights can still be achieved by applying the bevels to low-poly models as normal maps like other fine detail.

Procedural Modeling

While standard modeling and sculpting techniques work well for man-made objects as well as humans, animals, and imaginary creatures, they are not the most efficient for terrains, rocks, and plants. The realism of natural elements depends not only on the amount of detail – It relies on a seemingly contradictory combination of randomness and order. Nature's chaos can only be successfully simulated by subjecting it to natural laws: the way that water, wind, and ice shape mountains and valleys; how plants of the same species share structural and visual characteristics; or the way vegetation grows only in specific areas on a terrain.

Procedural modeling methods provide a better alternative to standard modeling or sculpting. Using fractal displacement and algorithmic erosion, an artist can quickly generate vast, highly detailed terrain surfaces which would otherwise be extremely tedious to sculpt. And by using a combination of random generation with deterministic algorithms like L-systems, artists can "grow" an infinite number of plant variations form a single species, or a large set of individual rocks that share specific characteristics.

Terrain Modeling

The concept of heightmap-based terrain generation is simple: a polygonal plane is displaced with a black and white image (heightmap). Precise elevation data like the USGS DEM maps[1] has been used for many years to create real-world 3D terrain for a variety of applications, from geological survey presentations to architectural and environmental visualizations. But accurate scientific height maps are not necessarily the best solution for creative VFX and games scenarios that require the ability to design and generate terrains with specific features and detail from scratch. Game engines like Unreal have some basic tools for terrain creation, while specialized terrain generation apps like Vue, Terragen, World Machine, World Creator and Gaea offer extended functionality by using different combinations of one or more of the following modeling procedures:

FRACTAL-BASED ELEVATION

Many natural elements display fractal characteristics as the progressive recurrence of larger patterns in smaller ones. The key here is to introduce a random factor, usually by shifting the

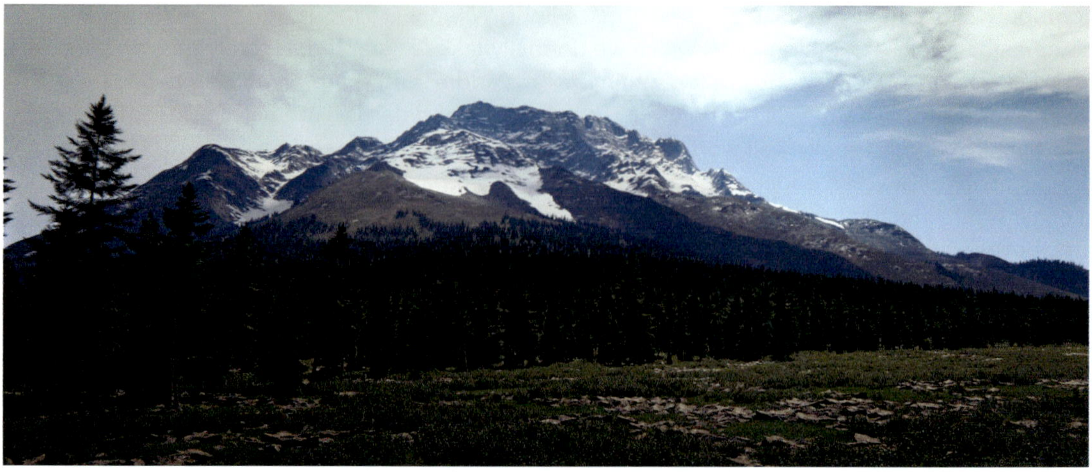

Mountain Environment by Caleb Spenrath. Gaea, Blender, Nuke.
© Caleb Spenrath.

center point arbitrarily through each subsequent subdivision. Perlin noise is the most used in fractal terrain functions (Voronoi is another common noise type). Attributes such as scale, amplitude, and roughness affect the overall look of the displaced terrain, while the number of subdivided iterations, or "octaves" affects the amount of detail (fewer iterations produce smoother surfaces with less small-scale features). Some terrain generation applications offer specialized fractal functions that better emulate the highly varied topology of large-scale terrains.

EROSION SIMULATION

Despite the complex and highly detailed look of iterative fractal functions, the resulting terrains still feel somewhat repetitive and mechanical. Real-world terrains are shaped by erosion – millions of years of water, wind, and ice that sculpt mountains and hills and give them their distinct, familiar look – the crisscrossed fluvial marks on the slopes, U-shaped glacial valleys, alluvial deposits at the foothills. What makes specialized terrain generation applications unique is the tools they provide for simulating natural erosion, which add a level of realism to fractal terrains that is extremely hard to recreate with standard modeling and sculpting techniques.

FREE SCULPTING AND TERRAIN FILTERS

While procedural modeling and erosion simulation are capable of producing a staggering amount of detail very quickly, they are still procedural – based on random functions or a set of rules – and offer little localized control over the shape of the terrain. The most powerful terrain generation applications offer the best of both worlds: the ability to use the power of procedural generation along with the hands-on sculpting. The artist can then choose to start a terrain by sculpting the general shape and features and then apply procedural displacement and erosion effects, or work the other way around, by first generating a procedural terrain and then sculpting specific detail. The new generation of terrain modeling software (like Gaea or *World Creator*) provides a wider toolset that include procedural filters for quickly adding unique terrain features like terraces and balconies, stratified rocks, sand dunes, plateaus, and scattered rocks.

Case Study: Erosion Algorithms in Gaea

Dax Pandhi is a CG environment artist, nature aficionado, and co-founder of QuadSpinner. Read his fascinating description of the R&D process behind Gaea, QuadSpinner's terrain generation software:

> QuadSpinner co-founder Cynthia Najim and I conducted research trips to the US Southwest and Pacific Northwest, Central America, Mediterranean, Iberian North Atlantic coast, and the deserts of Western India over the course of seven years. The mission was to understand geology in a new light and distill it into algorithmic form, striking a balance between geological realism and artistic requirements. Nature is beautifully complex. But when attempting to recreate it in algorithmic form, it becomes frighteningly challenging. Since our goal was not to replicate, but to approximate nature for artistic use, we began adapting some of the oldest tricks in the history of art. For example, a simple white dot on a pencil-shaded sphere hints at a smooth, reflective material. The position of this highlight is paramount to conveying the effect. Similarly, applying certain shapes and soil/rock modifications along specific curvatures, ridges, and deposits could convey a broad spectrum of terrain features. This was our way of translating immensely complex geology into 2.5D heightfields and polygonal meshes. This is a methodology we call geological minimalism.
>
> Back then, all procedural terrains were built with a shader-like model. You create noises, filter them to look like random geological features, and then erode. If modeling software had spheres, cubes, and teapots as primitives, we needed a similar language for making terrains. We started to create meaningful procedural shapes such as mountains, faults, valleys, ice floes, mesas – essential geological building blocks. Industry wide, digital terrains were fatigued from the overuse of fluvial erosion. If a terrain did not look realistic enough, artists tended to throw more erosion at it. To combat this, we started to craft new algorithms for alluvial deposits, terrain folding, stratification, anastomosis, and a range of other processes. Only by applying multiple processes would terrains look more realistic and diverse.
>
> Our new R&D Lead Daniil Kamperov augmented our work with his groundbreaking research on hydraulic and thermal erosion. Unlike other models during that period, this erosion model was resolution-independent, struck a new balance between erosion and deposition, and was both particle and grid based. In practical terms, artists would be able to modify erosion properties to give the terrain unique personalities and escape the limitations of previous models. Over time, this research has expanded to include snow simulation, hydrological features, and other natural phenomena.
>
> We completed our toolset with a set of simulations called Data Maps. They analyze terrains – even static terrains such as DEM data – and simulate natural properties such as water flows, rainfall velocity, vegetation growth patterns, and even quasi-random geological shapes for both processing and texture propagation. With this workflow, color production is enforced through systematic elements rather than on slope, randomness, and flow lines.

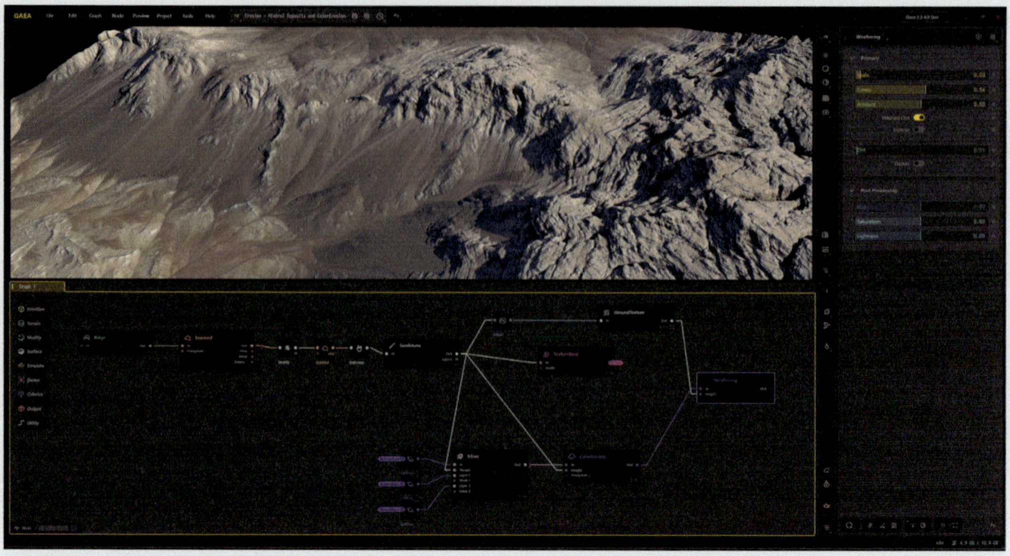

Applying various erosion algorithms in *Gaea*.

© Dax Pandhi.

Plant Modeling

Plants have a complex structure, but this structure follows a set of rules that is specific for each species. Individual plants of the same species share the same characteristics and visual traits, yet a certain factor of randomness within those rules makes each individual plant uniquely different. In 1968, Botanist Astrid Lindenmayer developed a system that defines the parent-children relationship that characterizes plant structures and can be used to model their growth. This system (known as the L System) is used today as the mathematical basis for procedurally generating plants. Plant modeling applications such as SpeedTree, Plant Factory, or the venerable Xfrog let artists define a set of parameters like length, diameter, curvature, and angle through the recursive structure of a plant – starting from the trunk (or stem), through the main branches, to smaller branches, twigs, and leaves. Each subset inherits characteristics from its parent while introducing new ones, and for each parameter, a certain margin of randomness can be set. This provides not only a quick and efficient method for creating a wide range of plants, it also gives the artist the option to "roll the dice" and let the random deviations kick in, which produces an infinite number of plant models of a single species, each uniquely different. These can then be exported to the main 3D application for populating scenes.

The leading specialized plant creation applications offer a complete toolset that includes not just core L System modeling but also shading, texturing, and mapping tools, various modifiers, poly count and LOD control, and specialized animation tools for wind, growth, and even physical reaction to objects. Applications like SpeedTree became ubiquitous in the games, VFX, and visualization industries because they offer a one-stop shop for 3D vegetation. However, some generalized 3D applications like Houdini provide L System tools that allow stripped-down plant modeling within the core application. A similar approach to plant modeling (though not necessarily with L-systems) can be used to quickly generate libraries of stones, rocks, bricks, minerals, or even animals and humans.

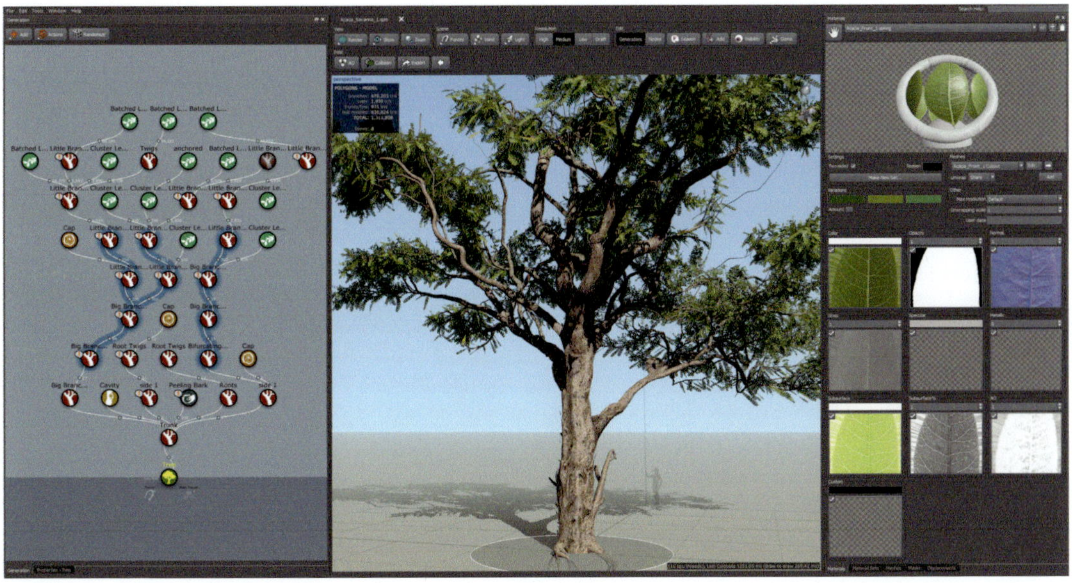

An Acacia tree model in SpeedTree. Notice the complex procedural hierarchy on the left, which includes nodes for trunk, branches, twigs, leaves and special features like knots, and the PBR leaf texture set on the right.

Note

1 https://www.usgs.gov/core-science-systems/ngp/3dep/about-3dep-products-services.

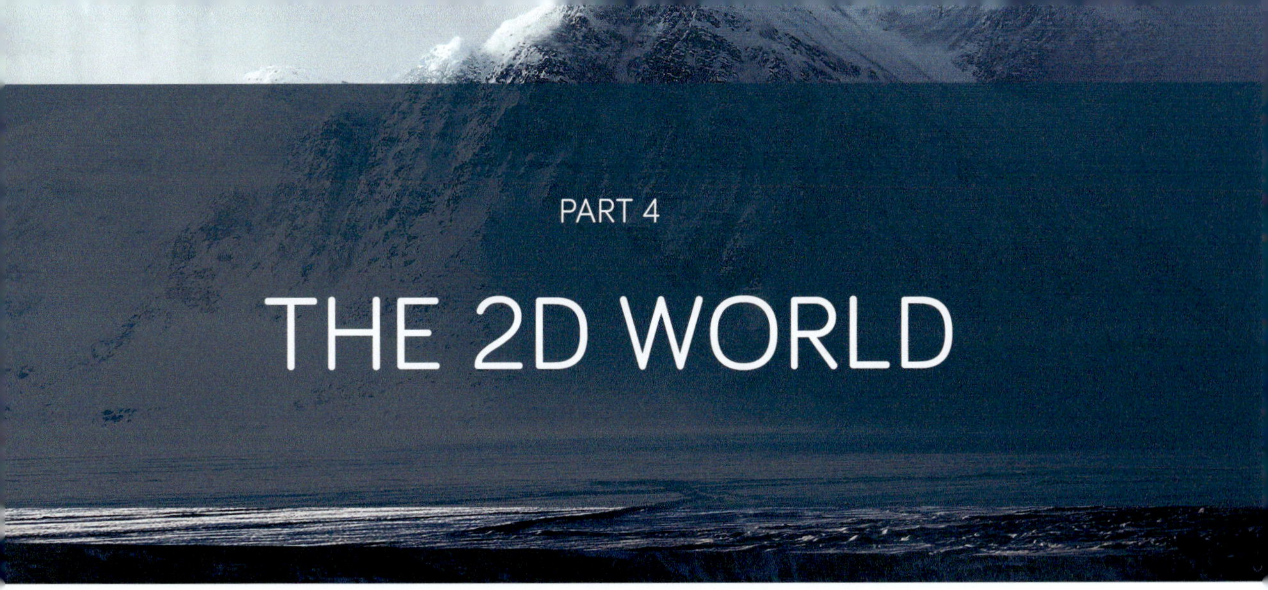

PART 4

THE 2D WORLD

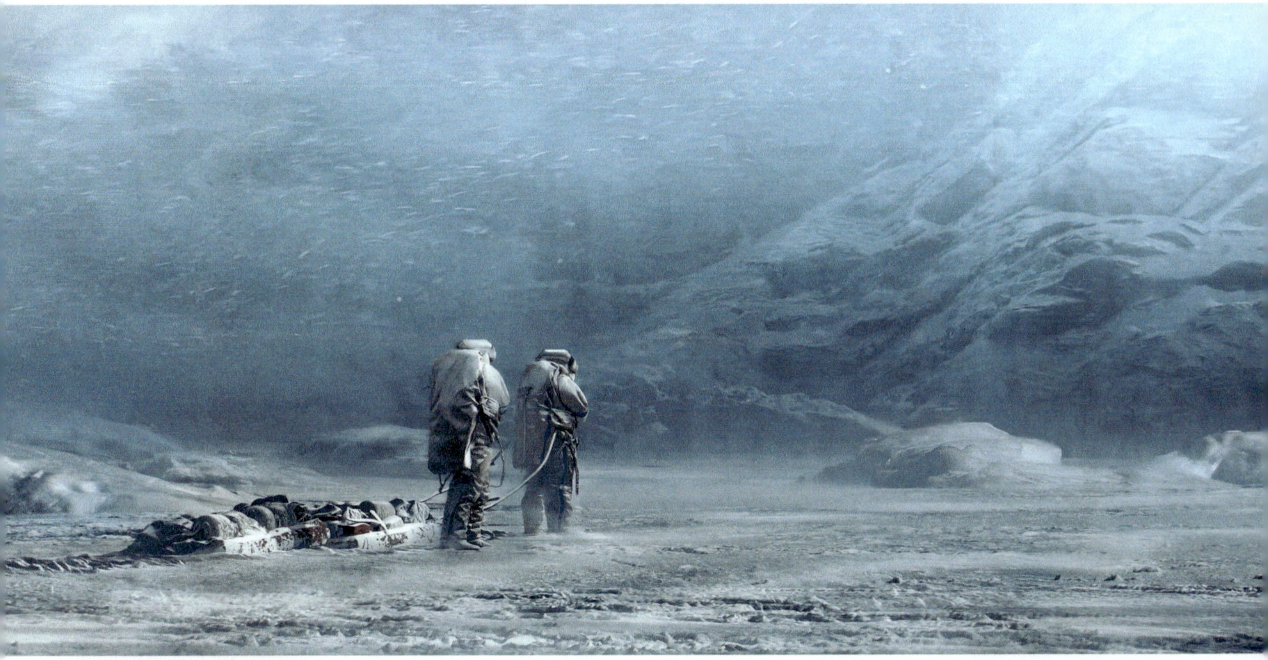

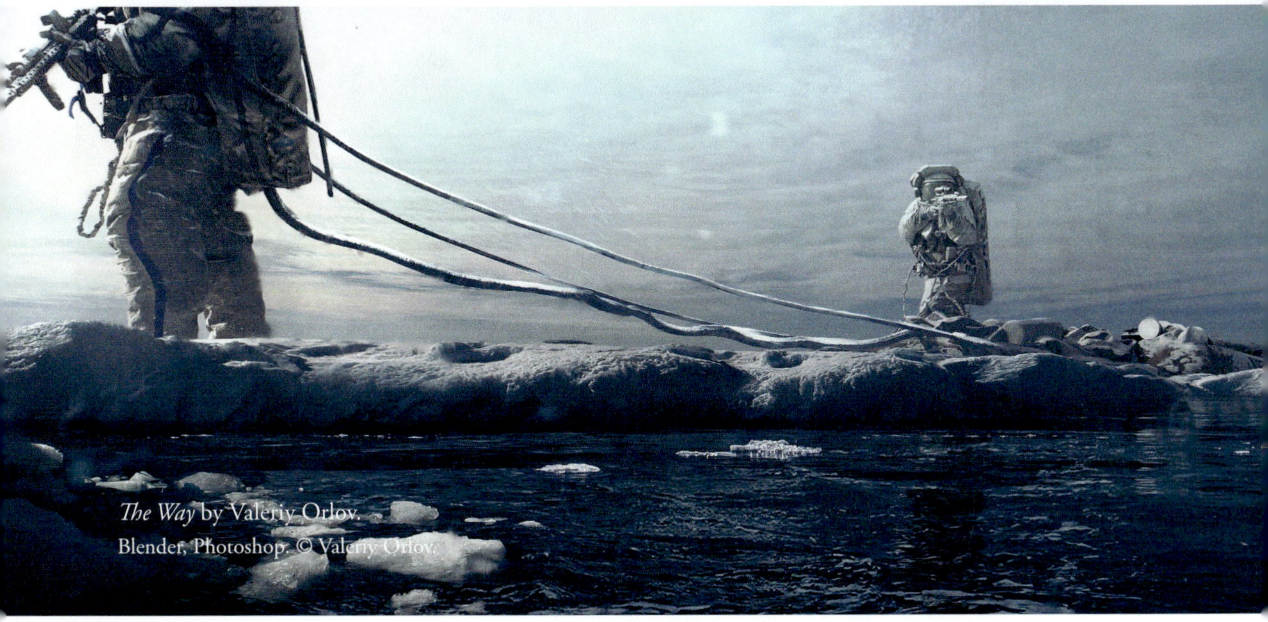

The Way by Valeriy Orlov.
Blender, Photoshop. © Valeriy Orlov.

2D workflows are undoubtedly more prominent in visual effects, which rely heavily on compositing and matte painting. Many of the most common bread-and-butter VFX shots start and end with compositing, and do not involve any type of CG work. The same can be said about matte paintings – a substantial chunk is done solely in 2D. Visualizations are certainly more 3D-centric, but many architectural models are composited into photographed environments, and 2D processing is used to add finishing touches to product renders and clips. Games and VR are based on 3D real-time rendering, yet despite the seemingly pure 3D nature of these fields, compositing techniques are increasingly being used to augment the game's cinematic experience, and 2D matte paintings are created for distant backgrounds. So, while this part of the book applies first and foremost to VFX artists, many of the subjects discussed here are useful to visualization and games artists as well.

Compared to 3D, the 2D workflow may appear extremely limiting – there is no lighting, shading, texturing, or modeling, and transformations are restricted to just two axes. But the notion that 2D is limited contradicts hundreds of years of prodigious artwork by numerous artists who managed to successfully convey depth, light, texture, atmosphere, and scale with nothing more than a brush and a canvas. They did it (and still do) through masterful use of color, and a deep understanding of three-dimensional space. For digital artists, the 2D process is essentially the same – in compositing and matte painting, photorealism is achieved through meticulous color work and a three-dimensional approach to a two-dimensional workspace.

The first advice I give to my students is that mastery of color is the key to successful compositing and matte painting. The second advice is to never, ever, think of compositing or image manipulation as a process of stacking layers, as "A over B", or "layer 5 over layer 4". Every image or piece of footage represents a three-dimensional world, and every 2D element that is added to that world should be approached as a three-dimensional entity placed within a three-dimensional space. In essence, the key to photoreal 2D is to never think of it as 2D: the seemingly unattainable

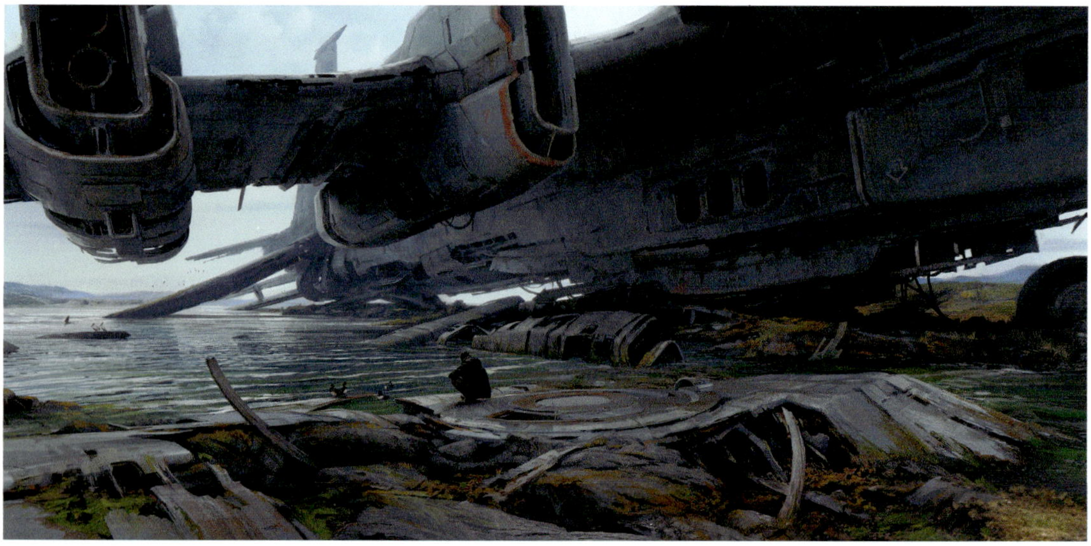

Quiet Coast by Sergey Vasnev.

Blender, Photoshop. © Sergey Vasnev.

third dimension should always be accounted for through the thoughtful use of scale, perspective, atmosphere, and depth of field.

Working in 2D has some significant benefits for a photoreal workflow. First and foremost, it is faster: compositing, matte painting, and image manipulation are not nearly as computationally intensive as 3D modeling, lighting, and offline rendering. The ability to view the end result in real-time as the artist interacts with the scene makes continuous tweaking and fine-tuning a much easier and intuitive process. This is why the final lighting and color adjustments of CG renders, as well as the addition of lens characteristics like depth of field are often left for the compositing stage. However, the most important advantage of 2D is the benefit of working with photos and footage. The inherently photoreal quality of such material allows compositors and matte painters to manipulate realism (or create realism from scratch) without the need to mimic or simulate reality by relying on mathematical rendering algorithms. The detail, the imperfections, the wear and tear, the subtle shading, the warmth of a physical lens – all the aspects that are difficult to produce in CG are already embedded in footage. The challenge, nevertheless, is in combining different photographic elements: it is akin to combining puzzle pieces from multiple unrelated puzzles into one, coherent image.

Chapter 15

Integrating 2D Elements

Color Matching

The ability to tweak the colors of an element to make it "sit" well within a new environment is one of the most crucial skills for compositors and matte painters, and arguably one of the toughest to master. Say you composite the same tree element over five different backgrounds. The tree's leaves will remain green, but it will be a different green in each of the five cases. How different? That depends on a variety of factors. Back in Chapter 3, I presented my six-layer approach as a way of simplifying the process of color matching. The green of the leaves exists as an absolute color only in a virtual world with no external influences. In reality, that green is already affected by the lighting and the camera's own color interpretation. When extracted and placed over a different background (or, to word it more appropriately, within a new environment), that green needs to be adjusted with respect to that environment's particular lighting, atmosphere, and camera color interpretation. The color operations I described in Chapter 3 (gain, gamma, lift, offset, saturation) are the essential tools in this process – mastering them and understanding the factors that affect color are the keys to achieving seamless integration and photoreal results.

There is no single methodology for color matching, and it is more a matter of personal preference. Some compositors start by matching the black and white points, others start with the brightness or the hue. Some use value sampling and others rely solely on their eyes. I like to start with the big brush strokes – adjust the gain and the gamma for general brightness, contrast, and overall tint, then proceed to finer adjustments for blacks and highlights, and finally (if needed), tweak specific ranges like shadows, midtones, and highlights, or use a hue correct tool to modify specific hues. I like to rely on my eyes more than on sampling or automated tools, because I believe that color matching is context-sensitive and relative, rather than absolute science. Then again, this is merely a personal preference, and the process depends on individual experience. In the end, the goal is to be able to quickly and easily match colors to achieve seamless integration.

I often give my students two versions of the same photo, an original and a color-corrected one, and ask them to adjust the colors of the original version to match the second as closely as possible. Depending on how the colors have been tweaked, this can be an easy introductory exercise or a mind-twisting puzzle. It may initially seem like a rather mechanical drill, but so is practicing scales for a pianist or doing jumping jacks for a soccer player. Color matching is an acquired skill, and the more you practice it, the more you train your eye to see color precisely and strengthen your grip on color tools and operations.

DOI: 10.4324/9781003590293-20

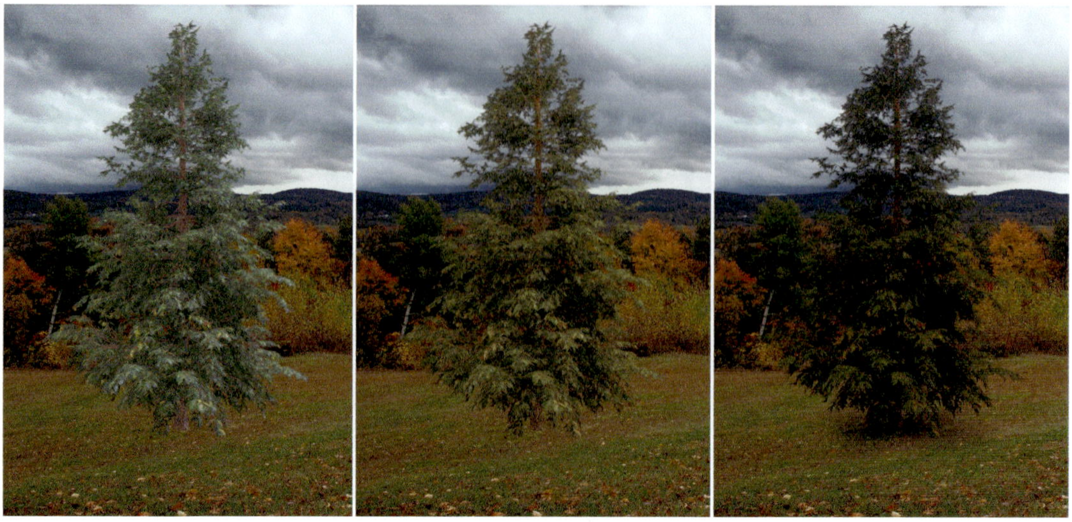

Three stages of color-correcting a tree element to better integrate with the footage.

Edges: Problems and Solutions

Unlike CG elements, which are normally rendered on a clean black background, every 2D element needs to be cut out or extracted from its original background before it is composited unto a different footage. If all elements had sharp, well-defined edges, this process would have been rather straight-forward. But that is hardly ever the case: edges often comprise of complex sub-pixel detail (hair, fur, fabrics, tiny irregularities), or are fuzzy and semi-transparent because of translucency, defocus, or motion blur. Such edges make the extraction process challenging even if the original background is a perfectly clean and well-lit green screen. And extraction is only half of the challenge – integrating the elements into their new environment becomes tricky when edges are soft. The reason lies in the fact that any edge that is not 100% opaque carries with it some of the original background. This is obviously a problem if that background is busy and detailed, but even clean backgrounds (like a green screen) can be problematic: chances are that the luminance level of the original background does not exactly match that of the new background. This difference in luminance shows up as dark or bright edges around the composited element, killing the believability of the comp.

This problem presents a tricky compositing dilemma: whether to preserve the detail and softness of the edges and risk ugly bright/dark outlines, or to tighten the extraction and end up with unnaturally sharp and harsh edges. Both options can damage photorealism: unnaturally sharp edges or dark or bright fringes around edges inevitably lead to composites that "look Photoshopped". The goal is to preserve as much of the edge detail and its original softness while also treating any bright or dark outline and making sure that the semi-transparent edges blend seamlessly with the new background. Easier said than done, but crucial for seamless integration. In this section I will discuss some key methods that can help achieve photoreal integration of extracted 2D elements.

Matching Background Luminosity

If you know in advance what type of environment the element will eventually be composited into, you can match luminance values already at the shooting or sourcing stage, eliminating edge issues down the line. For example, when I am on set shooting a green screen element for a nighttime shot, I ask the cinematographer to reduce the lighting on the green screen and take it down a notch or two below optimal level. While it is important to keep the green screen bright enough for proper chroma keying, slightly dimming the lighting in such a case can mitigate unnatural bright edges (or light spills) when the element is composited over a darker background. A similar principle can be used when selecting elements for a matte painting – those with backgrounds of similar luminosity to the target background will transfer and integrate much better and will require much less edge work. For example, a tree extracted from a bright sky background and composited onto the dark forest environment is bound to show unwanted bright edges. In this case, it makes sense to look for tree elements that have a similar, darker background.

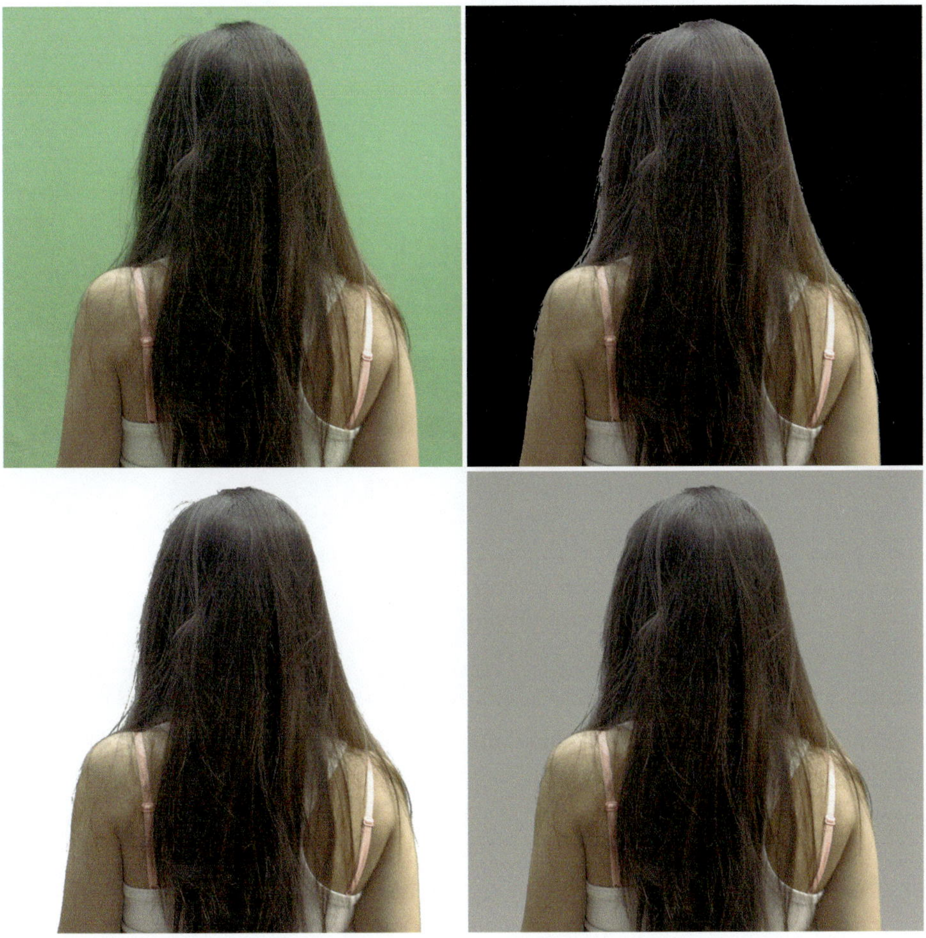

Bright edges around the hair are visible when the background is darker than the green screen (top right), while dark edges form along the arms when it is brighter (bottom left). A background with a similar luminance to the green screen produces the best results (bottom right).

Footage courtesy of fxphd/Eduardo Abon.

Spill Suppression

Green and blue screens must of course be highly saturated to serve their purpose, but this does not help with edge integration. Semi-transparent edges carry not just the background's luminosity, but also its hue (in this case, extremely unnatural green or blue). To be clear, spill suppression is required not just for edges – areas on the subject that receive bounce light from the screen or specular reflections are also susceptible to spill. But soft edges often require a separate spill suppression process, because they reveal the actual screen, not just its reflection.

A wide variety of despill tools are available for compositors, and many have controls that allow fine-tuning the spill sensitivity and output color. Some despill algorithms go even further: Nuke's IBK algorithm, for example, uses a dedicated screen color node that, when set up correctly, creates a clean version of the green/blue screen. This is used as a color "roadmap" for subtracting green spill. It is a more accurate method because it takes into consideration the different color values on a typical green screen (green and blue screens rarely have uniform color, due to wrinkles, smudges, and uneven lighting). Spill suppression is a relatively straight-forward aspect of keying and integration, as long as it is used judiciously and with attention to the results. For example, heavy despill of grainy footage can create "black holes" and increased noise where green pixels turn black by the process (in which case, it might be necessary to degrain the footage prior to despill and then re-grain it).

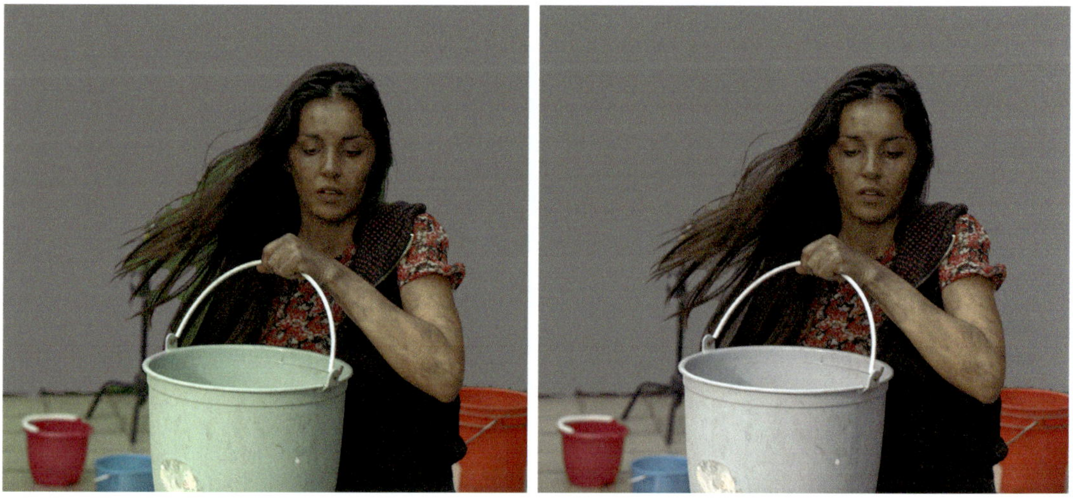

Spill suppression removes unnatural green spill from wispy areas like the hair or reflective areas like the floor. Care must be taken, however, not to despill areas that are naturally green, like the bucket in this shot.

Footage courtesy of fxphd/Eduardo Abon.

Edge and Core Extractions

Junior compositors often attempt to generate a single, perfect extraction matte. This is rarely the best approach (unless the element has perfectly sharp, solid edges) because of the constant tug-o-war between achieving a clean matte with no holes or noise and preserving soft and semi-transparent edges. It is much more effective to create two separate extractions – a hard-edged, solid matte for the core of the element, and another, softer extraction solely for the edges. If the element is a person, for example, then the core matte ensures there are no unwanted holes in the body and face, while the softer semi-transparent matte preserves hair and fabric detail, as well as any motion-blurred or defocused edges.

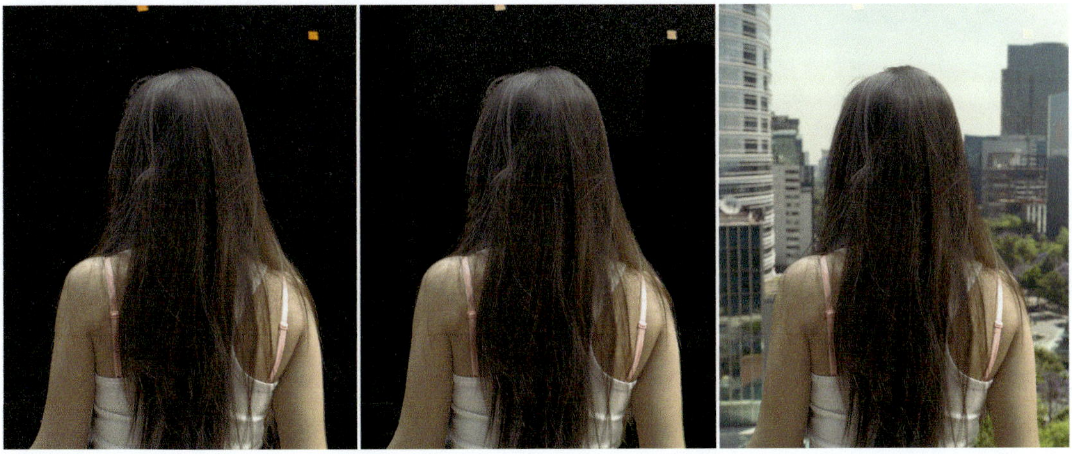

Nuke's IBK extraction algorithm does not use pre-multiplication, and generates noticeable background noise (left). But it uses the background luminosity and hue (middle) to camouflage the noise for a clean background (right).

Footage courtesy of fxphd/Eduardo Abon.

While a core matte is easier to extract with most keying tools (even the elementary ones found in editing software), edge mattes are trickier, and require skill, effort, and good keying tools. Some extraction algorithms like Nuke's IBK are especially suited for edge mattes because they do not use pre-multiplication, which can shrink the matte and make the edges crunchy. Instead, they "camouflage" the noise and extraction artifacts outside the matte borders by matching their luminosity and hue to the background. This method is especially suitable for integrating semi-transparent edges, and dynamically adapts the color of the edges to changing colors in the background.

Edge Reconstruction

When an element is heavily motion-blurred or defocused, it is vital to preserve its soft edges. Failure to do so results in a visibly awkward mismatch between the element's blurry detail and its unnatural hard edges. In addition, the lack of soft motion-blurred edges causes fast-moving elements to strobe. The problem, however, is that extended areas of heavy motion blur or defocus are so transparent and wispy that they contain too much of the background. This is of course undesirable with a green or blue screen, and especially tricky if there's detail in the background. The latter situation is a typical issue with rotoscoping: attempts to include motion-blurred and defocused edges within the roto shape generally fail, because there is no way to separate the background from the element in such semi-transparent areas. For that reason, enabling automatic motion blur in the roto shape usually does not provide good results.

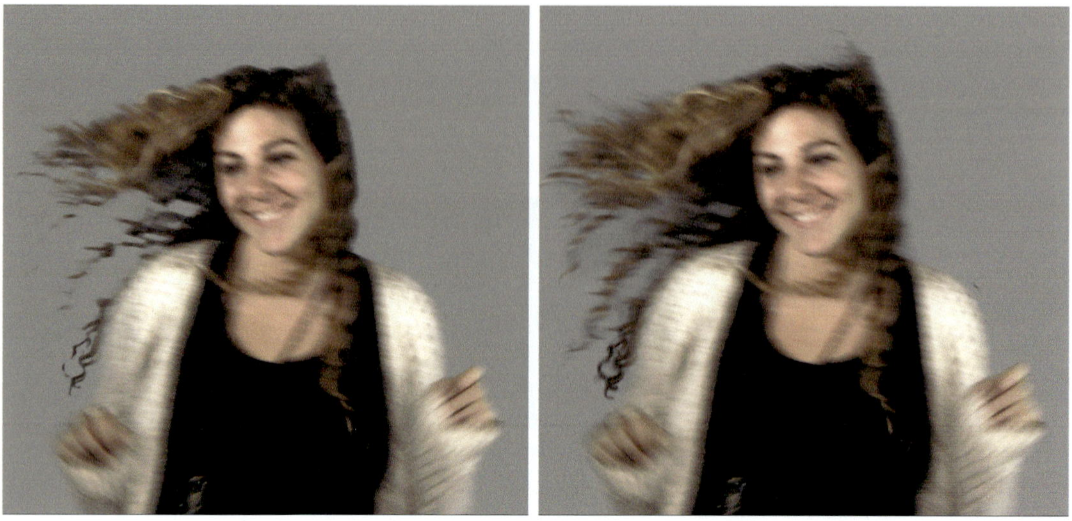

Initial extraction fails to pull the fine motion-blurred hair edges. With some additional edge reconstruction, they feel softer and more natural.

Often, the best approach is to not even try to extract such areas, but instead to recreate them artificially. With this method, roto shapes or extractions follow the last solid pixels along the edges to create a tighter matte. Then a copy of the extracted element is dilated, blurred, and placed underneath itself, to replicate the missing soft edges. Directional blur (or vector blur) can simulate the typical streaking effect of motion blur. Using the original footage as reference, the compositor can recreate soft and semi-transparent edges through a clean process that does not carry any of the original background and blends seamlessly with the new environment.

Matte-Less Extractions

The notion of compositing an element without a proper alpha matte may sound odd at first, but think of it this way: it's really all about color. When you composite an element over a background you are in fact changing the color of the background pixels. The matte is merely used to define where those color change happens, not to delineate any physical "thickness". But when it comes to thin and wispy sub-pixel detail like hair, fur, or grass, standard extraction techniques often result in a chunky, sticky look. Instead of containing the color channels within an area defined by an alpha channel, it is sometimes better to modulate the background color with the foreground color directly, using either an additive (brightening) or a multiply (darkening) operation. This method requires enough contrast between the original background and the detail you want to preserve. With dark hair over a bright green screen, the color channel with the most contrast should be isolated, then the whites should be brightened just enough so that non-hair pixels reach a value of one (and are clamped to one to prevent artifacts). When the target background is multiplied by this image, the darker hair values darken the background, while white areas (no hair) do not affect it. Sub-pixel detail can thus be added to the background, bypassing the normal chroma extraction and pre-multiplied alpha channel. The method for bright blond hair over a darker blue screen is similar, but this time non-hair areas are darkened down to zero, and the image is then added to the background, brightening only areas where hair exists. These methods are normally used as a complementary process to standard keying and roto, but their contribution to the photorealism of the comp is often substantial, as they help mitigate the cutout "Photoshopped" look by bringing back the very fine detail.

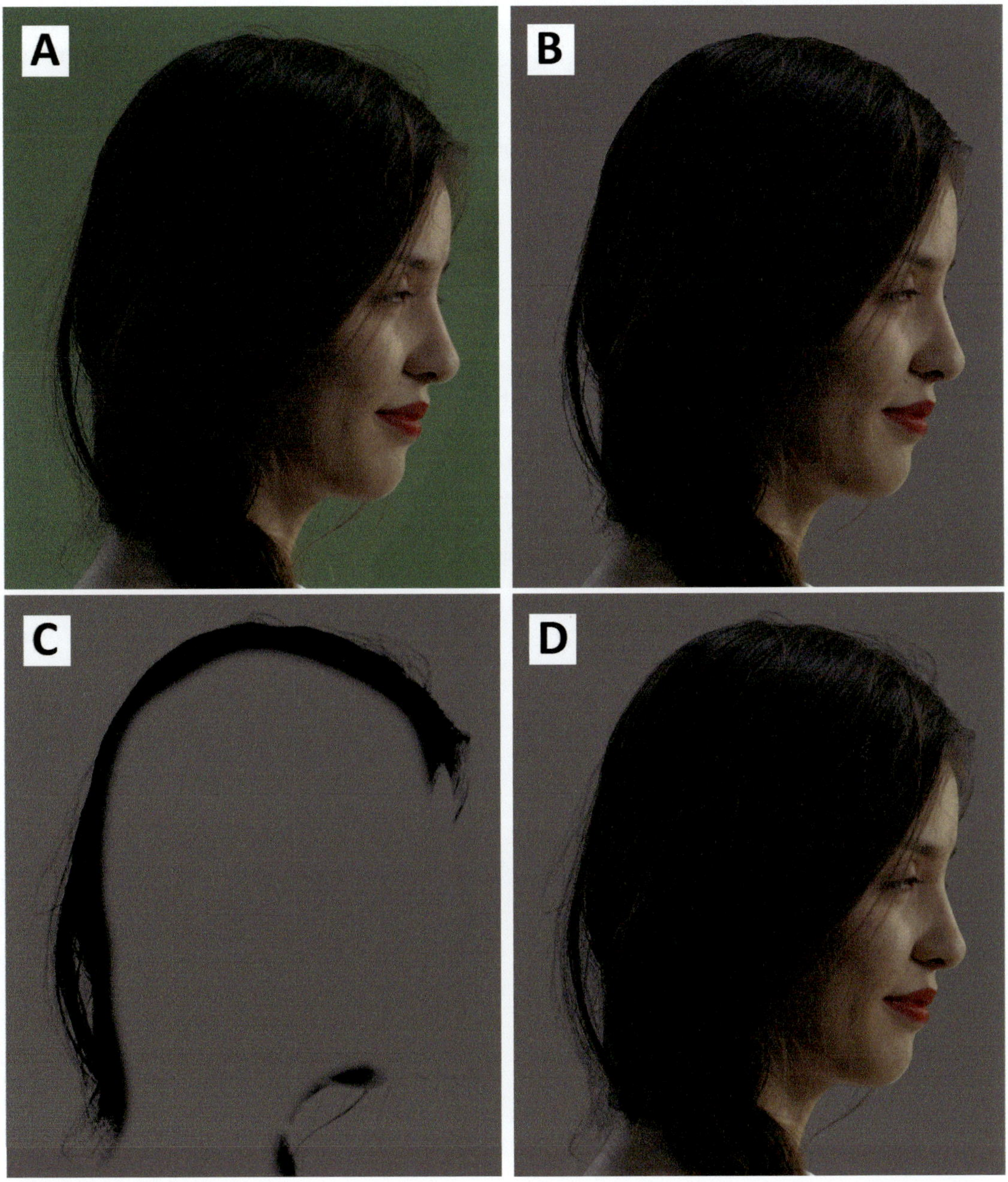

The wispier parts of the hair (A) are lost in the extraction process (B). Instead of pulling a hair matte, the background is multiplied by the darker hair (D), and the extraction is comped on top, with the detail now preserved (D).

Footage courtesy of fxphd/Eduardo Abon.

Edge Blur and Edge Color

A slight blur to the edges of an extracted element usually helps alleviate noise and improve the blending. An edge blur affects both the alpha and color channels, which softens harsh transitions and adds a photographic touch (lenses tend to slightly soften boundaries). Edge blur is most effective when used very subtly – depending on the resolution of the image, a blur value of one to three pixels is usually enough (higher values are useful if the element is out of focus). It is helpful to reference the footage for the amount of softness of real edges at a comparable focus level. Since an edge blur operation uses a mathematical edge-detect algorithm, some edge blur tools also offer edge color tweaks. Nuke's edge blur filter, for example, includes brightness and tint controls, which can be used as an effective last-resort remedy for bright or dark edges. Because edge blurring affects the color channels, it also affects grain. Missing grain along the blurred edges can call unwanted attention, but this issue can be easily fixed by using the edge-detect matte to add back grain only where it is missing.

Light Wrap

When a subject is located in front of a light source or an area of substantial brightness (such as a window), a certain amount of light bleeds over the subject's edges. Light warp is unfortunately often misused by compositors as a remedy for bad edges. It is important to remember that in most cases, objects display a rather solid silhouette, even when backlit. Light bleeding is limited to extremely bright areas (mostly light sources). Exaggerated light warp can result in unnaturally fuzzy, ghost-like edges. Light warp is most effective and realistic when used sparingly and always in direct relation to areas of strong luminance behind the subject.

Chapter 16

Integrating CG Elements

In VFX, CG elements are usually composited into live-action footage in VFX. Similarly, architectural and product visualizations often combine CG elements with background photos or footage. Even when perfectly textured and lit, CG elements rarely drop right in the footage without the need for some fine-tuning. The compositing stage is where this happens, and in this chapter we will look at the most essential tools and techniques for integrating 3D with photographic footage.

Compositing with Render Passes

For photographers, the ability to separate a photo into different lighting passes or various components would have been a truly astonishing concept. Just imagine the incredible breadth of post-processing possibilities that photographers and cinematographers could have had, if only they could separate their footage into direct and indirect light, highlights, shadows, transparency, refraction and depth (or even by individual lights). While this is not (yet) possible in the photographic realm, it is practical reality in the CG world. Indeed, render passes are an integral part of the VFX (and visualization) workflow, because they provide a flexible, fast, and interactive way to adjust, improve, and integrate CG elements at the compositing stage, without the need to go back to time-consuming 3D rendering. Even the most perfectly lit renders need some adjustments at the finishing stage, and good strategies for using render passes help pave the way to photoreal results. Render passes (also called **AOVs**, **render elements**, and **render channels**) can be output as separate stills or image sequences, or combined into a single multi-channel EXR image or image sequence. Generally speaking, render passes fall into two main categories: **lighting passes** and **utility passes**.

Lighting Passes

The final, complete rendered image is typically called the **beauty pass** (or just "beauty"). That image encompasses the full accumulation of all light interactions in the scene. Those interactions can be separated and output as distinct lighting passes, each representing a single component of the beauty render: diffuse, specular, transmission, subsurface scattering, and so forth. In addition, the influence of different lights in the scene can be separated into specific lighting layer passes. Most rendering applications provide the option to separate the direct and indirect aspects of

DOI: 10.4324/9781003590293-21

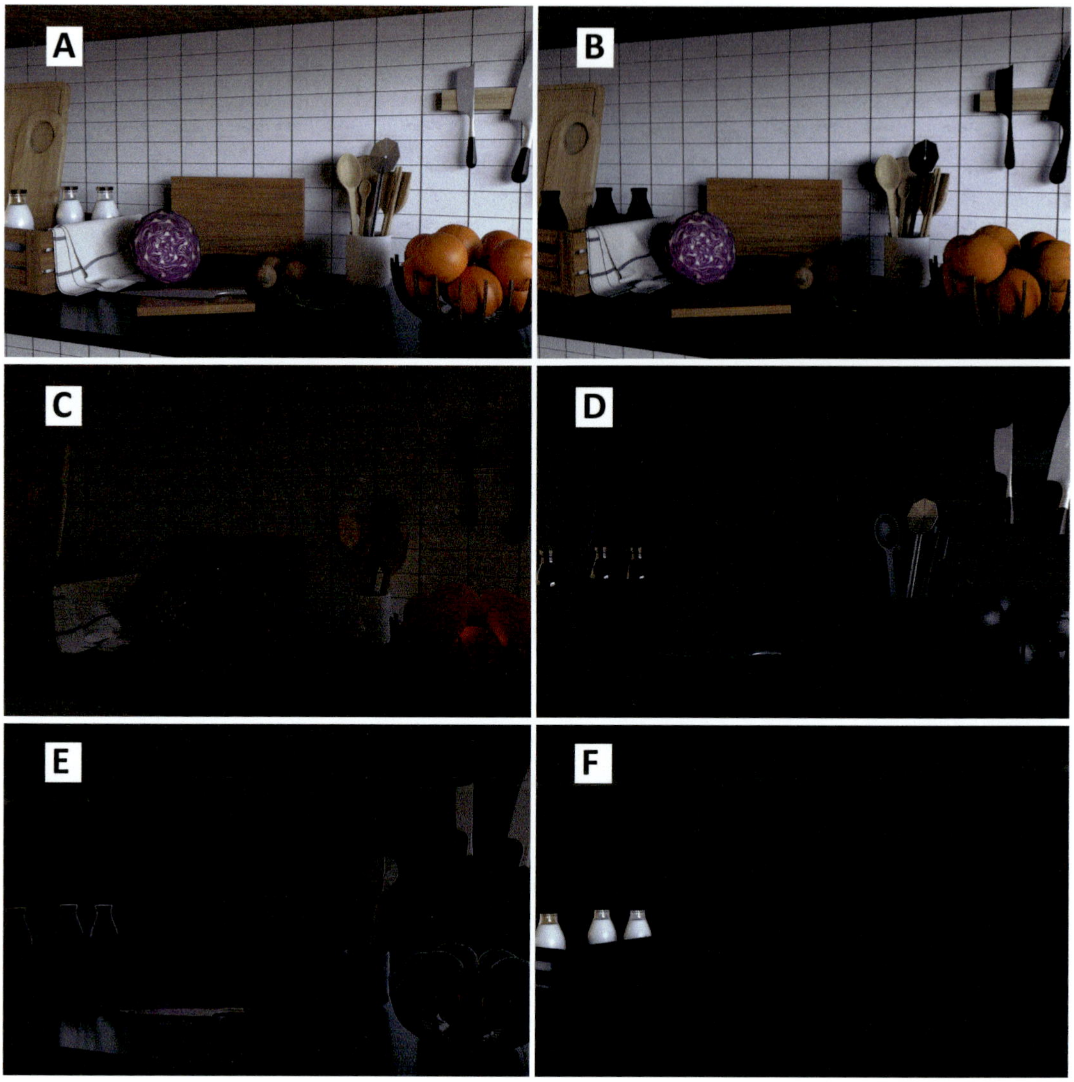

A: beauty, B: direct diffuse (notice how metallic surfaces turn black), C: indirect diffuse (notice the strong color bleeding between the oranges), D: direct specular (showing mostly the reflection of the window light), E: indirect specular (showing object inter-reflectivity), F: transmission (the milk bottles are the only transparent objects).

some components: a direct diffuse pass includes only the illumination of light sources, while indirect diffuse shows only illumination from reflected light. A direct specular pass shows the reflection of light sources (including sky and environment domes, which are considered lights), while indirect specular shows the reflection between objects. This type of separation is of course completely artificial – in the real world, there is no difference between reflecting the sky or a nearby tree – but it is nonetheless extremely helpful for achieving better real-world emulation.

There are two basic approaches for working with lighting passes in comp. The first is to start with the separate passes, then re-combine them by adding their values together (since light components accumulate, an additive operation of all the passes produces an exact replica of the beauty render). Like an audio mixing console, this approach gives the compositor full control over each component, its color and its relative amount in the "mix" and is suitable for situations where the original beauty needs to be substantially tweaked and modified. However, recreating the beauty by separating and recombining every single pass can be an overkill in situations where only minor tweaks are needed. In such cases, it is simpler to start with the beauty, and add (or reduce) only specific components. For example, increasing the specular level can be done by simply adding a bit of the direct specular pass to the beauty using an additive (plus) operation, while reducing specularity can be done with a difference operation.

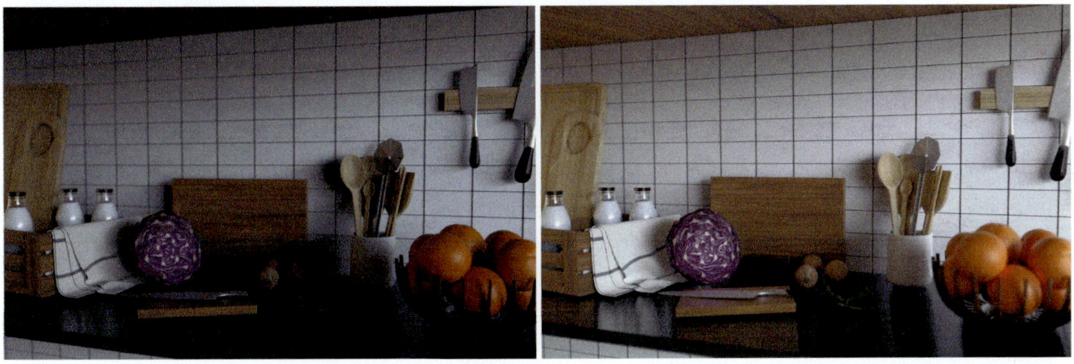

The raw render (left) is somewhat bland and murky, but by boosting the indirect light components, the lighting feels more natural and the image looks more pleasing (right).

Regardless of the approach used, it is important to remember that changing the balance between lighting components may affect the energy conservation principle of PBR. That is not necessarily bad – the goal is to make the image look convincing and sit well in the comp, even if it means breaking physical rules. But it is helpful to keep this in mind, especially with heavy modifications. I have often found myself tweaking render components ad nauseam, only to go back to the beauty at the end. Sometimes the well-balanced results of PBR work best unaltered.

In most cases, the use of the basic lighting passes is a straight-forward process of balancing the look of the rendered element: reducing the specular highlights a bit, tweaking the color of the diffuse refection, etc. Here are some tips for additional uses of lighting passes:

- Increasing the indirect diffuse presence produces a "poor man's subsurface scattering" effect, by reducing the contrast and softening the object's features. It can be used to subtly take the edge off of elements that feel too harsh and "CG".
- Although technically incorrect, boosting the indirect specular component is a sort of a "cheat" to make a surface feel more metallic, because it enhances the illusion of increased reflections at grazing angle.
- As mentioned in Chapter 11, indirect components (like indirect diffuse or indirect specular) are usually the main culprits in noisy renders. When re-rendering at higher sample settings is not an option, one way to reduce the noisiness is to separate the indirect passes, run them through a denoise or degrain algorithm and then plug them back in comp.
- Transparent objects are usually rendered with the intended background included for proper transmission and refraction. The problem is that at the compositing stage, the colors of the background may change (subtly or drastically), in which case the rendered

transparency will no longer match. By using the transmission pass, the compositor can adjust the transparency to match the modified background without affecting the other lighting components and surface properties.

- Separate per-pass color correction is an extremely effective way of grading CG elements in comp without generating a blanket effect on all the components that can hurt photo-realism. The classic example is changing the diffuse color without affecting the specular color. As explained in Chapter 9, specular reflections are not tinted in dielectric materials, so keeping the specular component unaffected while grading the diffuse maintains the correct look of the material.

Utility Passes

Unlike lighting passes, utility render passes do not represent any of the visual content of the beauty pass. Rather, the pixel data of utility passes provides additional information that can be used to manipulate the rendered image or diagnose render issues. The most common utility passes are **Z depth** and **facing normal**, but different render engines allow for a wide variety of utility passes, including fully customizable ones. Utility passes are indispensable tools for compositing CG elements and enhancing photorealism. Here is a brief look at the role of the most common ones:

SHADOW MATTE AND AMBIENT OCCLUSION

While shadows are calculated as part of the rendering process, they only apply to the existing 3D geometry in the scene. Rendered elements that are composited into footage need to interact with the 2D environment in the footage, and shadows are a primary factor in that interaction. This is where shadow mattes and ambient occlusion (AO) passes come in handy – they help bridge the gap between 3D rendering and 2D compositing. For example, if you need to composite a CG object on a photo of a room, you must create the illusion that the chair is actually standing on the (photographed) floor. This requires contact shadows underneath the chair's legs. Depending on the room's lighting, the chair may also need to cast shadows on the floor (and possibly on other elements in the footage).

To create these shadows, the CG artist needs to model the interacting elements in the footage (like the floor and nearby objects) as accurately as possible. These are added to the rendering scene, but only as shadow casting/receiving objects. They are not visible in the beauty render, but they are active in the AO and shadow matte passes. Ambient occlusion is perfect for contact and proximity shadows because it is completely independent of the actual lights in the scene. By multiplying the background image/footage with the AO pass in comp, the areas of contact or proximity (which have values lower than 1) darken the image, while non-affected areas (fully white at 1) remain unchanged. When the CG element is comped over, all areas of occlusion are properly darkened underneath (or behind) the element.

When the scene includes lights that also cast defined shadows, a shadow matte pass is used in addition to the ambient occlusion pass. Unlike an AO pass, a shadows matte pass is derived directly from the specific lighting of the CG scene, and represents the softness and intensity of each shadow in relation to the occluded light. This allows the compositor to set a global shadow grade for the scene, and then use the shadow matte as a mask for applying this grade to the footage, resulting in shadows that have the correct intensity and softness.

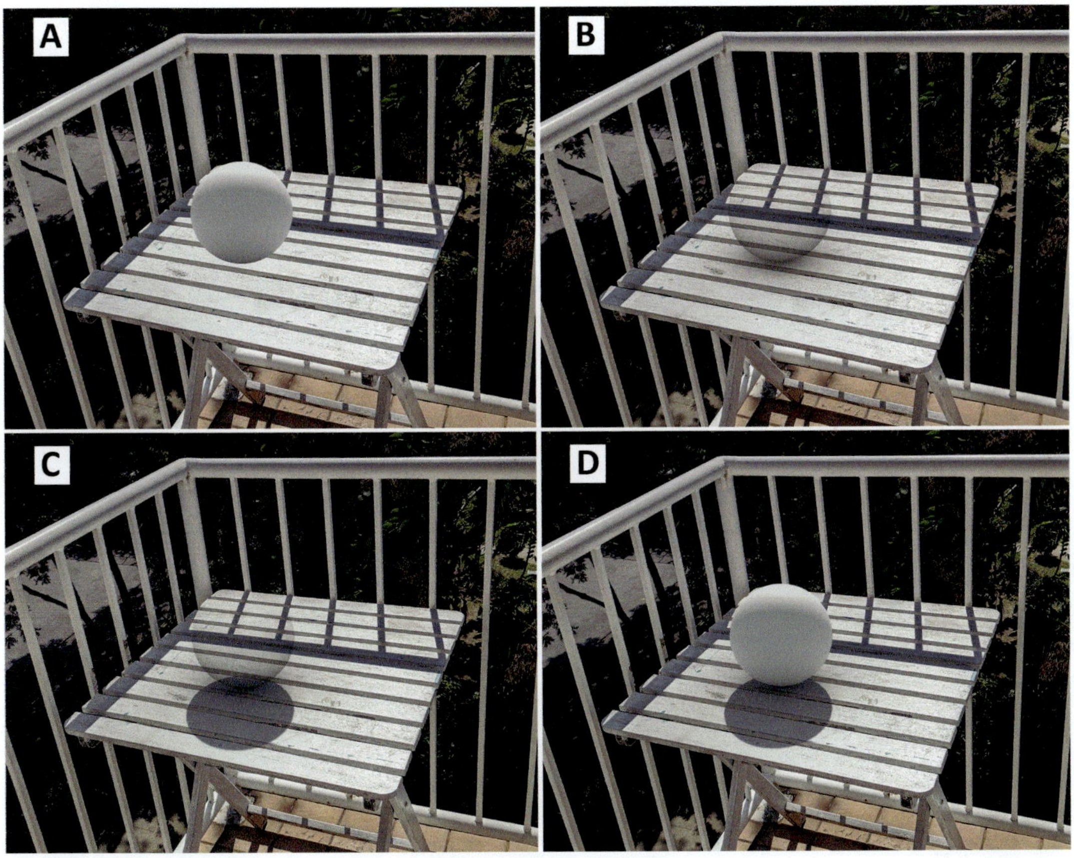

A: a CG sphere is composited over a photographed table, but is clearly missing a shadow and contact shadows. B: an AO pass is multiplied with the image to darken the areas of contact. C: a shadow matte pass is used to create the shadow through color grading, D: the sphere is added back.

POINT POSITION AND NORMAL PASSES

These two standard utility passes are used as 3D to 2D translators, and together they form an accurate two-dimensional map of the three-dimensional geometry. The **Point Position pass** (aka **Position pass** or **P pass**) denotes the location of every surface point in actual 3D space, using the RGB channels to represent the XYZ position values. Because the value range is huge and precision is paramount, position passes only work in a 16- or 32-bits floating point format such as EXR. The **Facing Normal pass** (aka **Normal pass** or **N pass**) shows the facing direction of each point in the 3D geometry. Like the position pass, it uses the RGB channels, but with a range of −1 to 1. For example, in a CG model of a room, the facing values of any point on the floor would be 0,1,0 (the floor surface is facing straight up on the Y axis), while a point on the right side wall would be −1,0,0 (facing left). Note that the facing direction is in relation to the world 3D coordinates, not the camera.

Modern compositing applications offer the ability to work in 3D space, but they are not built to handle heavy geometry sets. The position pass is used together to create a 3D **dense point cloud** that accurately represents the scene's geometry within the compositing application, directly from the rendered image and without the need to actually import any geometry. This provides an extremely useful reference for compositing additional elements in 3D space. For example, imagine a shot with a CG character running on a dirt road, where one of the compositor's tasks is to add 2D elements of dust as it is kicked off by the character's feet. In a 2D only workflow, the compositor can only guess where each foot lands when trying to place and track the elements. But with an animated point cloud derived from the position pass, the compositor can clearly see where the feet land in 3D space and place the 2D dust elements on cards in 3D space to match those positions.

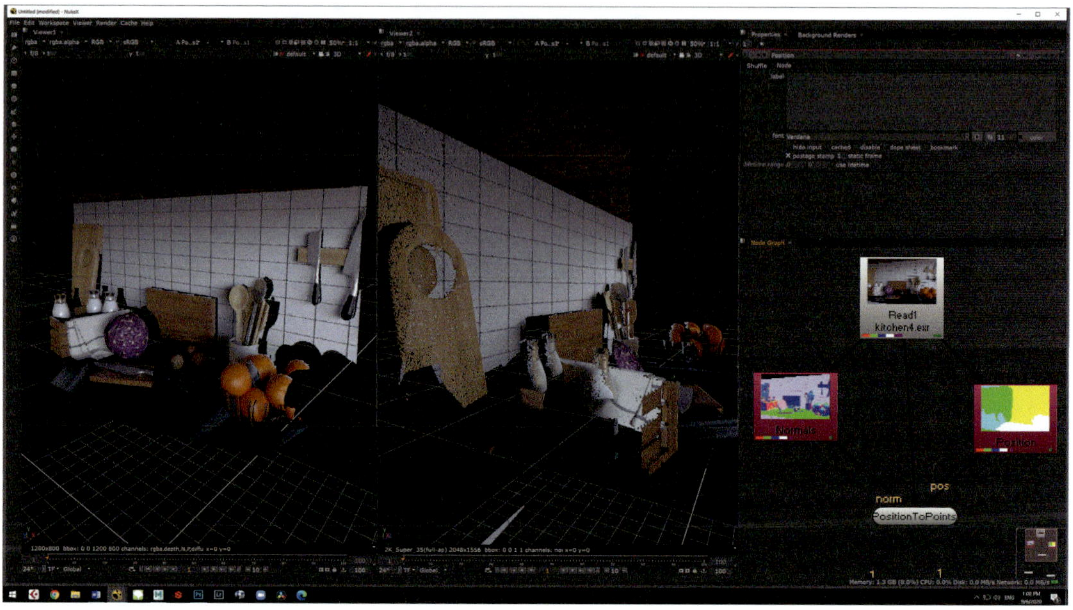

A Nuke screen grab, showing how the position pass is used in to generate a dense point cloud of the 3D scene. The normal pass is also factored in, to enable relighting in Nuke.

Position and normal passes can also be used for projecting textures onto rendered geometry at the comp stage, or to make three-dimensional selections within the render. Some compositing applications even offer relighting tools that are solely based on the information of the position and normal passes. This hybrid method requires placing CG lights in 3D space and using the compositing software's own 3D renderer. As a cheap and quick 2D alternative, the normal pass can be used to increase or reduce the light in the rendered images based on direction. For example, to increase top lighting overall, the Y channel of the normal pass can be used as a mask for brightening only up-facing areas in the render.

MATTE AND ID PASSES

One of the clearest advantages of working with CG renders is the ease of isolating specific elements or areas for separate treatment. Mattes can be generated at render time for individual objects (or a combination of objects), as well as materials or shaders. Colored mattes only work well as pure red, green, or blue, because mixed colors cannot be accurately isolated in compositing. A single matte pass can therefore have up to three separate mattes, one in each RGB channel. However, new tools like the excellent Cryptomatte plugin make the process of rendering mattes and selecting them in comp much more efficient. Cryptomatte automatically assigns a unique ID to every object and/or every material in the scene (along with a unique color), which allow the compositor to easily isolate any object or material (or any combination thereof) with a simple point-and-click process, and eliminates the need to render multiple matte passes.

Z DEPTH

Z depth is an important utility pass, as it serves two essential purposes: applying accurate depth of field in 2D and applying depth-based atmospheric effects (see Chapters 17 and 18). Despite its important role in compositing CG elements, the depth pass has some notable limitations:

- Most current render engines output a non-normalized depth pass, which can represent any distance to infinity. While this is generally advantageous, it does pause some difficulties when a non-normalized depth pass is used as a mask for atmospheric depth effects (especially when dealing with small distance differences) and may force the compositor to increase the contrast in the depth pass by pushing the black and white points.
- Transmission/refraction cannot be represented in a depth pass. All transparent or semi-transparent surfaces appear as fully opaque, and any see-through element is not visible in the depth pass. This may cause serious issues with refracted objects. For example, an out-of-focus element may suddenly jump into focus when it moves behind a glass object, because at that moment it simply disappears form the depth pass.
- Volumetric CG elements such as smoke and fog are similarly problematic, since the depth pass cannot accurately show multiple layers of semi-transparent pixels (some advanced render engines provide volumetric depth options, though these are not ideal).
- Reflected elements are not represented, so it is impossible, for example, to use the depth pass to focus on a figure reflected in a mirror (only the mirror itself is visible in the depth pass).

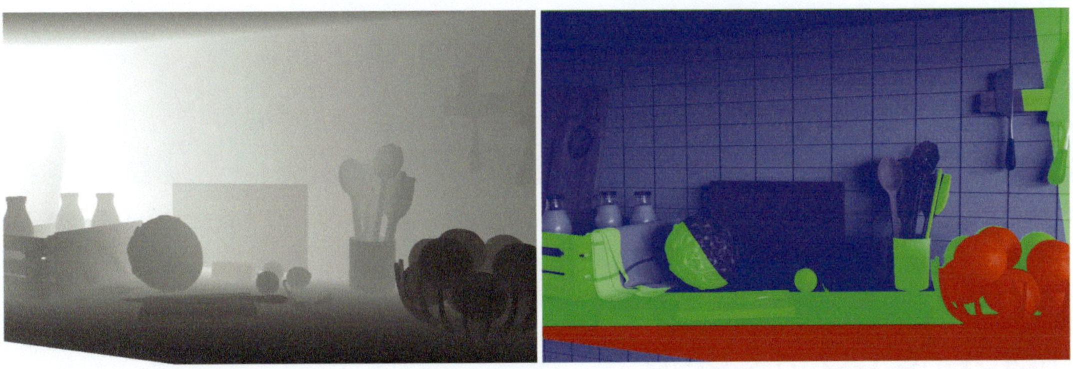

Left: a non-normalized depth pass (color-corrected to show the values). Right: Z-defocus in Nuke, displaying the depth of field settings (green is in focus, red and blue out of focus).

Deep Compositing

Most renderers can output **deep data** in a deepEXR format. Unlike the standard depth pass, which can only represent surfaces that are visible to the camera, deep data provides an array of multiple values per pixel at varying depths, and thus includes surfaces that are hidden behind foreground objects. Deep compositing provides an entirely new level of functionality because the rendered elements are treated as three-dimensional entities rather than 2D layers. When you layer standard 2D elements, the order of the layering defines which element goes in front. With deep compositing, the layering order does not matter at all, because the elements are automatically arranged at the correct depth in 3D space. Standard 2D operations like transform, crop and even color correction can be done along the Z axis in addition to X and Y. Deep compositing allows true 3D depth-based layering which is especially useful with complex scattered elements like particles or volumetric smoke, because the deep data automatically sorts which parts of one element go in front or behind other elements in the composite. Deep compositing can significantly enhance the process of compositing complex CG scenes, but because deep data can only be generated through a 3D rendering process (not live-action footage), and deepEXR file sizes are substantially larger than regular EXRs, it is not used as frequently in bread-and-butter VFX compositing work.

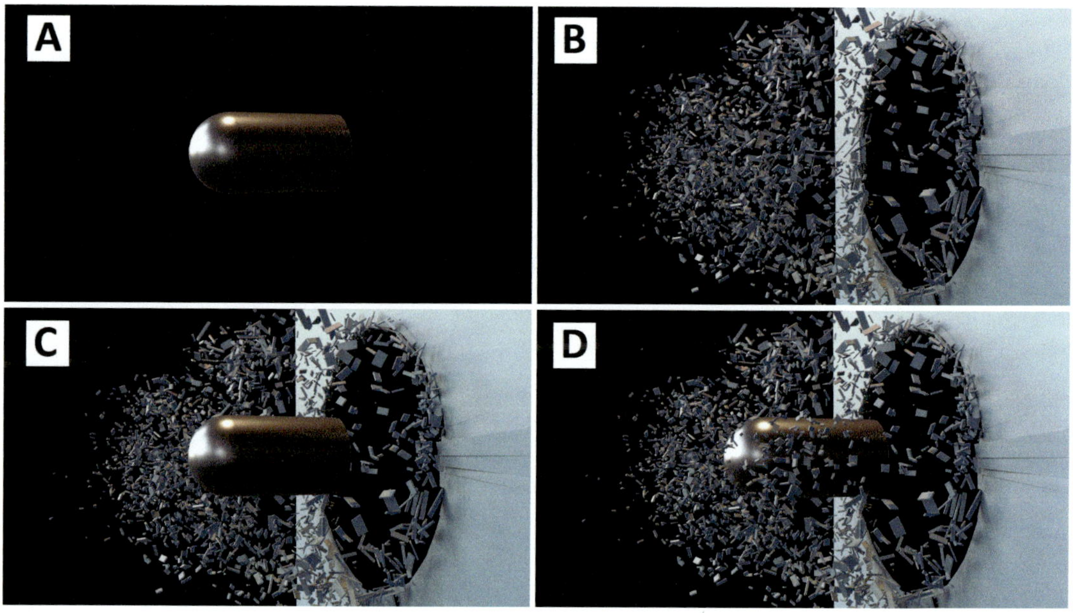

When the bullet element (A) is composited over the shattered surface element (B), the bullet is simply on top. But with deep compositing, the bullet is automatically placed in the correct depth.

Improving CG in Comp

Computer-generated imagery is purely mathematic, and thus inherently perfect. Such perfection often stands in the way of photorealism. In Part 3 we discussed ways of adding complexity and randomness in the modeling, texturing, and shading processes. Such methods are of course necessary to create plausible CG scenes. But the 2D process provides an opportunity for additional treatments that can reduce the "curse of perfection" of CG renders and help their integration with live-action footage. One thing to consider is the fact that CG is rendered through a virtual camera, not a physical one. The light that hits the virtual camera's "sensor" does not travel through layers of glass. Emulating specific lens characteristics like flares, defocus, or chromatic aberrations will be discussed in Chapter 18. In the following short section, I will offer some simple, basic "CG sweetening" tips that can work for almost any scenario.

Render Resolution

The logical assumption is that the render resolution needs to match the final output or source footage resolution. Sometimes it is necessary to render at the full resolution (or even higher). For example, if a scene contains a lot of thin straight lines (like the wires of a suspension bridge), high resolution is necessary to avoid jagged, aliased edges. But in many cases, rendering at a lower resolution is not only a time-saver, it actually helps the CG look more realistic. CG renders are almost always too sharp compared to photographed footage, and very often need some softening to sit well in the comp. Rendering at 80% or 75% of the full resolution, and then resizing back to 100% with a cubic or quadratic filter produces the same effect as softening, with the additional bonus of cutting down render times and file size. The exact render resolution reduction can be decided on a per-case basis, by rendering a test frame at full resolution and seeing how much softening is needed in comp, then reducing the render size to match.

Hazing Effect

This effect has nothing to do with depth-based atmospheric hazing (to be covered in Chapter 17). Rather, I am referring here to a very subtle, almost imperceptible effect. It is easily achieved by blurring the CG element a little (10–40 pixels, depending on the resolution), and then merging back that blurred version of the element over itself, while setting the mix (opacity) just high enough to slightly soften/smear the features of the CG render. It is a very simple yet effective trick to take the edge off of CG, as long as you keep the mix relatively low and avoid a cheesy "dreamy photo" look.

Highlights Bloom

As described in Chapter 10, highlights blooming is essentially a "mini-flare" that happens when very bright areas scatter slightly in the lens. The procedure is essentially the same as the hazing effect, but a max (maximum) operation is used instead of over or screen. This ensures that only the brightest areas in the render get hazed, mimicking real-world highlights bloom. This too works best as a subtle effect – lens manufacturers strive to minimize lens scattering, so overdone blooms tend to look distracting and unnatural.

Albedo Pass and Contrast

One recurring issue with CG is that the render looks terrific on its own, but somehow does not feel quite right in the context of the comp. Contrast, in particular, is elusive. Renders may look like they have just the right amount of contrast when viewed over clean black, but often feel much too punchy when integrated with the target footage. There are of course ways to adjust contrast through color operations like lift and gamma, but it is tricky to adjust lighting-related color without affecting texture-related color. In other words, flattening the contrast caused by the lighting also flattens the contract in the textures. The diffuse albedo pass is useful for this purpose because it contains all the texture color information minus the lighting and shading – in essence, it is a flattened version of the render. Blending the beauty with a bit of the albedo pass is an effective way of reducing light-shadow contrast without flattening any of the texture color detail.

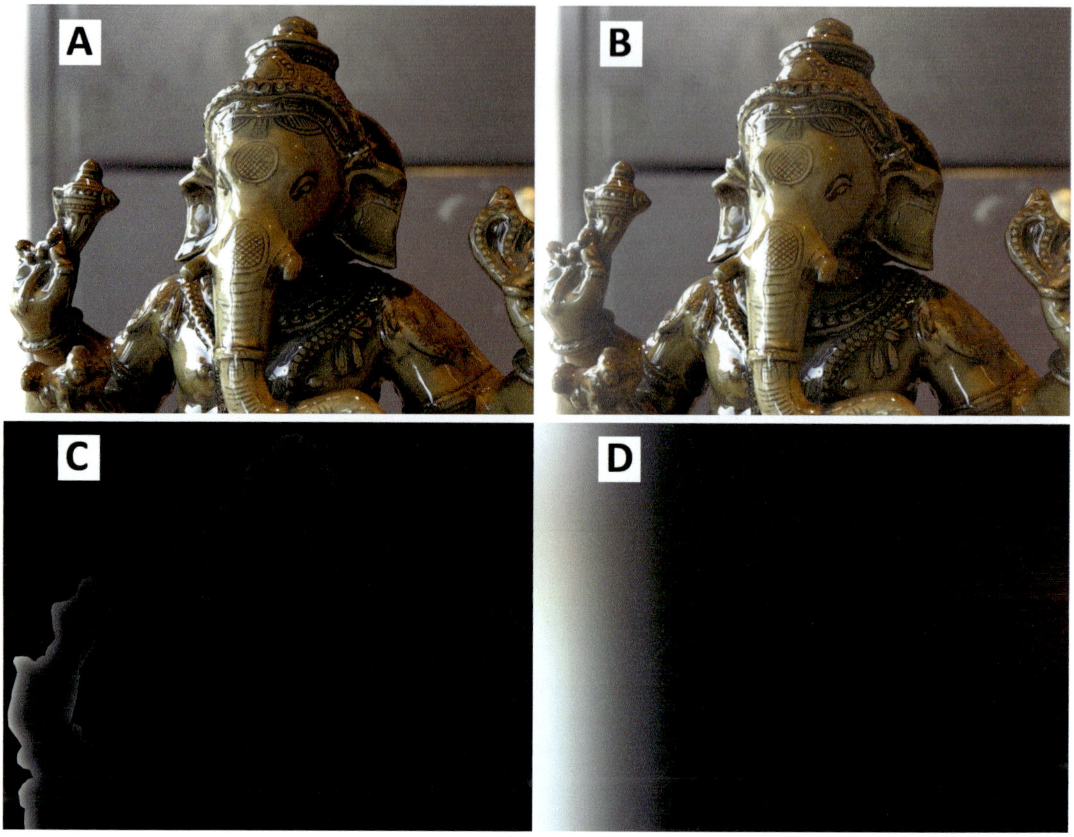

The CG render feels a touch too crunchy and clean over the background image (A). Adding some light wrap (C) and some light bleed/flaring (D), while also mixing in a bit of the diffuse albedo pass and adding subtle bloom, helps the element "sit" better in the comp (B).

Edge Treatment

CG elements are normally rendered over a black background (or over the target footage), and are thus free of the issues associated with extracted or rotoscoped elements, such as bright/dark edges, and green or noisy fringes. The relatively pain-free process of adding CG element to footage, however, does not mean that edge treatment is not necessary. Overly sharp, clean, or consistent edges and boundaries are often tell-tale signs of CG. Successful integration therefore requires one or more of the following treatments: subtle edge blur, blooms on rim highlights (making sure the blooming bleeds into the background), light warp, chromatic aberration (see Chapter 18), and, in some cases, breaking the edge consistency by subtly modulating the alpha with noise. Like everything else described in this section, subtle touches are usually more effective than heavy treatment – good CG renders usually require minimum comp intervention, just enough to break the curse of perfection.

Case Study: Hybrid 2D/3D Matte Painting

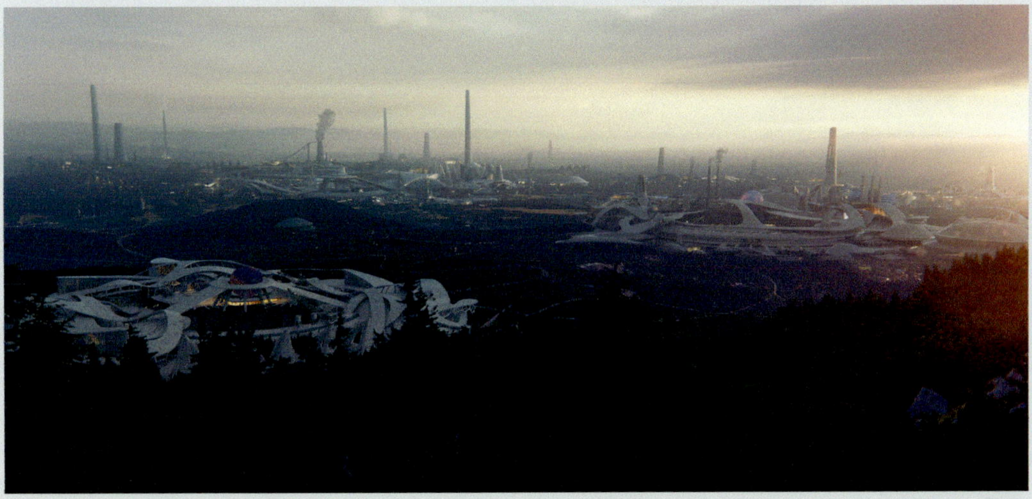

Ekio Utopia by Irena Šmitáková.

Ekio Utopia is a great example for a successful blend of 3D elements and 2D work. Matte painting supervisor Irena Šmitáková describes her process in creating this artwork: "the inspiration came to me during a trip to the Krkonoše mountains. As I stood in the valley, the beautiful landscape made me wonder: "what will this valley look like in the future?" My creative process started with extensive research into visual references. I gathered images of futuristic sci-fi facilities, industrial utopias, and aerial cityscapes to inspire the look and feel of the piece. Next, I organized my ideas using EdrawMind (formerly MindMap). Once I had a solid concept, I revisited the photos I took during that trip and selected a specific shot of the valley to serve as the foundation for the artwork."

"To better visualize the scene in 3D, I created a quick sketch in Photoshop", continues Irena," While I sometimes sketch on paper, I find Photoshop more flexible for this kind of work. After finalizing the sketch, I began blocking out the 3D models. I used a KitBash kit for various 3D pieces but modelled the main base from scratch. I layered, scaled, and rearranged individual pieces from the kit, often stumbling upon the best results by happy accident. I rendered the elements in Arnold, using my photo from the trip as an image plane for aligning the camera, lighting the scene with an HDRI and directional sunlight.

Once I was satisfied with the base render, I began the process of compositing. This involved masking the CG elements onto the original plate, removing unnecessary geometry, and integrating the roads into the landscape using a custom tree brush and alpha masking.

I also applied basic color grades to match the CG render and the photograph, ensuring a seamless blend. I used Cryptomatte to selectively adjust the lighting and shadows on each element, and painted in additional detail like interior lights, window glows, and subtle architectural features.

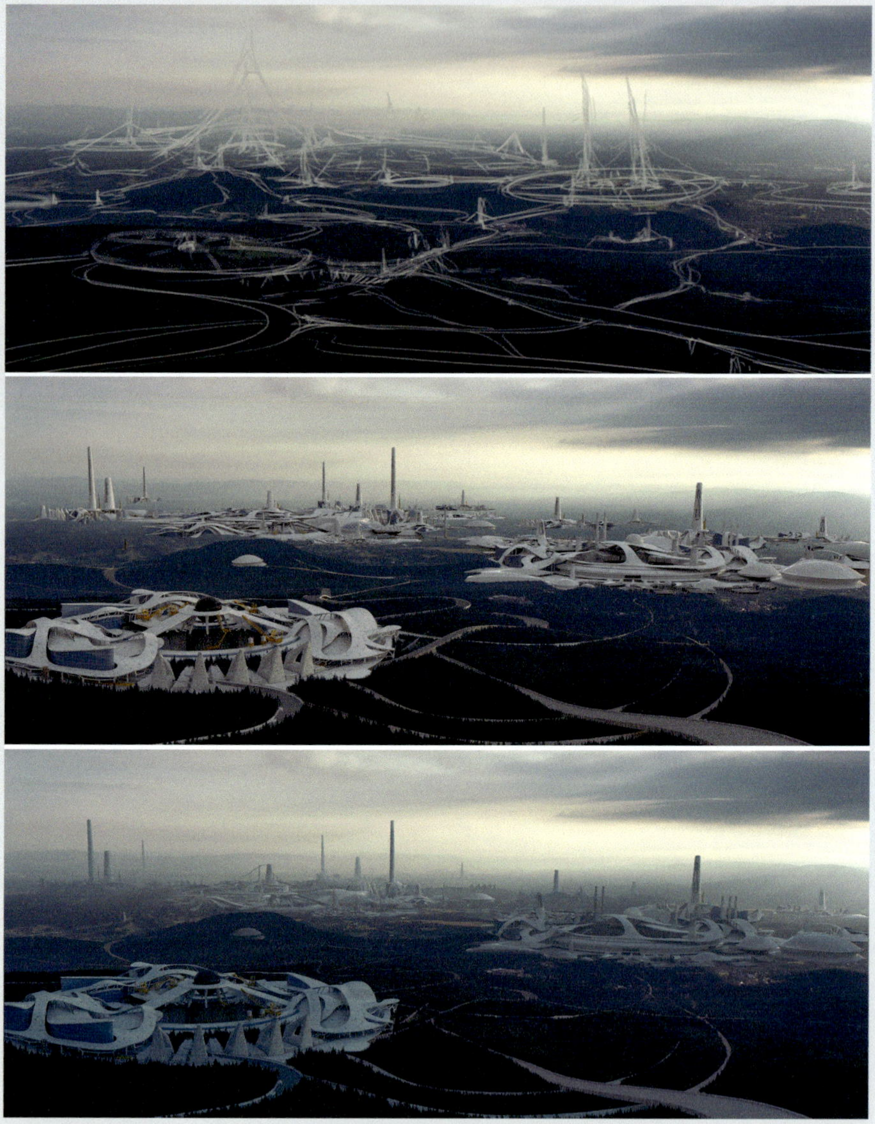

Three stages in a matte painting process: initial sketch (top), layout (middle), integration (bottom). © Irena Šmitáková.

To complete the scene, I added an extra layer of foreground trees, refined the existing trees in the front, and applied additional grades to convey a sunrise atmosphere. I also enhanced the sunlight on some of the buildings to generate more visual interest, added more atmosphere into the far background for depth, and added the sun flare and smoke elements. Finally, I added lens distortion and some grain to help merge all the elements. My last step is always the same: I step away from the artwork for a while and then return with fresh eyes for a final color grade in Photoshop or Lightroom, concludes Irena."

Chapter 17

Lighting in 2D

Relighting with Color

When working in 2D, artists have none of the convenience of CG lighting: they can't place lights at selected locations in the scene and control their intensity and color; There are no shaders or surfaces that can be set to react to light in specific ways; Accurate shadows and reflections are not generated automatically; And perhaps most importantly, lighting is already pre-baked in the image, and often needs to be undone before it can be modified – a challenging, sometimes even impossible, task. Just like a traditional painter, the only way a digital artist can create or change lighting in 2D is through the selective use of color. In matte painting, traditional techniques are used to create lighting by coloring areas of light and shadows, and in compositing, the paint brush is replaced with animated roto shapes when coloring is applied to moving and shifting elements.

The pre-baked lighting in the source material is a defining factor on how effectively that material can be relit in 2D. If an element is shot in direct sunlight, changing the direction or intensity of the light requires removing the baked light and shadows first – a difficult, or sometimes impossible challenge. Even the highest quality imagery may not contain the kind of super-wide dynamic range that allows grading highlights down and lifting shadows up to such an extent. In many cases, there is little or no information in the highlights or deep shadows, and extreme grading results in clamped, flat areas, or ugly artifacts. It is of course best to source material with lighting that already matches the target environment. But if no such material can be found or used, it is much preferable to start off with overcast/flatly lit material. As in texturing and 3D scanning, 2D material lends itself best to relighting through color when it is captured under overcast sky, or is evenly lit by non-directional, diffused lighting.

The Challenge of Reflections

Up to this point, we approached specularity as a single type of light interaction, whether it appears as simple specs of highlights, or discrete mirror-like reflections. In 2D, this distinction does matter. Our visual perception is not particularly sensitive to the top of the color range and we are pretty lenient toward misplaced or inaccurate specular highlights. So, without the need to represent a specific reflected image, generic highlights can be painted in with relative ease – even a few scattered daubs are often all it takes to hint at reflectivity (this is clearly manifested in numerous

DOI: 10.4324/9781003590293-22

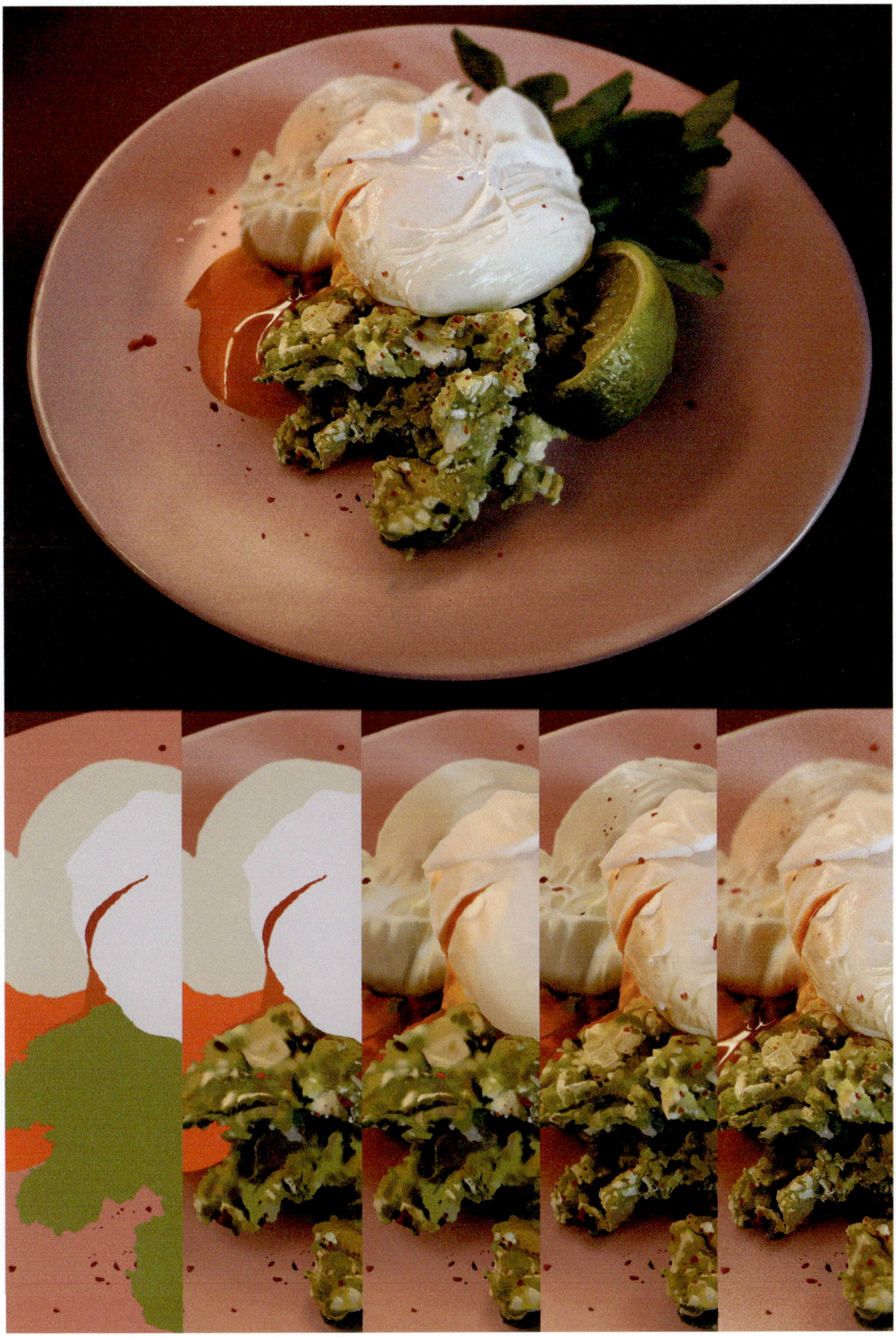

It took Luke Panayiotou a week to paint this image, using just a single brush in Photoshop. Every aspect of photorealism in this work is achieved solely through color.

traditional paintings). That said, film and video add the factor of motion – the effect of specularity is lost if the painted spots do not move in tandem with the movement of the object, light, or camera. Without proper movement, static highlights read as texture rather than specular reflection.

Clear reflections on smooth surfaces like windows, mirrors, water, or polished metal represent a much greater challenge in 2D and deserve a closer look, especially since "faking" reflections is a common compositing and matte painting task. This task can be tricky, mainly because what is visible in the reflection is not necessarily seen by the camera. In other words, **the reflected image represents the point of view of the mirror**, which can be different (or in fact the complete reverse) of the camera's point of view. Let us examine two VFX scenarios:

Scenario 1: a matte painting of a serene mountain lake. Such a setting obviously requires a reflection of the environment in the water. It seems like the reflection can be easily created by grabbing the background above the lake's distant shore and flipping it 180 degrees down. But is this really an accurate method? Is the reflection of the lake a simple flipped image of the environment? While the camera's line of sight is more or less horizontal, the lake is "seeing" the environment vertically from the bottom up. This 90-degree difference in POV hardly affects the appearance of the distant elements, but anything close to the water edge may look different in the reflection than in the camera. A deer standing on the lake shore is viewed face-on from the camera, but the reflection is "seeing" the deer from the bottom up, showing more of its underside. It is of course difficult to recreate this change in perspective in a 2D matte painting workflow without resorting to hand painting or 3D. Yet for most practical purposes, this level of accuracy is not needed. The viewer is unlikely to notice anything wrong in the reflection, and the simple flip can work just fine. The discrepancy becomes noticeable only in close-up views. Where there is distinct focus on elements close to the reflecting surface.

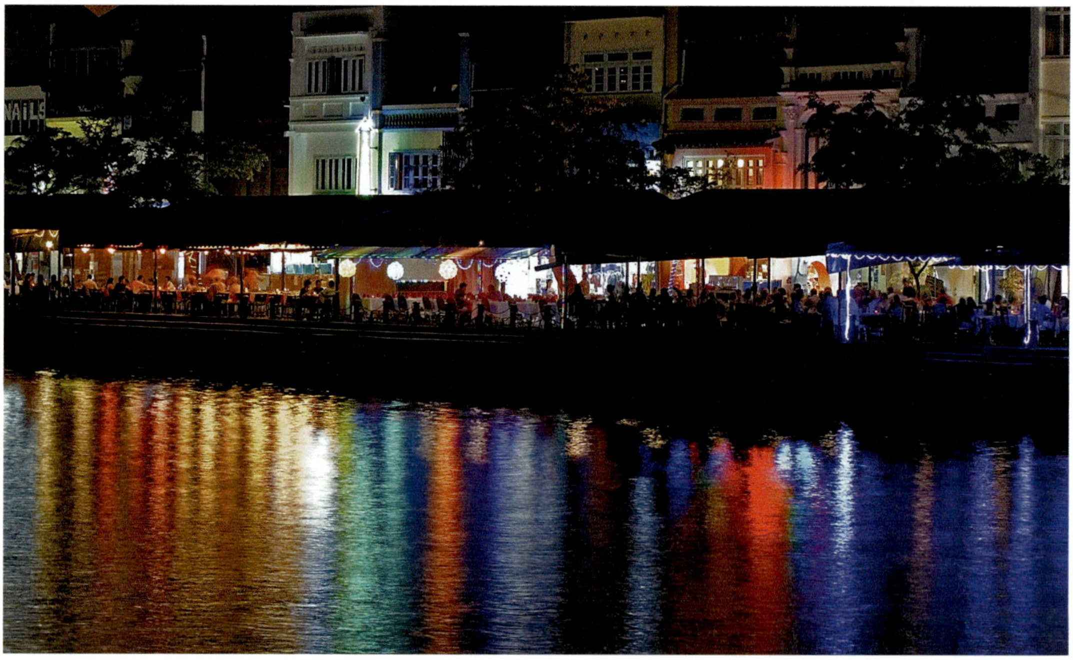

What the reflection "sees" is not necessarily what the camera sees. From the camera's point of view, most of the colorful lights are hidden behind the tarp roofs. But the water is "seeing" those lights fully from below, reflecting a view that is hidden to the camera.

Scenario 2: in an interior set on a soundstage, an over-the-shoulder shot of an actor standing by a window, looking at the "view" outside. A green screen is placed behind the window, to allow adding the exterior in post. While the actor is keyed out and the green screen replaced, a proper sense of window glass can only be achieved through reflections. The problem is that the camera is shooting the actor's **back**, while the mirror is reflecting the actor's **face**. In a traditional painting, the artist will simply paint the reflection. But in a photoreal VFX situation, the compositor must use real footage of the actor from the reflection's point of view.

The simplest solution would be to have a physical glass on the set window. Even though the green screen will be removed later and replaced by an exterior view, the compositor could still use luma-keying techniques to extract the reflection and screen it back on the added BG. This is a common practice, but not always feasible: often the glass reflects unwanted elements like the camera, the crew, or an unbuilt section of the set. Another option is to shoot a separate "witness camera" take from the direction of the window toward the actor's face. This ensures a clean footage but works only if the actor is not moving much (syncing could be a problem). The point here is that not every reflection can be created from scratch at the VFX phase. Sometimes the necessary material must be shot on the set or location.

Fresnel Effect

When adding reflections in 2D, the artist needs to be aware of the Fresnel effect discussed in Chapter 9, especially on non-metallic surfaces. The reflection should be stronger at shallow viewing angles, and gradually disappear as the viewing angle increases. In the lake example, the added 2D reflection would look rather unnatural if it were consistently strong. Rather, it should gradually fade out toward the camera, in correlation with the change in viewing angle (Translated to 2D, this fade happens from "top to bottom"). The same principle applies to any horizontal reflective surface (say, a shiny floor or table surface), and to curved surfaces where the reflection is stronger at grazing angles (on a sphere, reflection would be stronger toward the sides).

Breaking Up Reflections

As the irregularity of the surface becomes more pronounced, the reflected image loses its definition, and at a certain point only the areas of highest intensity and contrast (light sources, the sky) are visible as broken highlights. The lake surface is a good example: small ripples on the surface break the reflection to a certain extent, but the reflected image is still recognizable. The approach for creating such a reflection in 2D is to modulate or distort the reflected image with animated undulating noise. The noise pattern should match the scale and intensity of the ripples and should also match the perspective (ripples getting smaller with distance). However, when the waves are stronger, their height and complexity break the reflected image beyond recognition. In such a case, the reflected image should be blurred before being modulated. This ensures that only the strongest differences in illumination (like the sun) are noticeable in the broken reflection, but finer detail (like clouds in the sky) is not.

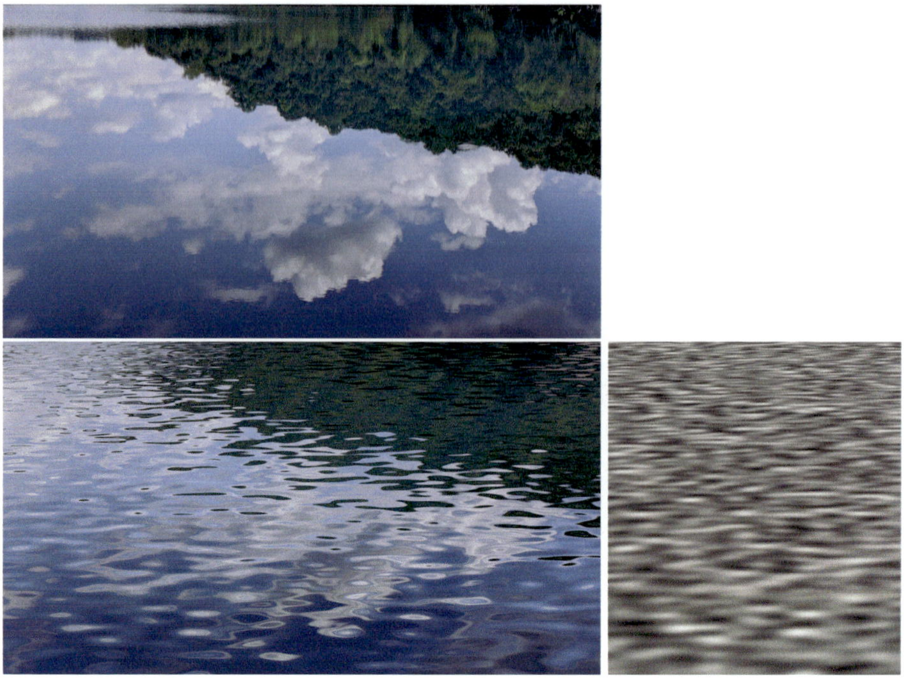

A simple noise pattern (right) is used to distort the reflection image to simulate a rougher water surface.

Creating Shadows

In Chapter 16 we looked at shadow mattes and ambient occlusion passes as tools for generating interactive shadows in comp. But these techniques only work with CG elements that are lit and rendered in 3D. When elements are extracted from 2D footage and comped into a new environment, shadows must be created solely in 2D. This is certainly more challenging, because the angle, perspective, falloff, and softness must be approximated and generated manually as in traditional art. The first step when compositing a 2D element into footage is to study the shadows in the footage and estimate the location and character of the light sources (even if they are not visible in the frame). This is easier to achieve in an outdoor daytime scene, as it only requires assessing the position of the sun, but can be much more difficult in a nighttime or interior footage with multiple light sources of varying characteristics.

Shape, Angle, and Stretching

The alpha channel of the comped element can be used as a starting shape for the shadow's silhouette, and can then be transformed, scaled, rotated, and skewed to achieve the correct perspective and shadow stretching. This is easier to do on a still frame, but much harder when the element is moving. For example, if the element is a green screen extraction of a walking actor, care must be taken when manipulating the shape of the shadow so that the shadow always sticks to each of the actor's feet whenever they touch the ground. It is usually more important to ensure that the shadow sticks to the contact areas then to emulate every small detail in the shadow's shape – we are much more susceptible to "floating" shadows then to missing detail.

The angle and length of the shadow need to correlate to the position of the light source as well as its size. For sunlight, the shadow angle and length can be assessed by gauging the position of the sun in an imaginary half dome and tracing a straight line from the sun's position through the subject at its point of contact with the ground. The angle and length should be kept consistent across the scene regardless of where the elements are. With any other light except the sun, the angle needs to change depending on the position of the subject relative to the light source. The straight line can be animated with its origin tracking the light and its pivot tracking the subject, and the result used as a rough guide to the shifting direction of the shadow.

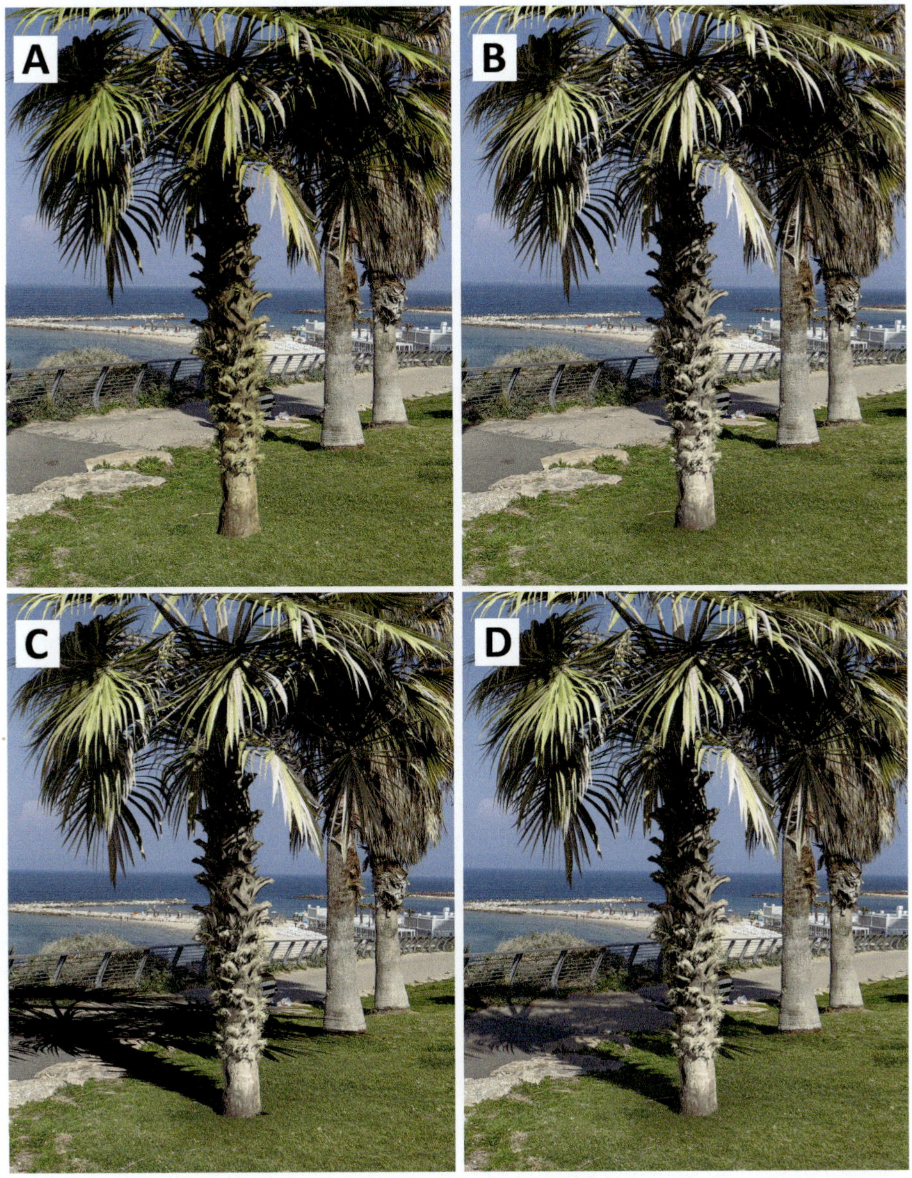

A: a palm tree is added to the foreground, B: color is adjusted to match the environment, the bottom of the trunk is darkened a bit and a very subtle proximity shadow added to the ground, C: the tree's alpha is skewed, rotated, and scaled to create the shadow, D: shadow color and softness adjusted.

Shadow Softness and Falloff

As shown in Chapter 8, the softness of the shadow (or the relative size of its penumbra) depends on a combination of several factors. While physically-based renderers faithfully calculate shadow softness, it is usually not necessary to consider every actor when creating shadows in 2D. Matching the overall shadow softness and falloff to existing shadows in the footage is usually enough. If the footage does not offer sufficient reference, consider the scale of the light(s) in the scene. Large lights produce softer shadows, while small, focused lights produce sharp shadows with a narrow penumbra. Overall shadow softness can be achieved by blurring the shadow's matte using local masks. When the light is coming at a steep angle, the penumbra is fan-shaped, and the falloff also happens gradually along the sides, not just the length.

Breaking Up Shadows

The most challenging aspect of creating 2D shadows is matching their shape and deformation to the underlying environment. Easy on a flat, even ground, harder when the shadow falls on an irregular surface or other objects. Simple cheats can be used to break shadows in a convincing way. For example, on a relatively flat but irregular ground like dirt, pebbles, grass, or water, it is often quite effective to break up the shadow matte with some simple noise. Although it is not an accurate representation of the surface, the illusion is usually quite sufficient. This type of treatment is of course not a solution for large surface irregularity (for example, stairs) or when the shadow happens over distinct objects. In such cases, the roto shape of the shadow must be animated to reflect the wide deformations in the shadow's shape, a challenging process, especially when both the occluding and occluded surfaces are moving or deforming.

Shadow Color

Junior compositors often make the mistake of overlaying the shadow as if it was a separate element. This forces a consistent color, when in fact the shadow's color depends on the color of the surface. Color-correcting the footage through a matte is a preferable method, as it preserves the underlying hue. However, simply reducing brightness (gaining down) may push the black values too low, so it is usually necessary to gain down while also slightly lifting the blacks and reducing saturation. The shadow area should be tinted toward the hue of the remaining light/s. In a daylight environment, this would correspond to the sky color, while in nighttime or interior scenes the hue would match the combined color of the remaining lights (such shadows are rarely as cool as daytime exterior). Shadow tinting should be subtle – the slight bluishness of shadowed areas on a clear day is nowhere near as saturated as the blue sky itself. Since our eyes are sensitive to black mismatches, it is important to check the blacks in the created shadows against existing shadows in the footage.

Contact and Proximity Shadows

AO and shadow passes help generate contact shadows for CG elements, nut 2D elements require compositors to manually create occlusion areas around the element's points of contact with the footage (or between two separate elements). Contact and proximity shadows can be created by using the element's alpha and slightly dilating and blurring it, or by drawing simple soft masks shapes around the areas. The softness and intensity must be animated to match the action. The closer the elements are to each other, the darker and sharper the shadow. For a walking actor, the ground contact shadow is darker and sharper when the foot is touching it, then gradually softer and lighter as the foot moves up. Similarly, if the composited actor is reaching out to touch an element in the footage, the shadow's intensity and sharpness increases as the hand gets closer to the element. Compositors and matte painters often focus on the obvious contact or near-contact areas and less so on the subtler proximity occlusion. For example, a very soft, very subtle darkening of a wider area underneath the walking actor can truly help the integration of the element in the environment. Such an almost imperceptible occlusion effect can go a long way in improving photorealism.

Case Study: Matte Painting Integration in Film

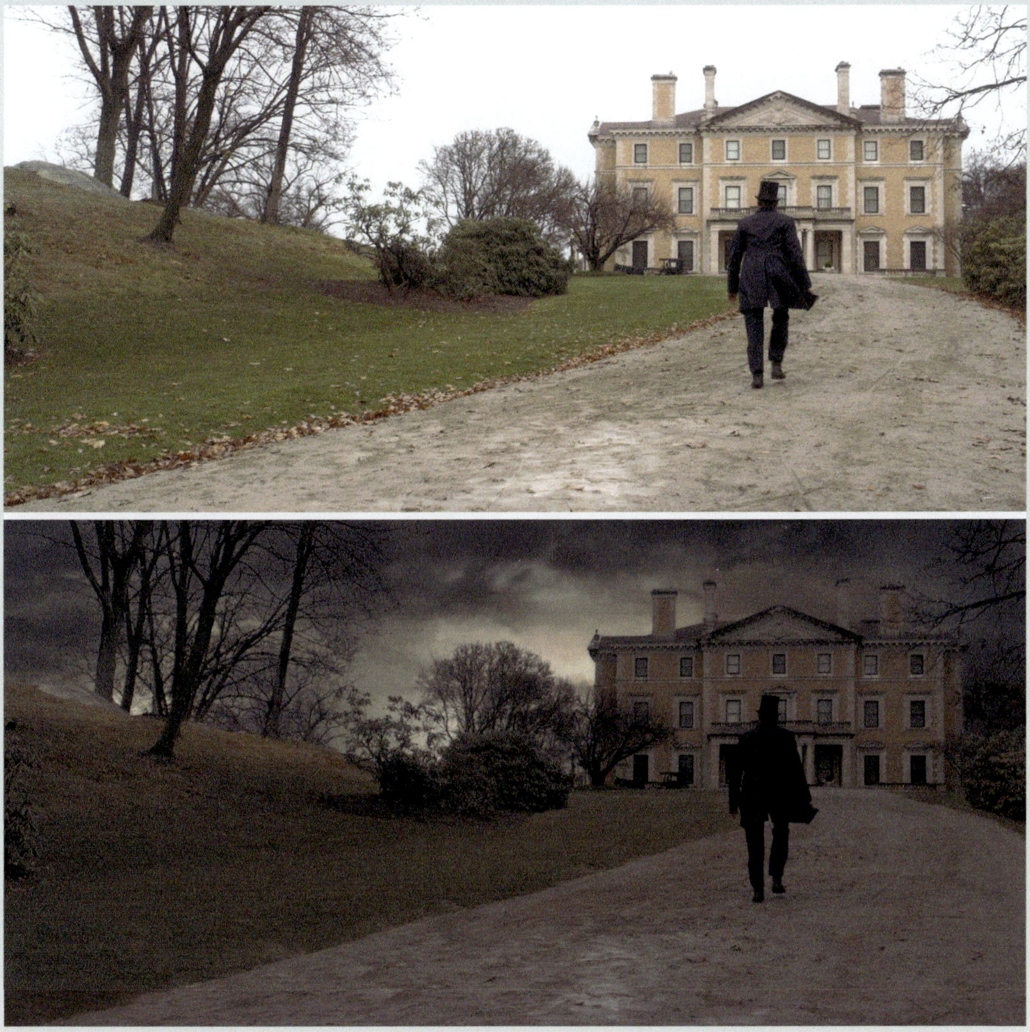

The Greatest Showman.

© Twentieth Century Fox. VFX by Brainstorm Digital.

Creating photoreal, detailed matte paintings is a challenge, but integrating them into the footage is often an additional challenge. Here are two examples from Brainstorm Digital's VFX work. The first, from *The Greatest Showman*, is a fall-to-winter conversion – done mostly by replacing adding a dramatic sky that changes the entire mood of the shot. The original sky was very bright and flat, and replacing it with a cloudy dark sky required substantial relighting of the shot. Some overall color corrections were used to bring down the tones to match the darkness of the sky, but the mansion façade was tricky: parts of it became too muddy and lost their definition because of the darkening, and additional matte painting was necessary to bring back the detail. Replacing a bright background with a much darker one inevitably causes issues of bright edges around wispy elements, and indeed, some tricky edge reconstruction work was necessary to treat the all those twigs and small tree branches.

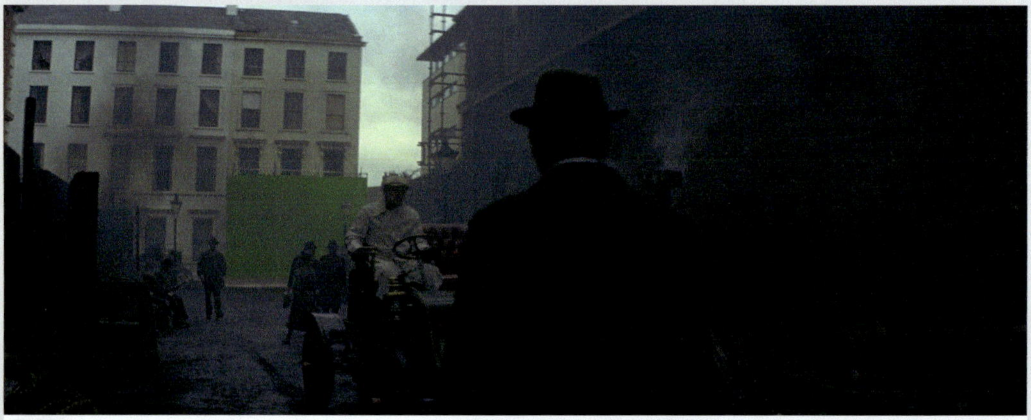

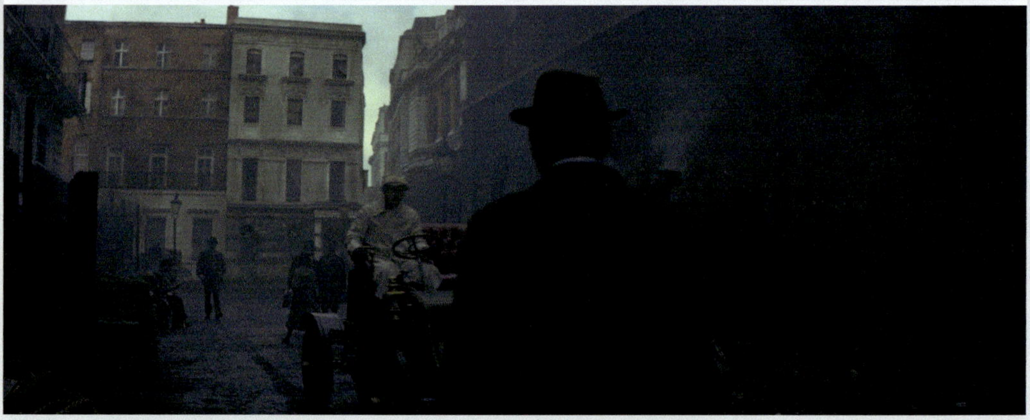

The Lost City of Z.

© Keep Your Head Company, Amazon Studios. VFX by Brainstorm Digital.

The second example is from James Gray's film *The Lost City of Z.* In this particular shot, explorer Percy Fawcett heads toward the Royal Geographic Society's headquarters. Gray wanted an exact replica of the historical RGS building on Savile Row, but the London parts were in fact shot in Belfast, so we explored several potential locations in Belfast. Eventually, we ended up choosing an intersection just off the crew's hotel, because of the similarity in layout and proportions to the real location. The green screen was intended to cover the all action, but cinematographer Darius Khondji ultimately went for a wider framing, and additional roto was necessary as some of the action went off the screen.

The movie was shot on film with an anamorphic lens, factors that make a difference when integrating matte painting elements: film has heavier and more pronounced grain than digital footage, which makes grain matching a crucial integration step. Film also has a subtle characteristic softness that is distinctly different from digital media. The anamorphic lens introduces a second layer of softness that is uneven across the frame, and, of course, a very pronounced distortion that must be matched (notice the e slanting of vertical lines). Yet the most challenging integration aspect in this shot was the dense atmosphere – several smoke machines were working hard on the set to emulate the thick, soot-filled air of early 20th century London. Adding or replacing buildings across the frame required meticulous matching of black levels, and adding back blowing smoke, either by keying it over the green screen or recreating it with external elements where it could not be keyed.

Atmospheric Depth

Photorealism strongly relies on an accurate depiction of depth. Our eyes are sensitive to depth mismatches, and digital imagery easily falls apart when aerial perspective is off or inconsistent. In visual effects, atmospheric depth is typically added at the compositing stage. Even if certain atmospheric elements such as fog or smoke are rendered in CG, the final adjustments are done in comp, with interactive control and in the context of the footage and/or final image. The method for generating atmospheric depth depends on the type of elements: for CG rendered elements, the Z depth pass provides an accurate per pixel representation of the distance form camera and serves as a base for all depth effects. When dealing with 2D elements from footage or still images, compositors and matte painters must rely on their visual skills and keen understanding of distance and depth. In lack of a depth pass, artists also need to manually separate or "slice" different depths in the images by painting masks or rotoscoping. Atmospheric depth is tied to accurate color grading, and the following color principles apply equally to 3D (Z depth pass) and 2D (manual depth separation) workflows.

Referencing the Footage

The first step when integrating an element is to assess its distance from the camera, based on its position in the image (how deep it sits within the footage), and then identify areas in the footage at the same depth as reference for the effect of the atmosphere on color. For example, when adding a building element to a photo of a city, existing buildings which sit at a similar depth in the footage can serve as an indication to the amount of hazing. Look at the blacks first – not just their value (how lifted they are), but also their specific RGB balance (it may be useful to temporarily raise the viewing gamma to see better into the blacks). Matching the subtle, almost imperceptible color detail in the low range is a key step in matching the overall aerial perspective and the effect of atmospheric elements like fog, dust, or smoke.

That said, there are times when the original footage does not provide the necessary depth clues. For example, when the elements are placed deeper than any existing object in the footage (adding mountains in the distance, or a plane high up in the sky), or when the entire scene is a CG

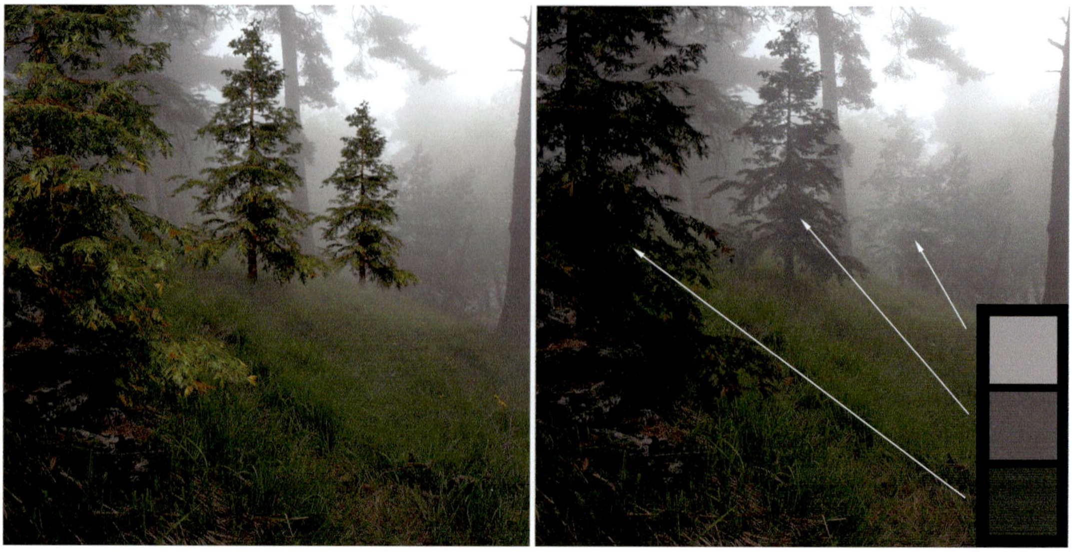

The same tree element is placed at three different depths (left). The substantial color changes must be applied to each tree to integrate it into the environment at that particular depth.

render or matte painting. In such cases, the sky is the best guide. Lifting the color values of distant elements toward the color of the sky is a basic way to simulate aerial perspective. As discussed in Chapter 6, humidity, dust, and other aerosols congregate in the lower atmosphere, which affects the sky color at the horizon. Since most distant elements sit near the horizon, these factors affect their color as well. Sometimes (especially with very large distant elements such as mountains) it is not enough to just push the blacks and midtones toward the horizon color. Atmospheric components like low-lying clouds or fog may partially or completely obstruct distant elements, and

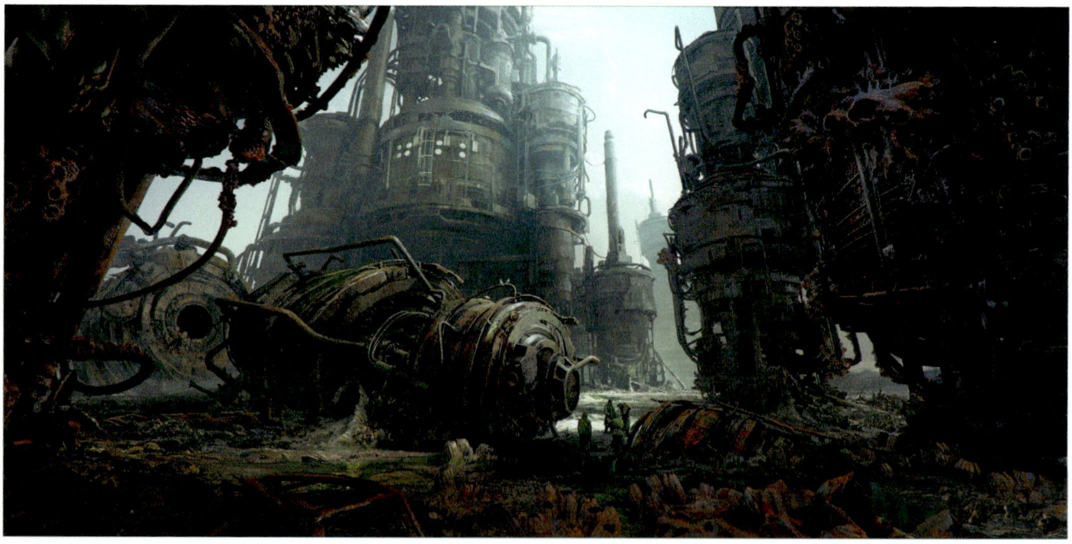

Abandoned Refinery by Sergey Vasnev.

Blender, Photoshop. © Sergey Vasnev.

it is important to show their interaction with the added elements. Low-lying fog, for example, can blanket the foothills at lower altitude, and then dissipate at higher altitude, with the peaks clearer than the base. It is important to treat the sky (whether it already exists in the footage or is added as a sky dome to a CG scene) as a true atmosphere – a three-dimensional volume that engulfs the scene and interacts with it, rather than a 2D background that simply sits behind the scene.

While lifting the low end for atmospheric depth, it is important not to brighten the higher midtones and highlights. Clean air usually leaves highlights intact, while haze may dim them, but highlights never get brighter because of the atmosphere. Therefore, color operations like lift or gamma are a good choice for depth grading. An effective technique for adding atmospheric depth is to grab the sky in the footage, blur it enough to erase any discernible detail, and then subtly screen it over the composited elements. It is important to use a screen merge operation and not an additive (A+B) operation. The screen operation (A+B-AB) protects the top range and prevents highlights from getting brighter, but effectively lifts the low end and lower midtones.

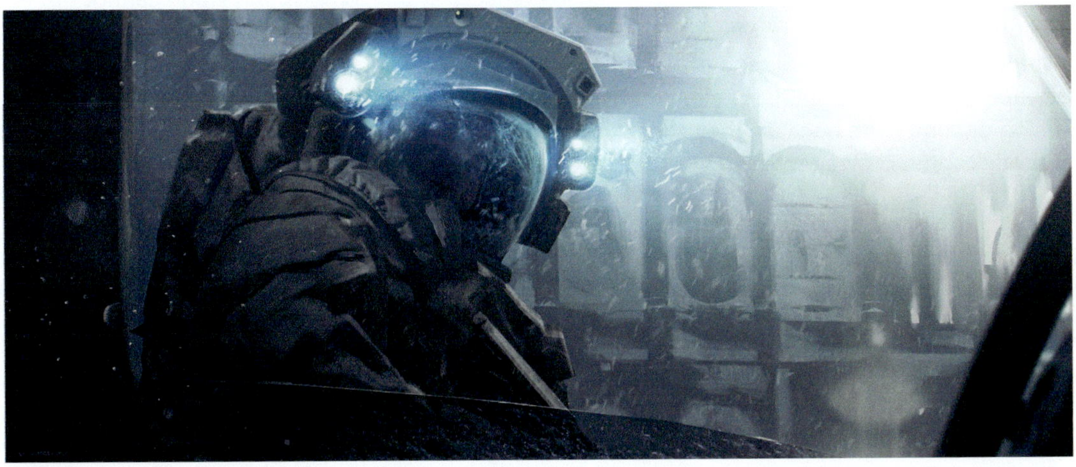

The Way by Val Orlov.

Blender, Photoshop. © Val Orlov.

Chapter 18

Lens and Camera Effects

The significance of adding lens and camera characteristics at the compositing stage cannot be overstated. This is not just the icing on the cake – this is the essence of photorealism. Even the most realistic CG render feels somewhat artificial without the appropriate touch of defocus, lens distortion, flares, and grain. Lens effects serve as a final glue that binds all the elements in a matte painting or composite because they strengthen the illusion that everything in the image was physically captured by a camera.

Defocus

Every CG render engine provides the ability to apply 3D depth of field as part of the render. This generally produces the most accurate defocus, without any of the issues associated with Z depth. However, 3D defocus is baked into the render, and cannot be changed or adjusted later in compositing. It also requires higher sampling rates and slows down the rendering. The 2D option is certainly more flexible – the focal point, depth of field, and defocus amount can all be tweaked and adjusted while the results are viewed in real time. As long as you are aware of the Z depth limitations discussed in Chapter 16 (in particular, issues with transparent and volumetric objects), then 2D defocus can be an efficient alternative to 3D depth of field for CG renders.

Compositing applications offer dedicated tools for applying accurate defocus based on the render's Z depth pass. Non-normalized passes work especially well since they cover an infinite range. The focal point is selected by sampling the Z depth value at the desired focus area, and then the depth of field and defocus intensity are adjusted. Defocus tools usually offer additional features for precisely controlling the look of the Bokeh, like setting the number of polygons (to match the number of aperture blades) and adjusting their relative curvature, adding typical highlights bloom with variable threshold, gamma shifting, or changing the defocus aspect ratio (for non-spherical anamorphic defocus). Such features allow the artist to emulate the specific characteristics of any lens and achieve seamless integration with photographed material.

When working with 2D elements, there is of course no Z depth pass. Applying convincing depth of field in a fully 2D workflow requires a good eye and a clear notion of where each element is

DOI: 10.4324/9781003590293-23

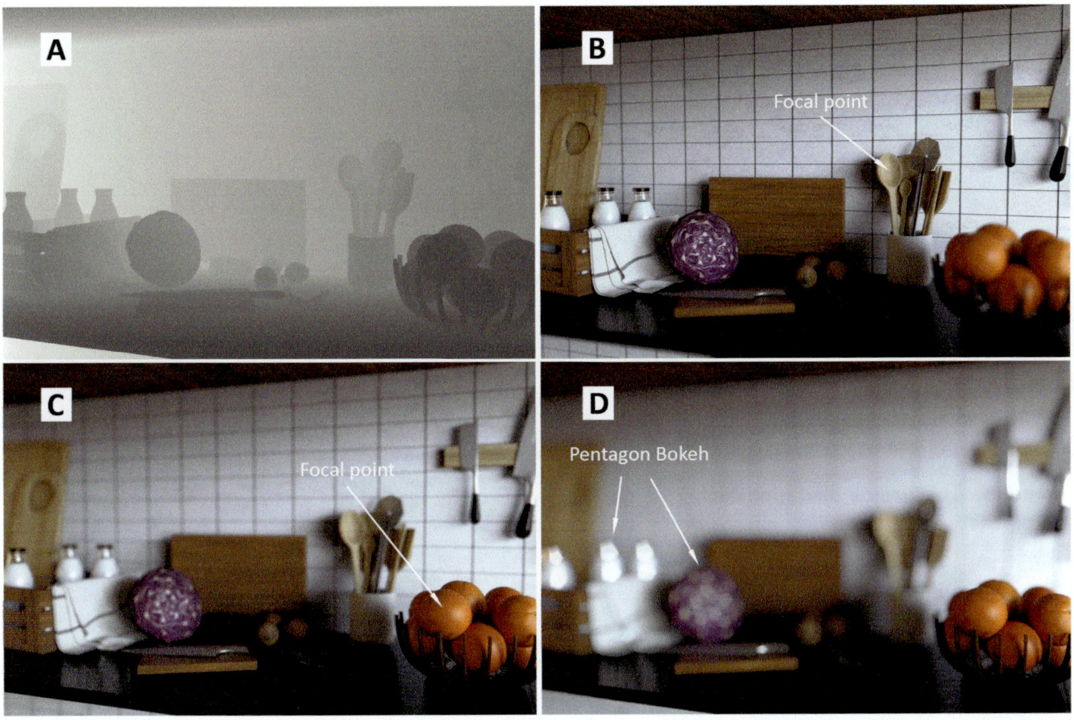

Tweaking defocus in comp. A: depth pass, B: focal point near the back wall, C: focal front moved to the front, D: exaggerated defocus with five-blade aperture, gamma correction, and highlight bloom.

placed in 3D space. In lack of a depth pass, it is certainly harder to create a smooth transition from in focus to out of focus. Focus level gradation can be somewhat simulated with soft masks, and although the result is not quite the same as true focal transition, it can nonetheless be an effective cheat. When integrating elements into live-action footage, focus level and Bokeh type should be matched to areas at a comparable depth.

Lens Distortion and Chromatic Aberration

Chromatic aberrations are important for integrating elements, and for adding a subtle sense of a lens medium to CG renders. The effect can be easily replicated by slightly offsetting one or two of the RGB channels. An offset of one to four pixels (depending on the resolution) is usually enough to create the typical fringing of chromatic aberration, while higher values are used more for pronounced stylized effects. Transverse aberration is stronger toward the fringes of the frame (see Chapter 10), and can be simulated by scaling the selected RGB channels from the center of the frame.

From a technical standpoint, **lens distortion** is an indispensable component of camera tracking. Distortion features vary greatly between different lenses, and although some of these variations are hardly visible to the eye, their precise calculation is crucial for 3D camera tracking. Some lens distortion tools offer elaborate distortion models that mimic real-world spherical and anamorphic

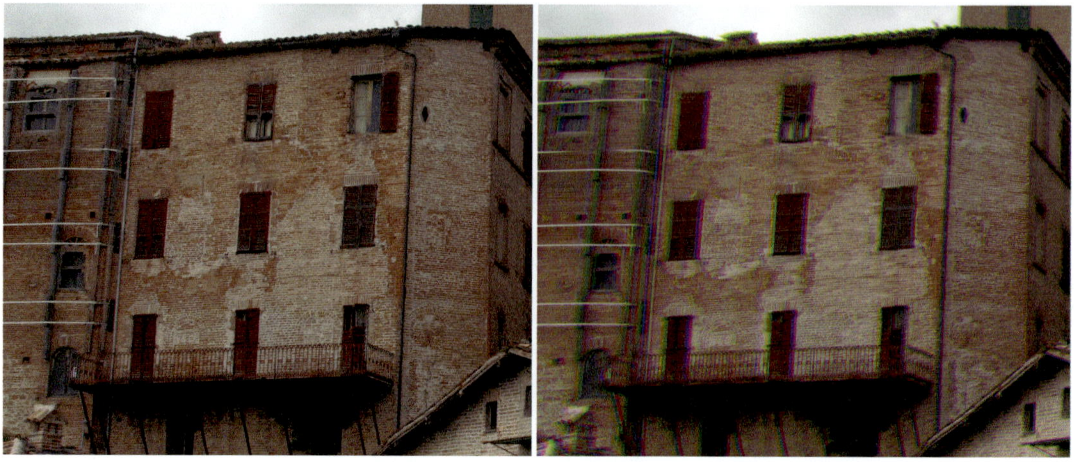

A pronounced chromatic aberration (right) is generated by offsetting the red and green channels in opposing directions.

lenses, as well as rectilinear and fish-eye projections. But beyond its technical importance, lens distortion can add a great deal of realism to fully CG games and even static renders, as long as it is kept within a plausible range that matches the context.

Lens Flares

Lens flares serve a triple purpose: they add visual drama, they give a sense of a physical lens, and they help tie together different elements in the comp. Flares are an important instrument for enhancing the photoreal aspect of digital imagery and serve as a very effective compositing "glue", because they happen over of all the elements in the image. The essential flares characteristics discussed in Chapter 10 need to be implemented in context of the shot. In particular, the movement and intensity fluctuations should be tied to the movement of the relevant light sources in or around the frame and random disturbances and imperfections must be added to avoid an overly clean and perfect look.

There are essentially two ways to create lens flares: procedural generation and real footage. Most compositing and photo-editing applications provide some procedural tools for designing lens flares, and there are also many third-party flare-creation plugins available. Procedural tools offer the advantage of accurate flare motion and intensity in relation to the light source, and full control over the shape and color of the various flare elements. However, procedural flares are too clean and perfect out of the box, and need to be further degraded and varied through noise or texture modulation. The extra effort can truly make a difference between a tacky effect and believable visuals. Real-world footage, on the other hand, has all the photoreal detail and imperfections built-in. There are quite a few stock libraries that include a wide selection of flares (ideally shot over a clean black background) that cover various lenses and lights. However, it can be difficult to match the motion and interaction between the footage of a full flare effect and a specific shot or CG render. Stock footage of single, isolated flare components is usually the most useful, because it allows the freedom to combine several elements and animate them in context of the particular shot (or use them to enhance procedurally-generated flares).

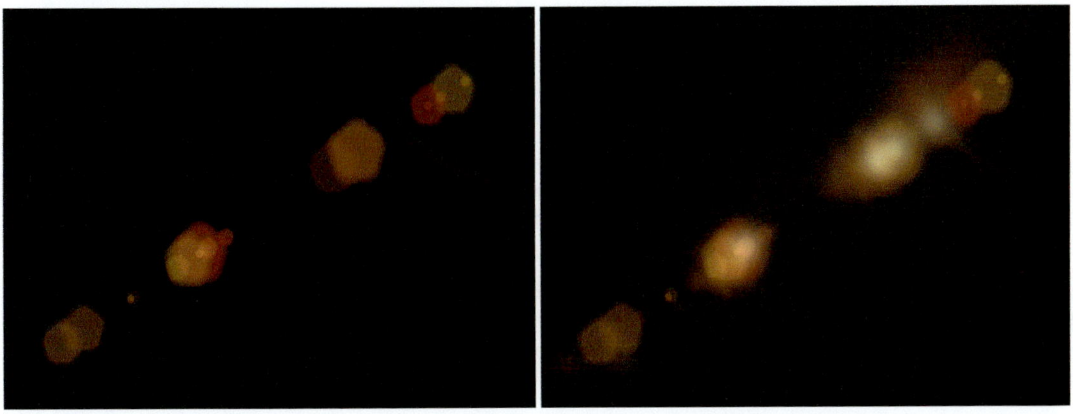

A procedurally-generated lens (left) looks a bit too clean and perfect. Blurring some parts, adding localized glows, and running the elements through a grunge map helps giving it a more natural look.

Motion Blur

3D motion blur is usually rendered as part of the CG scene. However, like defocus, it is baked into the render and cannot be further adjusted in comp, so sometimes it is more practical to add motion blur to CG renders at the compositing stage. This requires rendering a **motion vector** pass along with the beauty. Compositing tools like Nuke's **vector blur** node apply motion blur based on the motion vectors. The advantage of applying motion blur in 2D is not only shortened 3D render times but the ability to fine-tune the motion blur based on the context of the footage. However, when motion blur is applied to a complete render, bad edges and artifacts tend to form around areas of overlap between moving and static objects (or between objects moving in different directions and speeds). To avoid this, it's best to render the scene in separate layers.

Adding motion blur to 2D elements is of course limited to two axes, but thankfully, motion blur that happens on the Z axis (toward or away from the camera) is generally less noticeable. 2D vector information can be generated by tracking the motion of the 2D element. When tracking is not possible (or the results not satisfactory), motion blur can be faked with a directional blur. Of course, the exact amount and angle need to be keyframed manually to match the speed and direction of the object. As discussed in Chapter 15, extracted and rotoscoped edges often require added motion blur to match the original footage, and the same techniques (tracking or manual direction blurring) can be used for this type of edge treatment.

Grain

Grain is added last, on top of all the components, and its contribution to the final integrity of a composite image or shot is as crucial as it is subtle. The process of matching and adding grain often also requires removing any previous grain from different elements to prevent mismatched, double, or frozen grain. Grain and noise removal (**degraining** or **denoising**) is often a balancing act between removing as much of the grain as possible, and preserving sharpness and detail. Sometimes the best way to achieve an optimal balance is by experimenting with different tools and settings (In addition to native tools in compositing applications, there are plugins that use different algorithms for degraining and denoising). There are several techniques for re-applying grain to footage (such as degraining it first and then using a difference or minus merge operation to extract the grain), but visual grain matching and manual adjustments are often preferable. Grain size and softness should be matched and adjusted separately or each RGB channel, and both dark and bright areas should be examined. Grain in the blacks is particularly important, and most grain tools offer a separate control for that. When adding grain to a fully CG image or shot, the reference for grain should either be the surrounding shots (in VFX), or, in the context of visualization or games, by examining a desired camera or film stock characteristics.

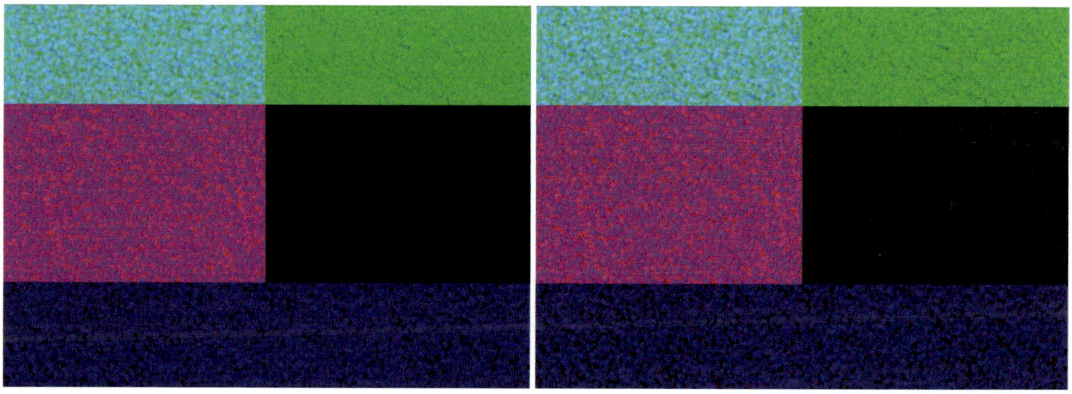

Artificially added grain does not show well in the blacks (left), yet in both digital and film mediums, grain is usually stronger in the blacks. It is therefore important to adjust the grain level in the blacks separately (right).

Chapter 19

Photorealism with Generative AI

Generative AI is used in VFX, visualization, and games in a variety of ways – from creating concept art and generating elements for matte painting, to modeling and texturing 3D assets – in any conceivable style. In this chapter, we'll focus on generative AI's role in producing photoreal content. Models that are trained mainly on photographic material can generate extremely realistic imagery, and in fact, can easily surpass anything created with CG or matte painting techniques. But achieving the desired creative and aesthetic goals with AI requires tackling an entirely new (and unique) set of challenges.

The process of working with generative AI essentially upends the foundations of CG and 2D. Much of what you've read in this book – the importance of detail, of imperfections, the role of lighting, atmosphere, lens aberrations – all of this can be easily generate with AI. Prompt an AI model to create a photo of, say, a dense jungle, and within seconds it will output an image lush with fine photorealistic detail, the kind of scene that would take hours (if not days) to painstakingly model, texture, and assemble in a CG workflow.

However, when precise control is needed, AI is at a disadvantage. A CG render engine takes a complex dataset – the 3D scene – and calculates the path of light rays within it to produce an image. Everything in that image is directly controlled by the artist, and every modification to the scene has a predictable outcome. The render engine does not really "create" anything – it is merely a powerful calculator at the service of the artist.

Generative AI models are trained on a large volume of tagged images. Through processes like diffusion (see later in the chapter), they form complex associative connections between words and visual content. When prompted to generate an image, they "create" it by rearranging pixels from a random noise pattern. Multiple factors, from the seed of the noise or the exact wording of the prompt to the model's particular training, affect the result. But while the artist can adjust many of these factors, the predictability of such adjustments is frustratingly low.

Moreover, the level of control and specificity that's at the core 2D and 3D workflows is severely limited with generative AI. We can't simply tell the model to "move this plant to the right" or "replace the texture on this rock" or "tilt the camera down a bit" (at least, not quite yet). Tweaks that are quick and easy in CG (like adjusting the sunlight angle or rotating an element) can be extremely tricky, if not impossible. To approach photorealistic generation with AI, we need to shed old axioms and think anew.

DOI: 10.4324/9781003590293-24

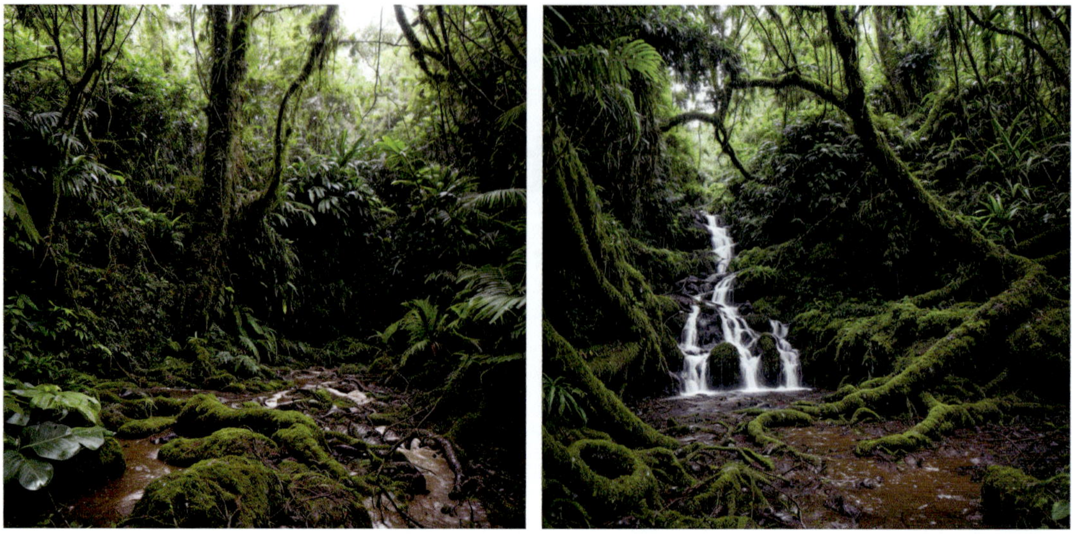

Two generated images of a lush tropical jungle. Prompt: *photo of a lush tropical jungle, moss, vines, muddy ground, photo, realistic*. Checkpoint model: Analog Madness.

It is tempting to imagine visual generative AI as a very strange and unusual artist – an artist that produces exceptional work at lightning speed and is extremely hard-working and eager to fulfill any request, but is also quite unpredictable and lacks essential communication skills. Ask this "artist" to change one thing, and it will change 15 others. Ask it to make flower red, and it will make the trees red as well. Ask for two people and it will give you three . . . and so on. But this analogy is not very helpful – AI, of course, is not a human. To truly get the best out of it, we need to understand its internal workings.

This chapter is a guide to handling generative AI's weaknesses and idiosyncrasies and explores various approaches and tools for achieving controlled results while guiding the model toward the desired creative goal. There are countless ways in which we can use the photoreal strength of AI to help with the creative process in VFX, visualization, and games. As the models themselves develop and become more powerful – so do the techniques for harnessing their power.

Generative AI for Video Versus Stills

Take the challenges AI image generation and multiply them by a hundredfold. Video generation is not only spatial, but also temporal. This added dimension complicates every single aspect of the process. Start with training, where every single image is replaced by multiple images – consider the toll this takes on storage, processing power, and time. Think of the relative abundance of still images on the internet as opposed to high-quality videos. And then, of course, the amount of detail that needs to be factored by the model as it generates a series of moving images: continuity, action plausibility, narrative, perspective, layering, and so on.

While most current AI video generators use similar diffusion models to image generators, the extra challenges make it harder to produce a believable, photoreal, and consistent video than it is to produce a still image. In addition, the processing power needed to generate multiple frames restricts users to cloud-based solutions, which can be costly.

But, as with everything AI, this is bound to change rapidly, and while the days of "creating a movie with a push of a button" are not quite here (despite the hype), the quality of AI generated videos is constantly on the rise, and we will likely see continuous improvements to both the video generation algorithms and the hardware used to run them. So, while this chapter focuses on image creation, the principles and workflows also apply to video generation.

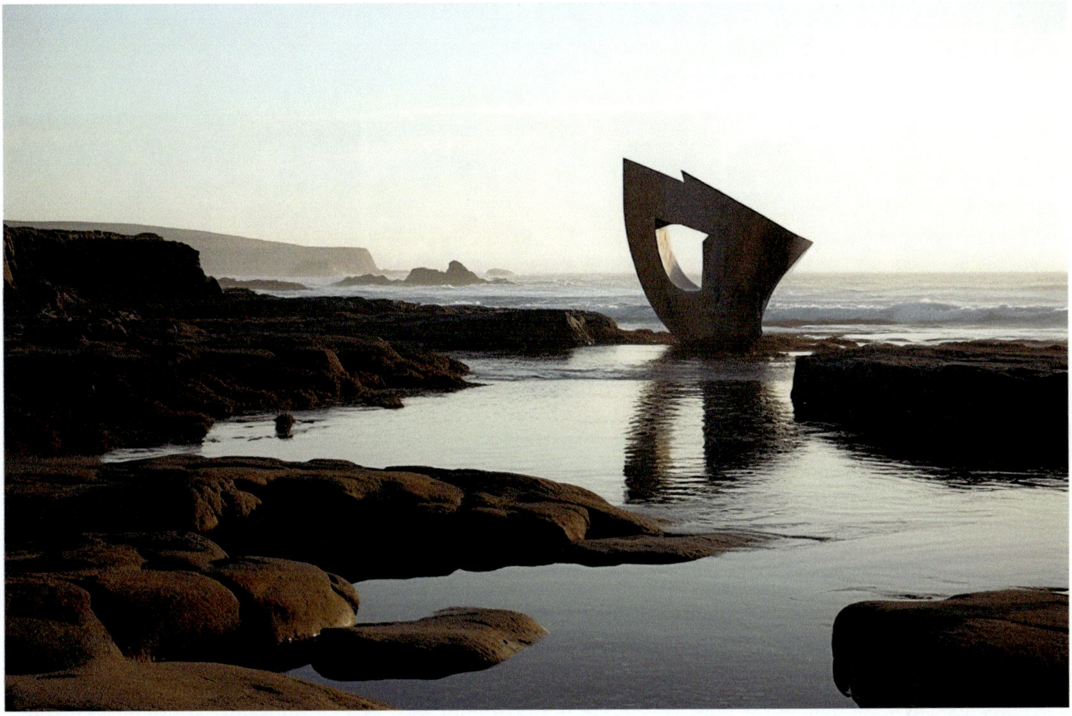

Oceanic Icons by Suburbia Studio.

Generated with Midjourney. © Suburbia Studio, © Julian Sadokha.

Oh, the Errors it Makes

There is a rather puzzling gap between AI's ability to generate impressively photoreal imagery and its tendency to commit gross errors in the simplest of details, the kind of errors that human artists are unlikely to commit. Here are a few common examples:

- AI models create stunningly realistic human faces, but consistently fail with hands.
- The prompt "a person looking at their reflection in the mirror" almost never works, as the reflection does not match the person.
- Text in the generated image (say, street signs or logos) is often plain gibberish.

These errors (and similar common errors with shadows, perspective and scale) all relate to the way generative models are trained. As mentioned before, the models are fed with large volumes of tagged images so that they can learn to associate words with specific visual elements. By "seeing" numerous photographs, those models acquire a good grasp of **what** photos look like, but no understanding of **why** they look like that.

Human faces are easy to identify from various angles and under different lighting conditions because the nose, mouth, and eyes all stay together and never switch places. But hands are much harder to decipher in photographs – fingers are often obscuring each other or are close enough so that they seem fused. On top of that, hands are much less prominent in photographs than faces. Since the AI algorithm has no understanding of human anatomy, it can only rely on its acquired visual "memory", and the results are often mangled hands with missing (or too many) fingers.

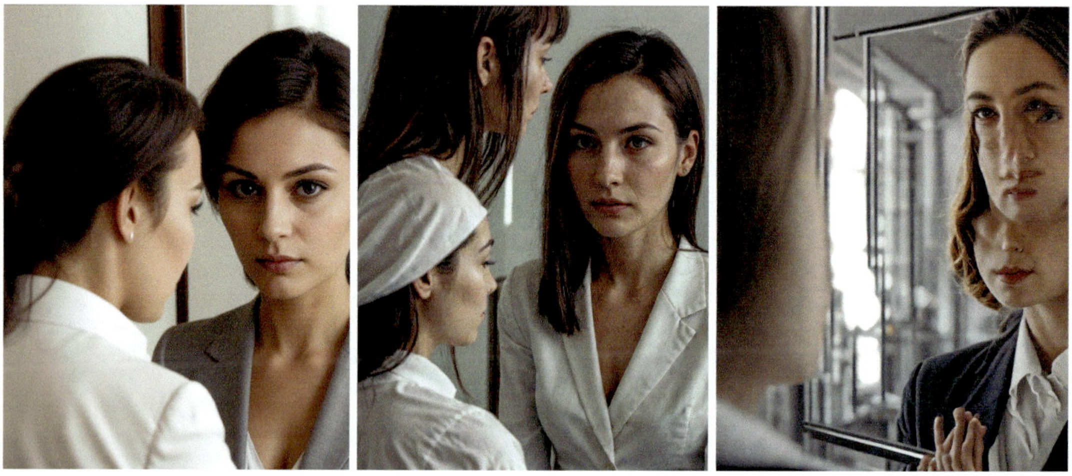

Three different checkpoint models (from left to right, CyberRealistic, Realistic Vision, SDXL Turbo) were tasked with this simple prompt: *close-up of woman in a business suit, looking at her own reflection in the mirror, photo, realistic.* All three failed spectacularly. As an aside, it is interesting to note how all three models generated similar-looking women, even though the only description in the prompt was "in a business suit". This says something about the built-in biases in generative AI models.

Likewise, AI can very effectively replicate intricate specular reflections on shiny surfaces like water, but it does so without understanding physics and does not trace light rays like a CG render engine. So, while it can create a convincingly photoreal image of a reflective lake (there are plenty of photos to train for that), it cannot accurately portray a specific subject and its own reflection. Similarly, while models can easily mimic text as seen in photographs, they do so with no understanding of what the symbols actually mean and thus generate random gibberish.

Are these limitations everlasting? Not at all. We are still at a nascent stage of visual generative AI and are likely to see new models with more complex neural networks that combine visual and contextual training, with better analytical capabilities. For example, models such as ChatGPT are already getting better at creating coherent text within generated images. But as AI keeps developing, it is good to be aware of its current limitations and find ways to avoid common pitfalls.

A Note about Stable Diffusion

In this book I strive to remain software-agnostic. 3D applications, for example, are very similar at their core, and one can easily apply the same knowledge and principles to any of them. But things are somewhat different with generative AI. Most current AI image and video generators (like MidJourney or Sora) are only available as subscription cloud services and offer limited control or customization. For professional digital artists who are used to applications like Houdini, Nuke, or Photoshop, the lack of controls and tweakable parameters feels quite restrictive (not to mention factoring in the cost per iteration).

Stable Diffusion, on the other hand, is a completely open-source diffusion model that's supported by a plethora of user-developed tools, from custom-trained models to algorithms that do anything from generating depth maps from photos, to isolating elements in images. Free UI applications like Auto1111, ComfyUI, or InvokeAI can be installed locally, and provide access to various parameters. The latter two even sport a flexible node-based workflow that should feel familiar to most digital artists. I think of Stable Diffusion's vibrant and ever-evolving universe as a fascinating prototype for future generative AI applications that offer the type of "under-the-hood" flexibility and control that serious artists need, and I explore its idioms and use examples from it in this chapter.

Undoubtedly, between now and the time you read this, many things would have already changed. AI is evolving and shifting very fast, companies and tools come and go, and open-source applications may not survive in a field that's increasingly dominated by big corporations. Still, I think that the principles and workflows discussed here are fundamental to creative AI workflows and are applicable regardless of the interface used.

Understanding the Diffusion Process

There are different types of models for generating graphic content. GAN (Generative Adversarial Networks) models, for example, use a two-player approach to training, where a "generator" attempts to create realistic images and a "discriminator" tries to distinguish fake from real. Autoregressive Models on the other hand, create images pixel by pixel, by predicting the probability of each new pixel based on the previous one. By far, the most widely used models for graphic content generation use the diffusion process.

Diffusion models learn to "imagine" an image from random noise through a reverse process in which images are gradually being destroyed. Starting with a clear tagged image, small amounts of Gaussian noise are iteratively added, and at each step the model attempts to identify the image, until the image is pure noise. In the reverse process, called denoising (more on denoising later in this chapter), the model attempts to reconstruct the image from varying levels of noise.

A diffusion model responds to a textual prompt by iteratively "rearranging" noise into a visually coherent image. In this process, the model associates between key words in the prompt and the numerous tagged images it has been trained on. In this section we will cover the main building blocks within the diffusion pipeline to better understand their respective function and how they affect the results.

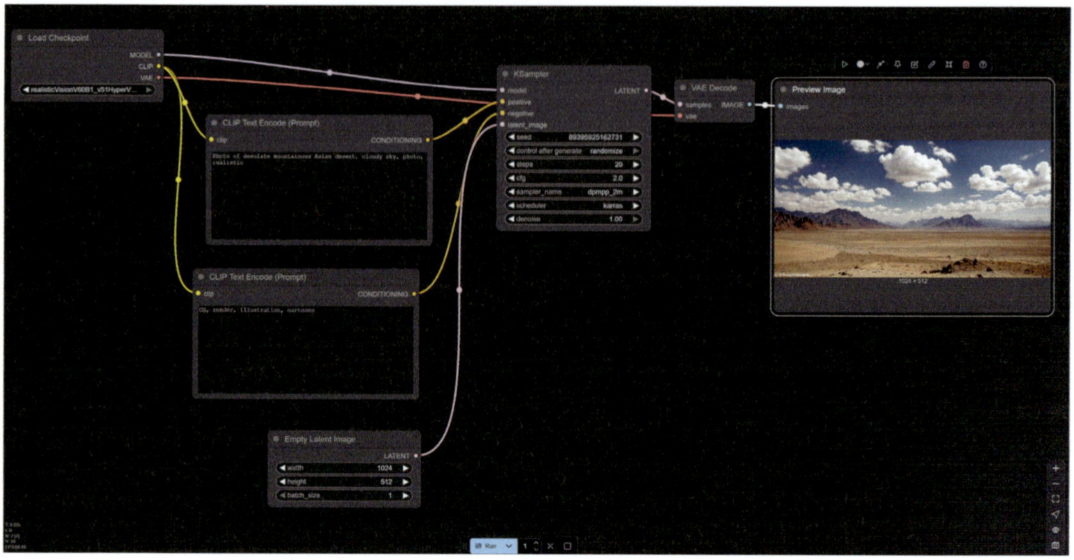

Basic workflow in ComfyUI, with the checkpoint model on the right, the positive and negative prompts, and the empty latent noise all feeding into the sampler.

The latent output is converted to pixel space by the VAE decoder and displayed as the generated image.

CHECKPOINT MODELS

Checkpoint models are the core (or "brain") of the diffusion pipeline, the part that encapsulates the knowledge gained from the training data and the calibrated weights applied in the training. These models are so named because of the "checkpoint" stages in the training process, where the model's output is checked, and weights are adjusted to control bias and reach the desired results. Stable Diffusion is an open-source diffusion model that serves as a base for numerous custom checkpoint models. These models are trained on more specific datasets and are usually stronger in a particular visual style (anime, cartoons, fantasy, photoreal, etc.). The same text prompt can thus produce widely different results depending on the checkpoint model used. Conversely, many models are merely tweaked or inbred versions of other models. It is therefore best to stick to a few models you know well rather than constantly switch models. Some of the best checkpoint models for photorealism are Realistic Vision, Juggernaut XL, and Analog Madness.

Can artists train their very own checkpoint model? Yes, in theory. But the process involves sourcing, processing, and tagging a massive volume of images. It also requires extensive GPU processing power and memory to handle extremely large datasets. Therefore, the vast resources required for effective checkpoint training are rather prohibitive for individual artists. Instead, LoRA models are a feasible alternative.

LORA MODELS

A LoRA (Low-Rank Adaptation of Large-Language Models) is a small "appendage" to an existing model that is focused on a narrow scope of training. LoRA modifiers can improve the main model's output in a specific area or expand its ability to generate specific visual elements or styles. Custom LoRA models require smaller datasets and are easier to train than checkpoint models. A quick search on the popular website CivitAI brings up a diverse selection of LoRAs: there is a

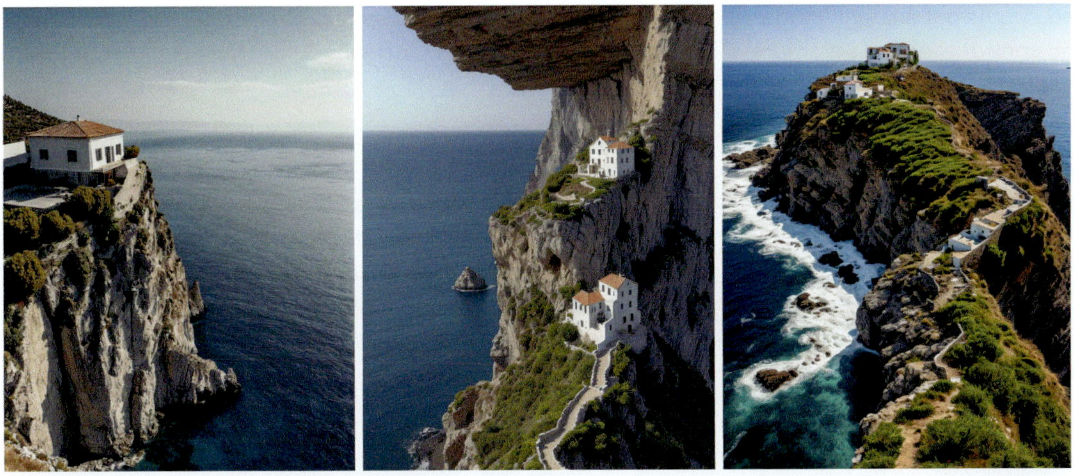

The same prompt (*photo of a lone house in Greece, perched on a steep cliff high above the sea*) and exact same settings were fed into three models to generate. Juggernaut XL (left) produced the most photoreal image by far, while DreamShaper 8 (center) generated a surreal fantasy look, and Deliberate struggled with perspective and scale. Checkpoint models differ not only in their training and biases, but also in the way different common settings affect their output.

LoRA for "person holding a camera and taking pictures", another one for creating "fairy mystic treehouse" and there's even a LoRA for generating images in a "Tron Legacy" style, just to give a few examples. From the photorealism standpoint, LoRAs are especially effective for filling in gaps where generalized checkpoint models lack sufficient training.

OTHER TYPES OF MODELS

There is a wide range of models that are not designed to generate the final image but are programmed instead to perform a dedicated task within the generative flow. The most common types are Upscale, ControlNet, and inpainting models – these will all be covered later in this chapter.

LATENT SPACE AND VAE

According to the Collins dictionary, the meaning of latent is "something which is hidden and not obvious at the moment, but which may develop further in the future". This describes latent space quire accurately – essentially a compressed, invisible space where most of the generative process happens. Input images need to be encoded first from pixel space into latent space, and the model's output must be decoded back into pixel space for us to view it as an image. The encoding and decoding operations are performed by a VAE (Variational Autoencoder) function supplied by the checkpoint model. Different models thus "translate" to and from pixel space somewhat differently.

DENOISER

You can think of the denoiser (also called Sampler) as the "render engine" of the generative AI workflow (though, of course, this has nothing to do with CG rendering). The denoiser processes the incoming random noise with the model data and the positive and negative textual prompts and churns out a latent image which is then decoded into pixel space. The denoiser is where you control essential parameters like number of steps, CFG, and scheduler type (see later in this chapter).

TEXT PROMPTS (CLIPS)

Text prompts are the fundamental communication interface between the artist and the machine. There are only two types of text prompts: positive and negative – detailing what should (and what shouldn't) be in the image. Writing effective text prompts is an essential skill in visual generative AI – beyond understanding syntax, order, and weight, the artist needs to "get into the mind" of the AI algorithm – even small changes in the prompt can have a substantial effect on the results. Text prompting will be discussed at length in this chapter.

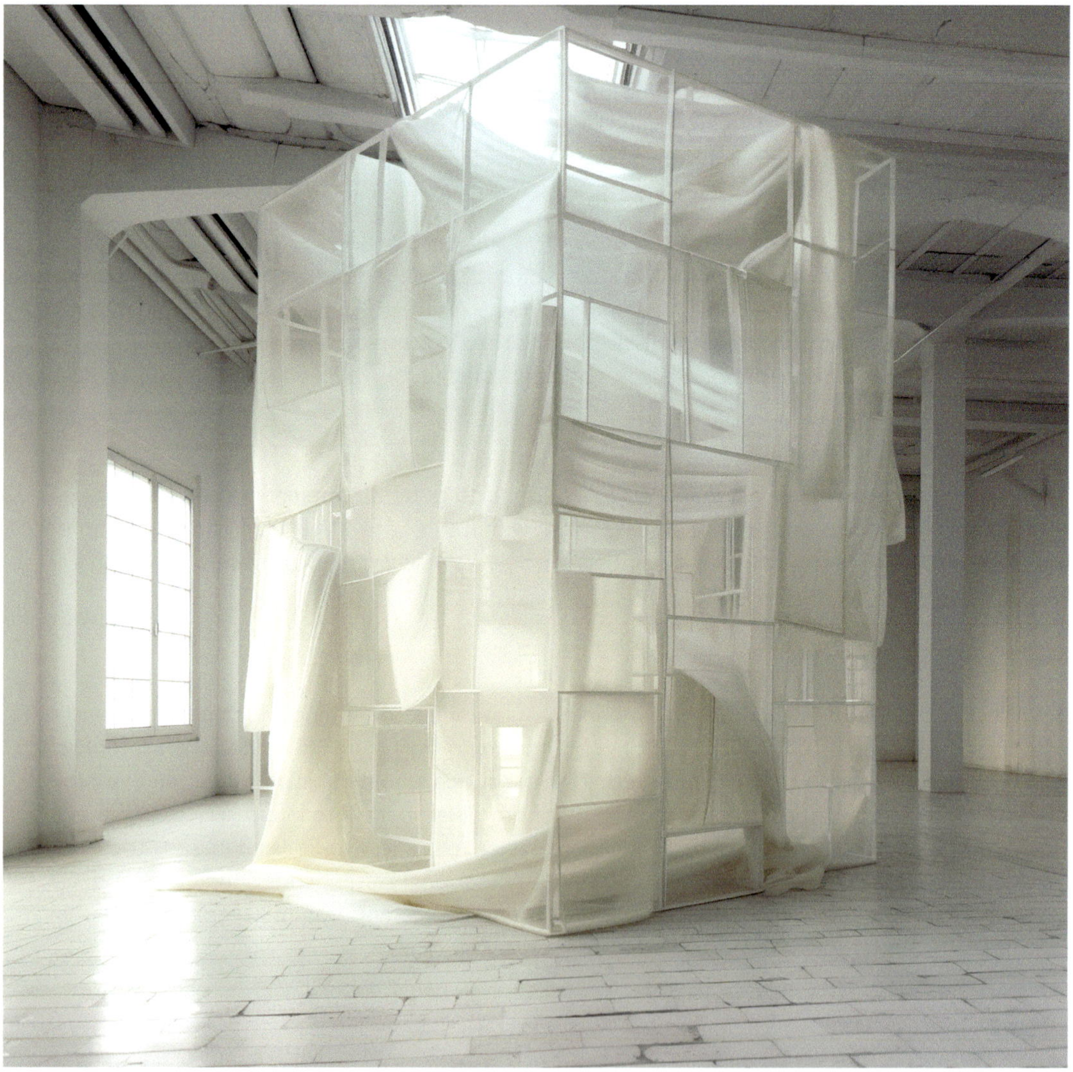

Light Whispers by Suburbia Studio.

Generated with Midjourney. © Suburbia Studio, © Julian Sadokha.

ASPECT RATIO

As digital artists, we think of aspect ratio as the "window" through which we view the scene. In CG, for example, the scene stays the same whether we use portrait or landscape aspect ratios. AI does not really "think" of framing this way. Rather, it strives to best fit the main subject in the prompt within the given aspect ratio. Prompting for "a single windmill in a field of tulips" would probably work just fine in a portrait format, but given a wide aspect ratio, it is very likely that the model will add another windmill (or more) in an attempt to fill the wide frame with the subject. Likewise, if the subject is a human figure, the model will often change its pose to sitting or lying to fit it within a landscape format.

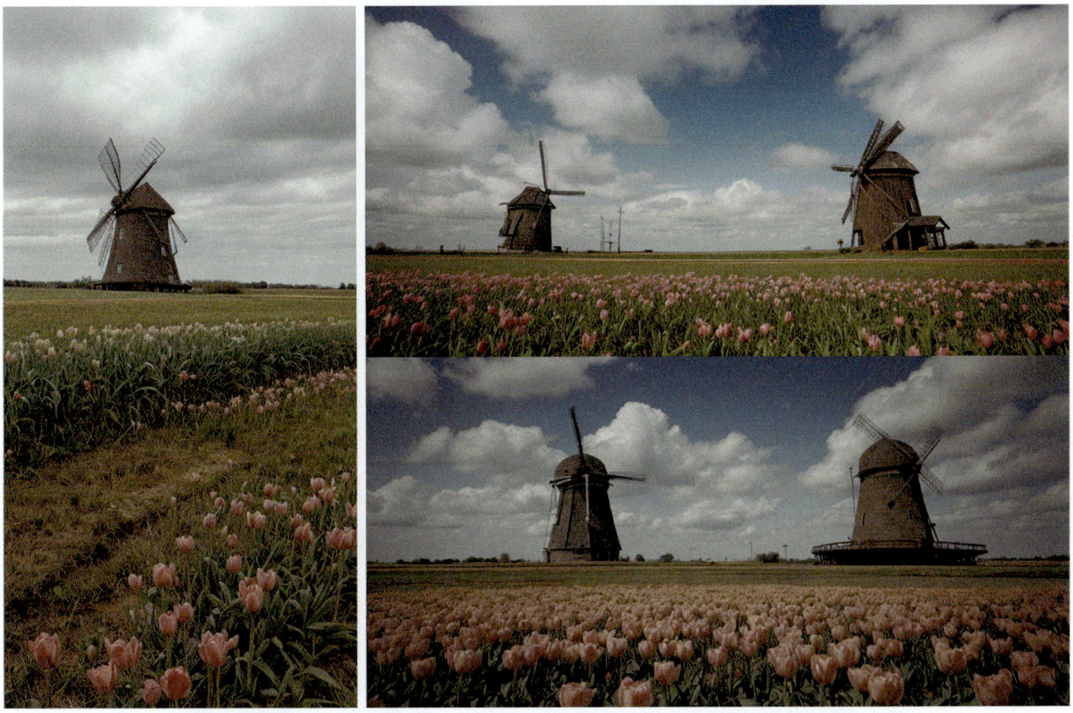

Prompting for a "single windmill" worked great in a narrow portrait format (left). But most attempts in a wide landscape format ended with multiple windmills, despite the prompt. Notice also the errors – the model is better at creating the structure than it is at generating the sails.

It is also important to note that most image-generation models are trained primarily on square format images, and although this is changing now with the newer video generators, models still struggle with the composition in wide aspect ratios. For example, they tend to place the subject at the center, and it is often difficult to place it near the edge of the frame without the guidance of a ControlNet (see later in the chapter). It is helpful to think of an aspect ratio as a substantial driving factor for the algorithm's creative output and choose the aspect ratio accordingly.

RESOLUTION

As in all digital art workflows, resolution affects the amount of detail in the image. But in generative AI, the number of pixels has a distinct effect on the realistic coherence of the generated image. AI models struggle with accuracy at lower resolutions: human faces are deformed or incomplete, objects lack defining features or appear distorted, and strange artifacts are common. There is a tricky balance to maintain here, because AI generation times grow exponentially with higher resolutions, as do the hardware requirements.

Granted, waiting a long time for a render is not a foreign concept to 3D artists. But in CG we have ways of visualizing the scene before hitting the render button – with generative AI, it's a dice throw, and AI workflows rely on choosing the best variation out of many. Additionally, an overall higher resolution is not necessarily a guarantee for better results, and the long waiting time may prove futile.

The most effective way to achieve high fidelity and detail without upscaling the entire image is to use inpainting techniques (see later in this chapter) to selectively increase the model's denoising

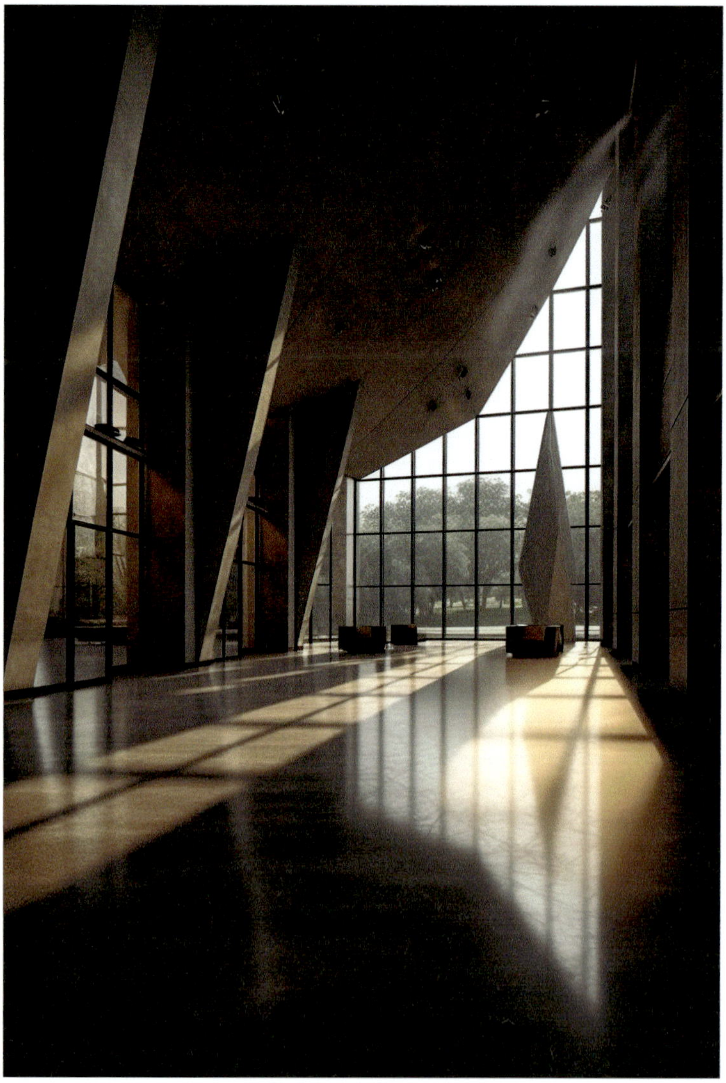

Tomorrow's Pavilion by Suburbia Studio.

Generated with Midjourney. © Suburbia Studio, © Julian Sadokha.

resolution in specific areas. However, if there is a need to increase the overall resolution of the image, latent upscaling models can successfully add pixels and increase detail while preserving the original image, thus allowing the artist to perform the initial re-generation and adjustments at a lower resolution.

CFG

Some AI image generators like Stable Diffusion and DALL-E provide a prompt adherence control called CFG (Classifier-Free Guidance). At higher CFG settings, the algorithm attempts to follow the prompt as closely as possible. At lower settings, it takes more "creative" freedom. CFG setting, however, affects more than just prompt adherence. Higher values generate a more contrasted and saturated image, often with an exaggerated focus on the prompt keywords. Lower settings tend to produce more pleasing features but may "ignore" some parts of the prompt or produce a flat look with lower contrast and reduced saturation.

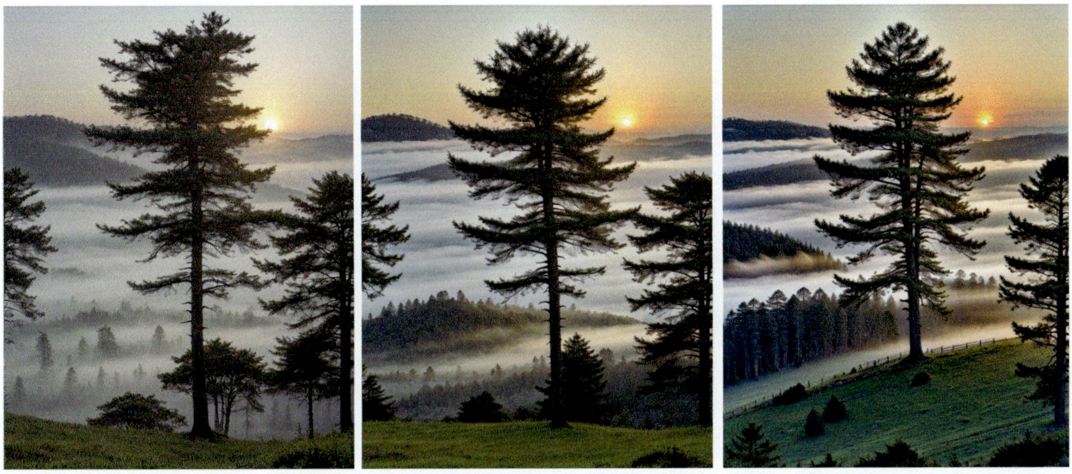

The same prompt (*photo of a tall, impressive pine tree on a hill, overlooking a verdant valley, patchy fog, sunset*) with the same noise seed is generated with a CFG of 2 (left), 5 (center), and 10 (right). Notice how the features become more pronounced and the colors more saturated as the look shifts from fairly realistic (albeit a bit flat) at the low end to almost cartoonish at the high end.

Tweaking the CFG control while maintaining the same noise seed is a common technique for finding the desired "sweet spot". Many AI images suffer from over-cranked CFG settings that make them look harsh, unnatural, or cartoonish. Photorealism is often easier to achieve with lower CFG settings, as contrast and saturation can be boosted as a post process. That said, the ideal CFG setting varies substantially depending on the checkpoint model, the subject of the image, and the cumulative effect of ControlNet inputs and other guidance features.

DENOISING STEPS AND SCHEDULERS

The quality and fidelity of the output image rely on the number of iterative denoising steps. Setting this value too low inevitably generates crude, low-quality images. But setting it too high can lead to long generation times with very little benefit, as additional denoising steps become progressively less effective. This is not unlike sampling in CG rendering – the optimal setting is a fine balance between quality and processing time and depends on the checkpoint model and other factors.

Schedulers are algorithms that control the iterative denoising process. The difference between the various scheduler types is rather obscure and requires some experimentation. Choosing the best scheduler depends on the checkpoint model and the desired results. For example, DPM2-Karras or DPM3-Karras tend to work well for generating photoreal imagery with models like Realistic Vision and Analog Madness, but other scheduler types like Euler and DPM Ancestral may work better in some scenarios, or with other models.

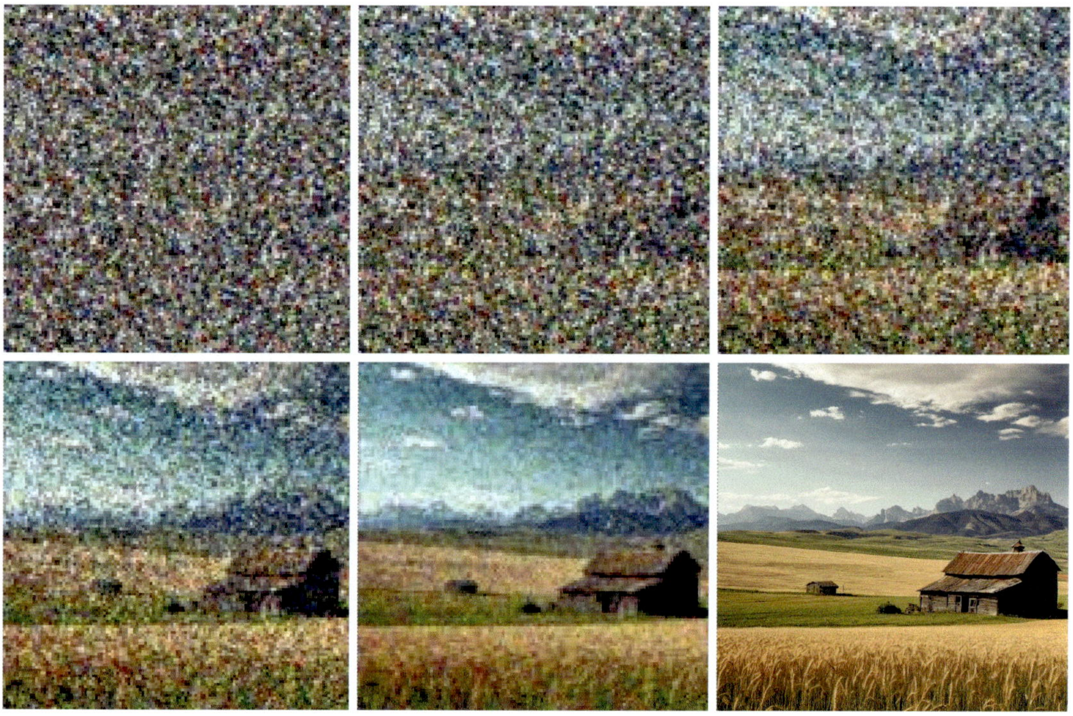

From pure noise to finished image, six steps of the denoising process.

denoising strength keeps the original image mostly intact, resulting in minimal changes. At a high denoising strength, most of the input image is converted into noise, and the results can be widely different from the original. Setting the right amount of denoising strength is a key technique for successful inpainting and upscaling (see later in this chapter).

The Art of Prompting

Digital artists who are used to 2D and 3D tools may find text prompting both fascinating and extremely frustrating. There is certainly magic in watching the computer generate your wishes out of thin air (or, more accurately, out of random noise), and the process is often creatively rewarding. But achieving specific results is challenging – AI is no genie in a bottle, nor is it a human artist – to write successful prompts we need to understand its "thinking process".

Let's look at this example prompt:

> A large red sun settles behind tall snowy mountains, painting the gray clouds with tiny specks of orange, while casting its diminishing light on a small green garden in a forgotten Tibetan monastery with white-washed walls.

This is a lovely, quasi-poetic prompt that can surely trigger the imagination of a creative human artist. But then again, AI is not human, and this prompt is unlikely to generate the desired results, because it suffers from several key problems:

- **Too wordy**: as previously noted, visual generative AI algorithms look for keywords ("snowy mountains", "clouds", "green garden", "Tibetan monastery"). Unnecessary verbs ("settles", "painting", "casting"), vague descriptions ("diminishing", "forgotten") or compound grammatical concepts ("while") contribute little to the success of the prompt, and may shift the emphasis unnecessarily. The guiding principle for choice of words is simple: can these words be easily portrayed graphically?
- **Wrong order**: the order of the subjects in the prompt matters, especially if no weights are specified (see in next section). In the previous example, it is clear to us humans that the temple and its garden are the main subjects of the image, but they appear last in the prompt, after the description of the sky and the background. Prompts should start with the main subject and foreground elements and then move on to the surrounding environment and background elements.
- **Confusing words: s**ince the AI algorithm is not trained to make sense of complex sentences, it can easily confuse the verb "painting" with the noun and assume that this is a stylistic request to make the image look like a painting. "Diminishing" could also be misinterpreted as "getting smaller" and wrongly applied to the scale of elements. Make sure to use concise nouns and adjectives and avoid verbs if possible.
- **Too many colors:** while it is natural to assume that providing detailed color descriptions can better guide the model toward the desired result, multiple colors in a prompt usually achieve the opposite. The prompt includes red, gray, green, and white, but red is mentioned first and will most likely be emphasized. For example, the walls of the temple may end up red even if they are specified as "white-washed" in the prompt. It is better to avoid multiple colors, especially if the color can be derived from the context. For example, "red" can be dropped, since the model will likely use this color in the context of a sunset.

With these four points in mind, here is a revised prompt:

> Old Tibetan monastery with a small garden, snowy mountains in background, partially-cloudy sky, sunset.

This prompt is much shorter, starts with the main subject and then moves to the background, sky, and time of day/lighting. It is formatted as a list separated by commas and avoids unnecessary words. "Partially-cloudy" is hyphenated to avoid any misinterpretation by the model. This is of course just a starting point – based on the first generated iterations, further adjustments, additions, and weights will probably be needed to fine-tune the results.

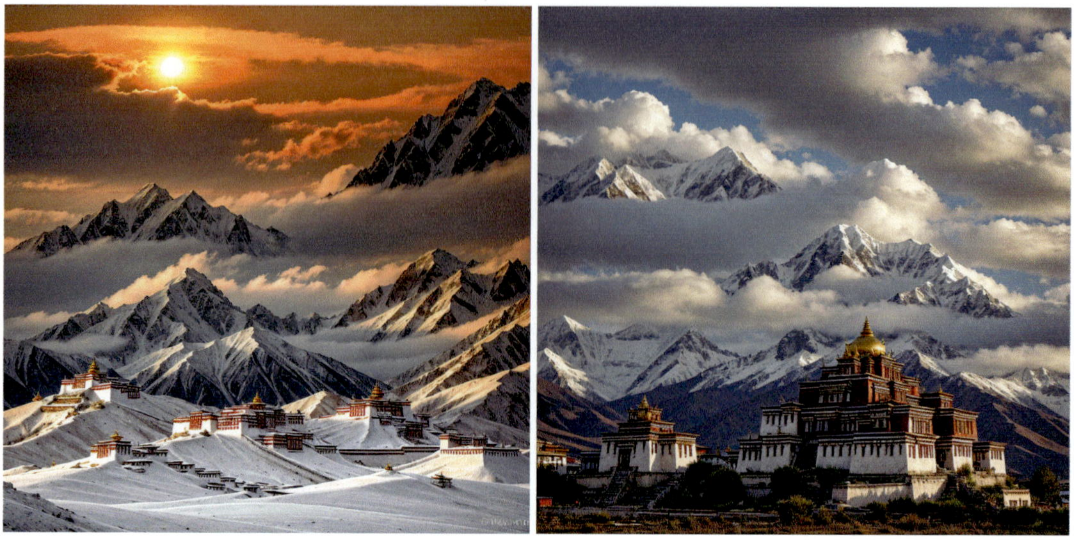

The original prompt (left) is confusing to the model, which overemphasizes the sun and mountains and completely ignores the garden. The revised prompt (right), though not without flaws, takes us much closer to the desired results and composition.

PROMPT WEIGHTS

Weights are an effective tool for balancing emphasis within the prompt. For example, if our Tibetan monastery prompt fails to produce the background mountains, we can increase their relative weight:

> Old Tibetan monastery with a small garden, (snowy mountains in background: 1.3), partially-cloudy sky, sunset.

If, on the under hand, the mountains are too prominent, their weight can be reduced:

> Old Tibetan monastery with a small garden, (snowy mountains in background: 0.8), partially-cloudy sky, sunset.

There are several common syntaxes to express weights – including using just parentheses or adding plus signs: *(((snowy mountains in background)))*, or *Snowy Mountains in Background+++*. Fine-tuning weights usually requires some trial and error, but in general, less is better – exaggerated weighting or too many weights in a single prompt may reduce their effectiveness. Also, weights and position within the prompt are somewhat interchangeable. Adding weights to a word at the end of the prompt can achieve a similar result as moving it to the start.

NEGATIVE PROMPTS

Negative prompts are essential for keeping the model away from unwanted results. Generalized negative prompts can be used to further push the model toward a desired style (for example, "CG, render, illustration, cartoons" to ensure a photographic look). They can also be used to keep the output quality high ("low quality", "blurry"), or to reduce the likelihood of common AI errors ("missing fingers", "wrong shadows", "watermark"). Specific negative prompts are particularly helpful for adding focus to the positive prompt and guiding the model toward the desired results. As with positive prompts, simplicity and clarity are key – overly restrictive negative prompts tend to become less effective.

PROMPTING FOR PHOTOREALISM

Checkpoint models are generally trained on a wide variety of visual styles, and even the ones that are biased toward photographic look (like Realistic Vision or Juggernaut) need some prompt pushing to produce photoreal results. Models "dig within their memory" for relevant references as they process the prompt, and we can nudge them in the right direction. So, in addition to such obvious keywords like "photo", "photograph" "realistic" and "photorealistic", we can specify, for example, "4K". This does not mean we're asking for a 4K image (the generated resolution is not defined in the prompt), instead, we are asking the model to favor those 4K images within its "memory", which can ensure higher quality and detail fidelity.

Likewise, using common lens aperture widths like f/2.8 or f/11 does not necessarily mean that the model will precisely emulate the depth of field of those apertures (generative models don't understand optics). Rather, it directs the model toward matching metadata specs in the training material, which can help achieve a more specific photographic look.

PROMPTING WITH AI

While it may sound like a doomsday concept out of a cheap sci-fi book, there's no reason not to use AI's help to write the prompts. In fact, this is a common technique for improving prompts – after all, AI know AI language best. For example, if you have a reference image, or any image that is close to what you're trying to generate, you can ask an AI model like Chat GPT to describe it. You can then use that description as a guide for the prompt.

SPLIT PROMPTS

One of the most powerful prompting techniques is to use two or more sets of prompts within the frame by assigning each set its own mask and restricting it to a particular area. This workflow provides a substantially higher level of specificity and control at the initial generation stage. Complex compositions and detailed visuals can be created by mixing masked prompts at varying strength levels. For example, one prompt can describe the background and cover the entire frame, while two other area-specific prompts establish key elements.

Inpainting

Of the various techniques discussed here, inpainting is the most intuitive and artist-friendly approach to guiding generative AI. The workflow is simple: start with a generated image, then regenerate smaller areas within the image to replace, modify, or fix specific parts. Inpainting is also a very efficient way of producing better detail and fidelity by regenerating smaller areas at higher resolution without rescaling the entire image.

On the surface, inpainting is similar to a basic compositing: the general inpainting area is defined by a bounding box and a mask. The model regenerates everything within the bounding box, but only the masked area is composited back on top of the original image. As expected, the mask's edge sharpness affects the transition area – soft mask edges are usually needed for a smooth, seamless blends. The bouncing can be used as a resolution multiplier by using the original resolution for the smaller area it defines, which is extremely effective for selectively adding or fixing detail.

The denoising strength parameter is key to successful inpainting. Lower levels are likely to generate content that matches the lighting and look of the underlying image and seems well-integrated. It is, however, harder to generate completely new detail at lower denoising strengths. Higher denoise settings work better for adding new detail or substantially modifying existing features in the image, but at the risk of losing integration and coherence.

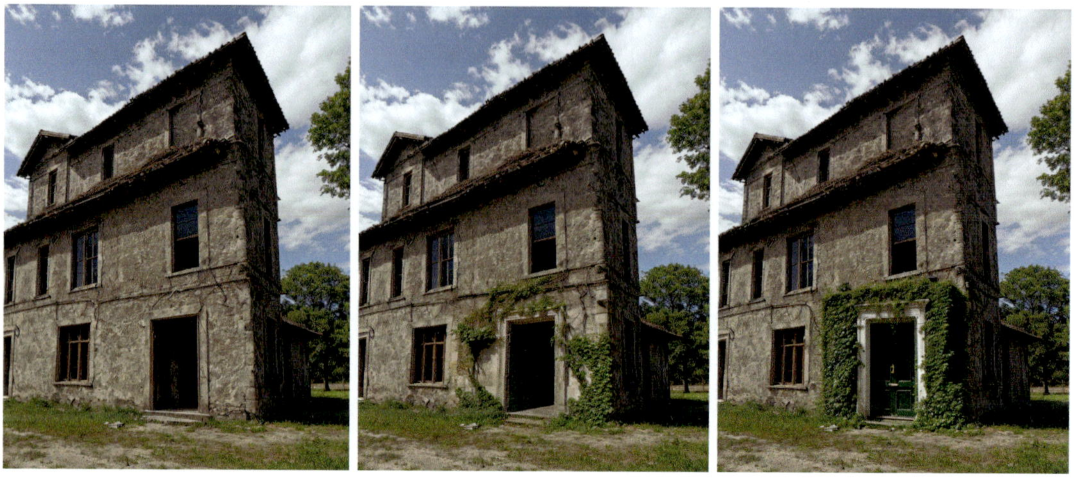

Inpainting dense ivy cover around the door area. With a denoising strength of 0.7 (center), the ivy is not very dense, but it's decently integrated and the structure remains mostly intact (though the model did add a green door). With a denoising strength of 0.9 (right), the ivy is dense but not well-integrated, and the entire entrance is regenerated with a wrong perspective.

INPAINTING PROMPTS AND ADDITIONAL GUIDANCE

It is usually not effective to reuse the original image prompt when inpainting. Instead, new prompts should be written (and adjusted) for each inpainting area with focus on the specific detail required. Additional inpainting guidance tools include drawing or sketching on the image (even simple lines or scribbles) – shapes and colors are factored in the image-to-image denoising process and are helpful along with the prompt.

Using ControlNet Models

ControlNet models are some of the most effective and accurate tools for guiding generative AI toward a desired result. There are two stages in the operation of a ControlNet: first, the ControlNet preprocessor generates a specific type of guidance information from a given image (edges, depth, normal, human pose, color, object identity). Then this information is used by the ControlNet processor to direct or constrain the generative model. ControlNet models can also be combined, and their relative weights adjusted.

ControlNet workflows are especially suited for professional digital artists. CG renders or 2D illustrations can be used as the ControlNet inputs – allowing for much tighter artistic control and specificity. In most cases, renders and drawings do not need to be fully fleshed out, detailed, or textured – just enough to define the placement, size, and shape of the elements in the scene. Sometimes, a simple illustration or a crude CG model is all it takes to effectively guide the generative algorithm toward the desired composition, camera angle, or action.

There are two points to keep in mind when working with ControlNet models. First, their ability to "understand" an image and generate data from it, while certainly impressive, is not failsafe. As you rely on a ControlNet to extract depth, detect edges, or identify objects in the image, you need to also check for errors, as this can negatively affect the result. The other factor to consider is the ControlNet's weight. Depending on the ControlNet type, higher values can push the generated image toward a flat and painterly look, or a CG render look. It's a fine balance between controlling the generative algorithm and giving it freedom to creatively add detail and preserve photorealism.

DEPTH CONTROLNET

The depth ControlNet is particularly suitable for laying out the composition and defining shapes and locations of elements within the frame, without implying any color, texture, or fine detail. There is a fair number of depth preprocessors that can "guess" depth and generate a depth map from an image (though not without errors, especially toward the deep background). However, if you are building a 3D scene as a guide, you can skip the depth preprocessor and use the Z depth pass generated by the CG renderer for ultimate accuracy. Most contemporary render engines output a non-normalized Z depth pass with extremely high color values, so some processing may be required to adjust it for a Stable Diffusion workflow.

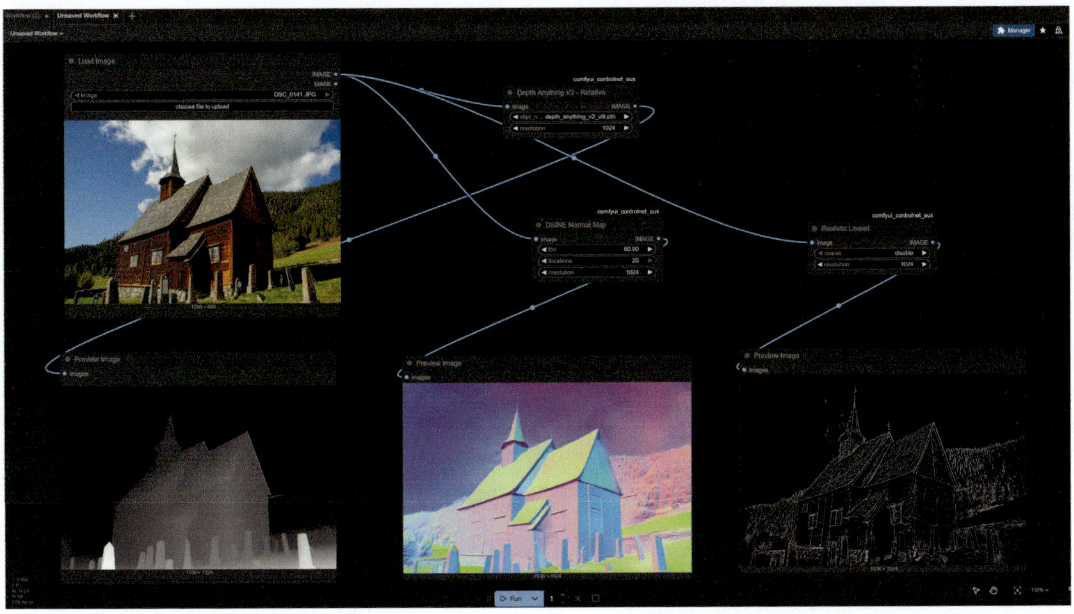

Preprocessors in action: depth, normal, and line art are extracted from a photo to be used as ControlNet guides.

NORMAL CONTROLNET

This ControlNet (sometimes referred to as lighting direction ControlNet) works particularly well coupled with a depth ControlNet. The normal map indicates the direction of the surfaces and thus helps further define detail that the depth map may not show. The preprocessor can generate a normal map from an image, but a CG generated normal map provides better accuracy, especially when coupled with a matching depth map.

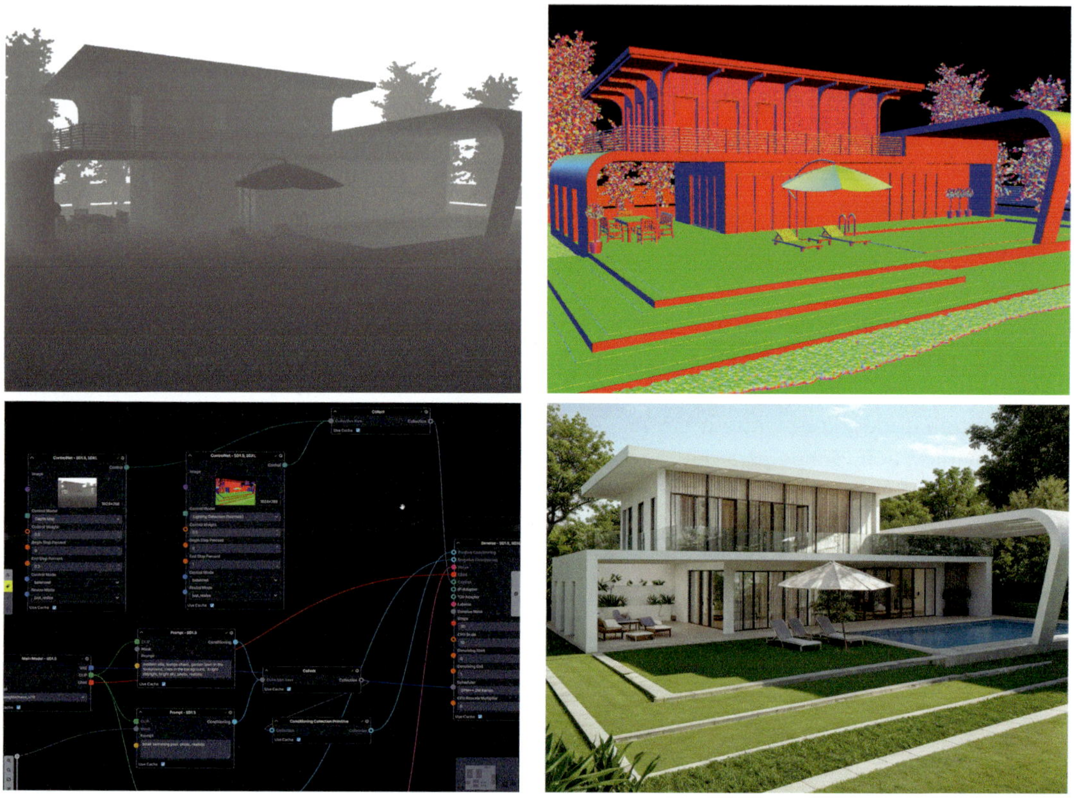

A 3D scene in Maya is used to guide the generative process: the depth pass (top left) and normal pass (top right) are fed into ControlNet nodes in InvokeAI (bottom left) to drive the generated image of a modern villa (bottom right).

CANNY

Canny is an edge-detect ControlNet that can be used to trace the outlines in an image. Low and high threshold control let you fine-tune the desired amount of edge detail and avoid overly noisy results. Canny is an effective guide, but should be applied very carefully, as higher weights tend to make the generated image flat and unnatural.

OPENPOSE

While AI models excel in generating photorealistic humans, they struggle when prompted to produce specific poses or actions. "Standing", "sitting", or "walking" usually works, but more specific requests, even as simple as pointing a finger, holding something, or looking in a certain direction, can be very difficult to achieve. On top of this, attempting to direct specific poses and actions for more than one person in an image or video can be extremely hard.

The OpenPose ControlNet preprocessor identifies human poses in a reference image and creates a simple line skeleton which is then used to drive the image generator. OpenPose is very effective in guiding AI models toward specific poses, down to individual fingers or facial expressions. But like any ControlNet, the OpenPose preprocessor can have a hard time figuring out the correct pose when the hands, feet or head in the image are obscured, or when the camera angle is very low or very high. It also struggles to decipher images with multiple people interacting, so it's better to use clean and uncluttered images (or CG renders) for pose references.

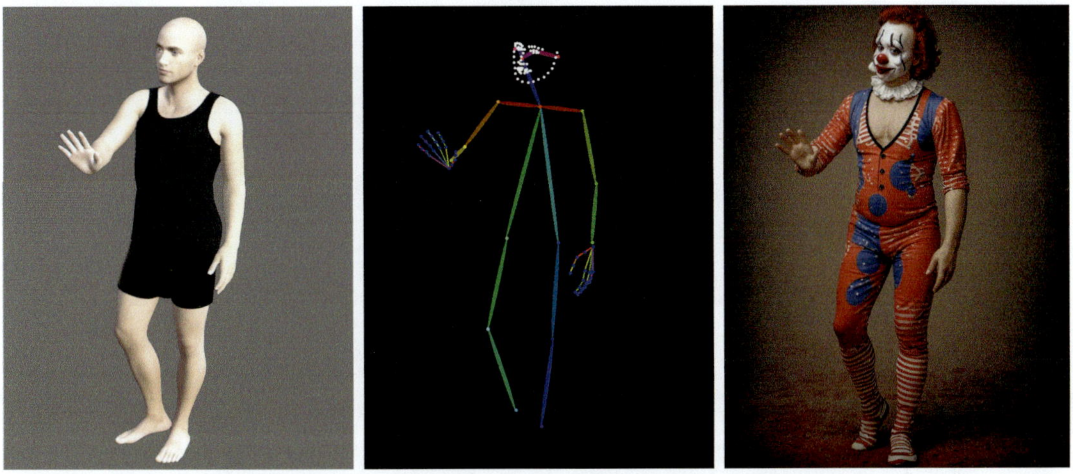

A simple render of a generic CG human (left) is fed into the OpenPose preprocessor (center), which then applies the pose to the generated clown (right).

IP ADAPTERS

Image Prompt (IP) adapters use an input image as an additional graphic prompt, along with the text prompt. IP adapters come in several different flavors. A style IP adapter can be used to transfer mood, color, lighting, or graphic style from a reference image. A composition IP adapter is used as a general loose guide for placement and scale (but not nearly as accurate as a depth, normal, or OpenPose ControlNet). Finally, some IP adapters are programmed for face transfers (an understandably popular use of IP adapters).

SEGMENTATION

Segmentation algorithms essentially reverse the text-to-image process and can identify various elements within an image. They are especially helpful for quickly generating masks for inpainting, area prompts, and other operations. As with all ControlNet preprocessing, the segmentation algorithm's accuracy depends on the complexity and clarity of the image, as well as the type of element it is asked to identify. Common, simple categories like "people", "cars", or "buildings" are more likely to be identified.

Case Study: Season Changes with ComfyUI and Nuke

Kevin Samar, a senior compositor, used ComfyUI and Nuke to apply season changes to a shot from a short movie he directed and produced. "When thinking about practical use cases for Stable Diffusion in narrative storytelling, a season change was one of the first ideas that came to mind, as it's something I've encountered many times in my career" says Kevin. "This panning landscape shot from the short film *Viva Extra* was a great candidate for the idea. It was captured on a 100mm lens, which minimizes parallax between depth layers and reduces the number of projections needed in comp".

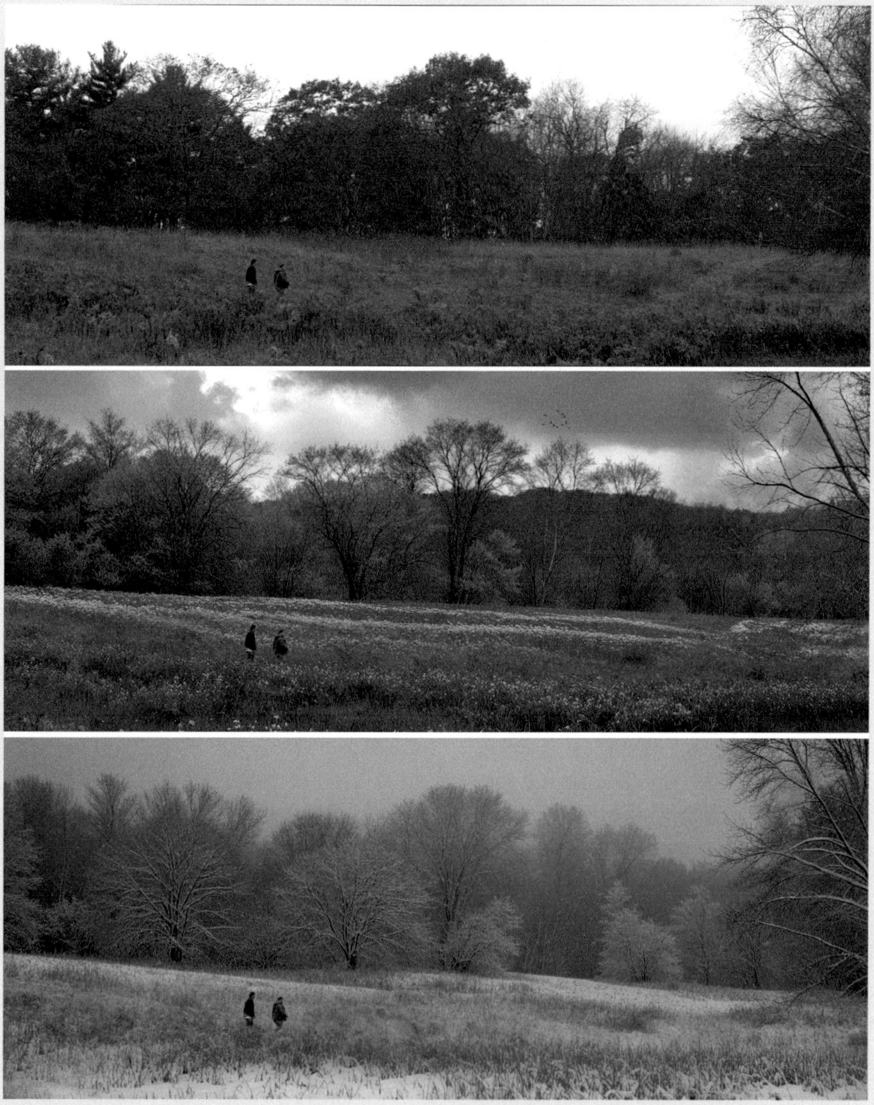

Same shot, three seasons. Nuke, ComfyUI.

I started by painting out the two characters walking through the frame. Since I planned to use image-to-image generation in ComfyUI, I didn't want the characters to affect the output. After painting them out, I rendered clean stills of the first and last frame. I used these stills as input images in ComfyUI, along with a depth-based ControlNet and the Stable Diffusion XL (SDXL) model. I kept the ControlNet weight value high to preserve the geometry and layout of the original scene.

I used simple prompts for each season. For example, the winter prompt was *winter scene, snowy trees, field, overcast, telephoto*. I ran multiple generations for each season, and then brought the best output images into Nuke, where I merged them using traditional compositing techniques and projected them over the original scene. I blended the projections, then reinserted the characters and added snowflakes using standard Nuke particle setups. I used a similar process for the spring version. Generative AI with ComfyUI is a perfect tool for compositors who want to have a bit more creative control on their images, concludes Kevin.

Epilogue

Where are we heading next? What does the future hold for photoreal visual content production? Will generative AI take over? Will skills like modeling, texturing, lighting, or compositing gradually become obsolete? This is certainly not the situation at the time of writing this text. Live-action visual effects still rely prominently on traditional 3D and 2D work. Video games still require skilled artists to build environments and animate characters. Visualization still needs a middle person to effectively relay the design. As I said earlier, we are not yet making movies at the push of a button. Not yet. But every day we see new AI tools, improved models, better results . . . no matter how you look at it – fascinating new frontier or a looming threat, AI is bound to change how we work as digital artists.

For thousands of years, humans created visual content by painting and sculpting. Then, in just a couple hundred years, a series of technological breakthroughs drastically changed how we generate imagery – photography first, then moving pictures, then computers, the internet, social networks . . . each of these changes required completely new skillsets and different learning paths. Old jobs were gone, new ones sprung up. But will this be the case with AI? If we believe the hype, then skills will not be required anymore, we'll just ask the genie in the bottle to realize our wishes. But if we look back at history, the changes are more nuanced. Yes, portrait painting as a day job became obsolete when photography took over. But painting did not go away, and even though we all have cameras in our phones, we are not all photographers. Cinema promised to be the ultimate story-telling medium, yet books are as popular as they ever were. New merges with the old, trends come and go – we have yet to see how this all develops, and my guess for the future is as good as yours.

But what about the essence of photorealism? Traditional art gradually shifted from figurative to abstract around the turn of the 20th century. Artists lost interest in recreating reality and instead searched for new avenues of expression. Will photorealism also go out of vogue at some point? Drawing and painting always served a much broader function than just pure art. Historical documentation, family portraits, scientific research, journalism, advertisement, entertainment – for most of these day-to-day uses, photography emerged as a better alternative. Similarly, photorealism serves a wide range of practical purposes, from visualizations to games. It is much less an artistic, aesthetic, or stylistic ideal as it is a functional necessity for immersive experiences. As such, it is very unlikely to lose its importance, but rather the opposite – the immersive experience is bound to be pushed even further.

DOI: 10.4324/9781003590293-25

But photorealism as we currently know it is merely an artificial emulation of photography, nothing more than 2D images on screens (even VR is essentially projected moving pictures). Will we ever shift into a new phase of artificial reality, one that is truly three-dimensional and involves the simulation of touch and smell? Such a shift is bound to overtake photorealism in entertainment and games, in the same way photography replaced painting in the late 19th century. The "photo" will then be taken out of "photorealism", because we will no longer emulate photography – we will emulate reality.

INDEX

Note: Locators in *italics* refer to figures.

For Product Safety Concerns and Information please contact our
EU representative GPSR@taylorandfrancis.com Taylor & Francis
Verlag GmbH, Kaufingerstraße 24, 80331 München, Germany